Tribology in Materials and Manufacturing - Wear, Friction and Lubrication

Edited by Amar Patnaik,
Tej Singh and Vikas Kukshal

Published in London, United Kingdom

IntechOpen

Supporting open minds since 2005

Tribology in Materials and Manufacturing - Wear, Friction and Lubrication
http://dx.doi.org/10.5772/intechopen.87674
Edited by Amar Patnaik, Tej Singh and Vikas Kukshal

Contributors
Bendaoud Nadia, Mehala Kadda, Yasunori Sakai, Auezhan Amanov, Debashis Puhan, Sumit Kumar Panja, Nguyen Van Minh, Hai Ninh Le, Huynh Nguyen, Alexander Sergeevich Kuzharov, Andrey Alexanderovich Kuzharov, Djilali Kaid Ameur, Brahim Tlili, Jeyaprakash Natarajan, Che-Hua Yang, Sabarinath Sankarannair, Avinash Ajith Nair, Benji Varghese Bijo, Harigovind Sureshkumar, Hareesh Kuttuvelil Das, Lorena Deleanu, Constantin Georgescu, Mihail Botan, Jun H. Zhong, Alexey Vereschaka, Sergey Grigoriev, Vladimir Tabakov, Mars Migranov, Nikolay Sitnikov, Filipp Milovich, Nikolay Andreev, Friedrich Franek, Ulrike Cihak-Bayr, Robin Jisa, Naofumi Hiraoka, Biswajit Swain, Rameswar Behera, Soumya Sanjeeb Moahapatra, Ajit Behera, George Tumanishvili, Tengiz Nadiradze, Giorgi Tumanishvili

First published in London, United Kingdom, 2021 by IntechOpen
IntechOpen is the global imprint of INTECHOPEN LIMITED, registered in England and Wales, registration number: 11086078, 5 Princes Gate Court, London, SW7 2QJ, United Kingdom
Printed in Croatia

British Library Cataloguing-in-Publication Data
A catalogue record for this book is available from the British Library

Additional hard and PDF copies can be obtained from orders@intechopen.com

Tribology in Materials and Manufacturing - Wear, Friction and Lubrication
Edited by Amar Patnaik, Tej Singh and Vikas Kukshal
p. cm.
Print ISBN 978-1-83880-557-9
Online ISBN 978-1-83880-575-3
eBook (PDF) ISBN 978-1-83881-899-9

For EU product safety concerns:
IN TECH d.o.o., Prolaz Marije Krucifikse Kozulić 3, 51000 Rijeka, Croatia, info@intechopen.com or visit our website at intechopen.com.

Meet the editors

Dr. Amar Patnaik is currently working as an associate professor in the Department of Mechanical Engineering, Malaviya National Institute of Technology, Jaipur, India. His area of specialization is composite materials and alloys. He has more than fourteen years of teaching experience and has taught a broad spectrum of courses related to mechanical engineering. Dr. Patnaik has guided twenty-one Ph.D. students successfully and several MTech students as well. He has published more than 250 research papers in international journals (SCI), written thirteen chapters in different books, and filed seven patents. Dr. Patnaik has delivered more than thirty invited guest lectures in different institutions and organizations. He is a lifetime member of the Tribology Society of India, Electron Microscope Society of India, and International Society for Technology in Education.

Tej Singh is an assistant professor in the Faculty of Informatics, Savaria Institute of Technology, Eötvös Loránd University, Szombathely, Hungary. He obtained his MTech in Nanoscience and Nanotechnology from Panjab University Chandigarh, India, and Ph.D. in Mechanical Engineering from the National Institute of Technology Hamirpur, India. His major research interests are composites, tribology, nanoparticle synthesis, multi criteria decision-making, and solar energy. He has more than 100 scientific publications in peer-reviewed journals and conference proceedings to his credit.

Dr. Vikas Kukshal is an assistant professor in the Department of Mechanical Engineering, National Institute of Technology, Uttarakhand, Srinagar (Garhwal), Uttarakhand, India. His area of specialization includes composite materials and alloys. He has more than nine years of teaching experience and has taught a broad spectrum of courses related to mechanical engineering. He has authored twenty-five articles published in reputed international and national journals and conferences and written four book chapters. He has supervised several MTech students and is presently supervising two Ph.D. students. He is also the reviewer of many referred international journals. He is a lifetime member of the Institution of Engineers (India), Indian Institute of Metals, and Tribology Society of India.

Contents

Preface

Tribology has garnered much research attention, especially regarding its role in advanced engineering materials. As such, this book presents current research on the tribology of advanced materials in bio-lubricants and engineering applications. There is great potential for using advanced materials in biomedicine and engineering by enhancing the current knowledge about composition, fabrication, properties, and performance of materials and providing solutions to many unaddressed issues in tribology. The chapters in this volume examine the effect of tribological properties on ionic liquids, polymers, interacting surfaces, and additive manufacturing.

Significant research efforts worldwide are exploring the various properties associated with materials and manufacturing. Researchers are collectively focused on minimizing wear by proper utilization of lubricants and lubrication mechanisms. This book discusses all applied and fundamental elements related to fabrication or synthesis, combination or composition, properties, characteristics and performance of engineering materials in the field of tribology. It also examines several critical issues and provides suggestions for future work in the field of tribological science.

I would like to thank IntechOpen and Ms. Sara Gojevic Zrnic for their invaluable input in the organizing and editing of the book. I also gratefully acknowledge those who invested their time and effort in compiling the material that made this book. Many authors who contributed their expertise and perspectives are clearly the backbone of this work and they deserve the lion's share of the commendations.

Dr. Amar Patnaik
Associate Professor
Malaviya National Institute of Technology Jaipur,
India

Tej Singh
Eötvös Loránd University,
Hungary

Vikas Kukshal
National Institute of Technology,
Uttarakhand, India

Section 1

Tribological Assessments

Chapter 1

Tribological Properties of Ionic Liquids

Sumit Kumar Panja

Abstract

Our main focus is to report the tribological properties of ionic liquids (ILs). Mainly, lubricating of ILs has been reported to understand the applicability of ionic liquids (ILs) in petroleum-based lubricant industry and energy conversion process as oil additive. The influence of counter parts of ILs on tribological property has been reported for designing efficient lubricating and oil-additive property of ILs. The effect of halogenated and nonhalogenated ILs on corrosion is also reported during tribological studies at different metal surface. Further, role of ILs as oil-additive has been discussed in terms of better tribological performance. Structure modification and role of anion on better performance of tribological property have been mentioned for enhancing effectiveness of lubricant and oil-additive properties. Origin of corrosion and thin film formation on metal surface are also discussed in detailed using different types of ILs and metal surfaces.

Keywords: ionic liquids (ILs), tribological properties, halogenated and nonhalogenated ILs, lubricant, corrosion

1. Introduction

Lubricants are very important materials for human and society due to their applications from "mobility" in ancient era to durability in modern times and then most recently in enhancement of "energy efficiency process". Petroleum-based lubricants are popular and used as the standard materials in transportation, manufacturing, and power generation industries etc. [1]. From economic point of view, 1.0–1.4% of a country's GDP may be achieved through lubrication R&D, which has provoked the relentless quest of advances in lubricants in order to increase both energy efficiency and durability [2]. Generally, commercial lubricant contains a combination of base oils and additives including antioxidants, detergents, dispersants, friction modifiers, antiwear and/or extreme-pressure additives, and viscosity modifiers.

As energy and environment play an important role in our life, there need for energy efficient systems, and utilization/conversion of energy in environmentally benign practices have been increasing immensely because of high volatility in fuel prices, stringent environmental regulations and global awareness on the sustainability of fuels. High fuel consumption is arisen due to high friction and wear in the transportation system during energy conversion process [3, 4]. Due to high friction and wear, failure of engine parts is often happened with large amount of discharge of partially oxidized fuels and greenhouse gas emission etc. For reducing the production of these hazardous materials, low friction and wear are required for energy conversion process. Lowering the friction and wear are important to reduce the production of hazardous materials during energy conversion process to the mating surfaces of the engine. Only an efficient lubricant can solve the problem related to energy conversion process and global awareness on the sustainability of fuels. Zinc dialkyldithiophosphate (ZDDP) is well-known as efficient antiwear and friction-reducing additive for iron-based components. Presently, it is observed that ZDDP is an efficient antiwear and friction-reducing additive but has shown toxic nature to aquatic wildlife, human-health issues and poisonous automotive exhaust gas as catalyst components.

Ionic liquids (ILs) have been known as new ionic materials and great important of applications in organic chemistry to as electrolytes in alternative energy generation/storage devices etc. (**Figure 1**). ILs have been known for their stability, well-established structural characterization and low viscosity etc. The choice of cation and anion is an important parameter for IL to determine the desirable physical properties. The tunable physical properties of the ILs make also an important material for the application in lubricant industries [5]. The length of side chain of the cation is responsible for making ILs as tailor-made lubricants and lubricant additives. Due to presence of unique physical and chemical properties of ILs, strong surface adsorption, high thermal stability, and low sensitivity in rheological behavior are observed compared to conventional oil lubricants. In early 2012, exploring the feasibility of ILs as lubricant additives was limited due to very low solubility in common nonpolar hydrocarbon lubricating oils [6–10]. The efficient oil-miscible ILs were discovered and reported as promising antiscuffing/antiwear functionalities [11, 12]. Since then, ILs is used as efficient lubricant additives in oil-based lubricant to increase both energy efficiency and durability due to improved solubility property [13, 14]. Hydrophobic cation or anions of ILs is responsible for showing good lubricant properties and making significantly stable thermo-oxidative materials.

Recently, ILs have been studied as versatile lubricants and lubricant additives for various engineering surfaces. The solid surfaces mediated thin films of ILs have shown more efficient lubricating properties compared to conventional non-polar hydrocarbon liquids due to presence of hydrophobic character, change of geometry of cation and charge characteristics of ILs. The dynamic conformation changes of cation and anion play important role to show the lower shear stress and friction than conventional non-polar molecular lubricant. ILs have also been studied as lubricating additives in water and lubricating oils due to their unique polar and

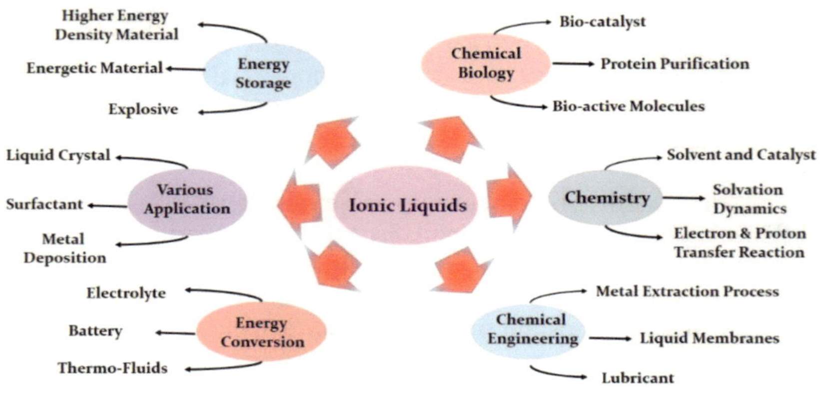

Figure 1.
Various application of ILs in different fields.

non-polar domain solutions and miscibility with polar and non-polar solvents. Now, ILs as lubricant and lubricant oil additives have become the new central research topic in lubrication processing.

This book chapter starts with the tribological performance of ILs as lubricant additives. The physicochemical properties of ILs have been correlated with their nature of cation and anion. Future research directions are also suggested at the end of this book chapter.

2. Tribological performance of halogenated ionic liquids

Generally, lubricants are used for extend the device life cycle and reduced parasitic energy loss by reducing friction. For these purposes, the lubricant must be high non-flammable and thermal stable with safer transportation and storage. ILs have shown interesting application in tribological studies due to their unique characteristic physical features [15]. It is also observed that addition of ILs to grease has shown substantially improved tribological performance. Similarly, IL-additive has shown to reduce more friction and wear compared to synthetic oil additives in base oil. Interestingly, imidazolium cation based ILs with long side-chain substituted cation and different anions have reduced more the friction and wear of steel-steel sliding pairs compared to base oil without additives. The excellent tribological properties of ILs as additives are due to their formation of physically adsorbed films and antiwear boundary film to reduce the friction and antiwear performance [16, 17].

The purity of IL is also key factor for improving wear and friction properties of ILs with additives. The highly purified IL has shown excellent friction reduction, antiwear performance and high load carrying capacity [18]. Further, lubricating performance of ILs depends on thermal stability, polarity, ability to form ordered adsorbed films and antiwear boundary film at the interface. Specially, polar nature of ILs can able to facilitate interactions in engineering surfaces forming the boundary thin film. The formation of unique protective thin film of ILs can able to avoid the direct contact between mating surfaces and is believed to be responsible for showing the antiwear property. ILs can provide an effective surface separative film at wide temperature ranges compared to conventional oils due to higher thermal stability. The area of functional fluids for lubricants and hydraulic oils is still under research and development.

Literature survey reveals that tribological study has been examined in ILs consisted of ammonium, phosphonium, pyrolidium, pyridinium, imidazolium cations as the cation and tetrafluoroborate (BF_4), hexafluorophosphate (PF_6), bis(trifluoromethanesulphonyl)imide (NTf_2), for the anion (**Figure 2**). On the other hand, ILs containing halogen exhibit have shown low friction and wear with good boundary lubrication properties.

Last one decade, several types of ILs like ammonium, phosphonium, pyridinium, imidazolium, etc. as cations and X^-, PF_6^-, $CF_3SO_3^-$, $(CF_3SO_2)_2N^-$ etc. as anions have been extensively studied as lubricant and lubricant additives for wide range of application in surface engineering. ILs have also exhibited structure dependent lubrication properties depending upon cations and anions [19–21].

The halogenated ILs are used over the steel surface for avoiding direct contact between tribo interfaces, consequently reduction in both friction and wear. During tribological test of BF_4^- anion based ILs, it is observed that the developing a tribo-thin film is composed of FeF_2 and B_2O_3 [22]. Phillips et al. have reported that BF_4^- anion based ILs can under go into several reaction with product of FeF_2, and lead to deduction of lubricant properties and corrosion of the substrate surface [22]. Metal fluorides (Like FeF_2) are formed on a boundary lubricating layer of

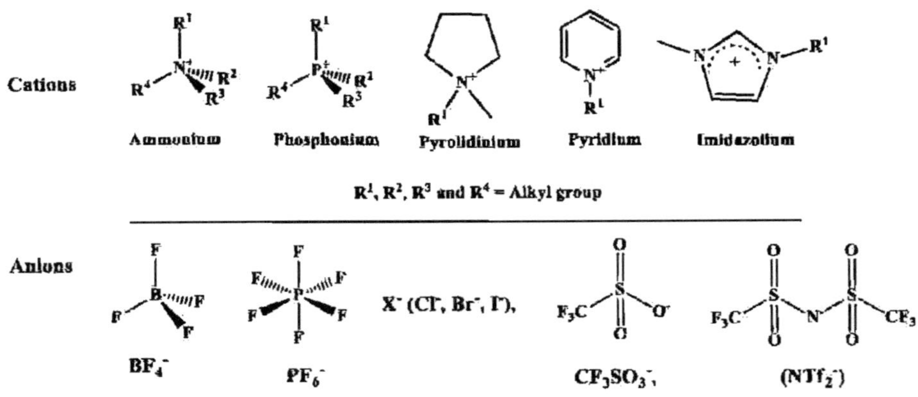

Figure 2.
Structures and abbreviations of cations and anions of the halogenated ILs used as lubricant additives.

friction surfaces by a tribochemical reaction. It is also known that ILs containing a halogen such as fluorine has been known to cause corrosion in steel aluminum alloy, bronze, and titanium alloy sliding materials [23, 24]. The corrosion of alloy sliding materials has been reported to be the formation of hydrogen fluoride (HF) due to the decomposition of halogenated ILs [25–28]. The formation of hydrogen fluoride is accelerated due to presence of water impurity in halogenated ILs. The change in color of the friction surface for steel bearings is observed using the hydrophobic IL as the lubricant in air at higher humidity [29]. The corrosion products are containing mainly metal fluorine and metal oxide on the surface which are experimentally verified [30].

After detailed investigation, halogenated ILs have hazardous and toxic effects to the environment and corrosive nature towards the engineering surfaces. The halogenated ILs can produce toxic and corrosive products after decomposition under different tribo-chemical reaction conditions for environment and the surface-engineering. High cost of halogens, particularly, fluorine-based precursors and disposal/discharge of halogenated ILs are big challenges for their penetration to the industrial applications.

Thus, halogen-free IL have been attracted more interest for developing the new type of lubricant for the energy efficient and environmentally-friendly processess.

3. Tribological performance of halogen-free ionic liquids

Accordingly, developing environmentally friendly ILs from renewable and biodegradable resources to diminish or avoid corrosion and toxicity has been becoming an inevitable strategy. A great effort has been devoted to searching for new halogen-free ILs. From literature surveys, halogen-free bioactive ILs such as saccharin [31–33], amino acid [34, 35] and ibuprofen [36, 37] ILs have been reported to replace traditional corrosive or hazardous halogenated ILs. Unfortunately, these halogen-free bioactive ILs are very poor thermal stability. However, low thermal stability and high cost of precursors cause less usable from application perspective. Interestingly, physicochemical properties and nontoxicity of these ILs can be regulated and customized by building precursor units from active pharmaceutical ingredients and biomass [38–40]. Literature survey reveals that tribological study of halogen-free ILs has been on the boundary lubricating capacity. Examined ILs are mainly consisted of ammonium, phosphonium, pyrolidium, pyridinium and imidazolium as the cation and phosphonate, dicyanamide, tricyanomethanide for anion (**Figures 3-5**). The halogen-free ILs have showed good tribological performance compared to synthetic lube oils.

Phosphonate-based halogen-free ILs has shown good thermal stability [41–43]. Phosphorus-containing ILs have been used for tribological study and shown effective lowering friction and wear reductions ability [12, 13, 44, 45]. The effectiveness of lowering friction and wear reductions ability, from high to low, was observed in phosphonium-phosphate, phosphonium-carboxylate, and phosphonium-sulfonate [46]. Experimental studies suggest that [$P_{8,8,8,8}$][DEHP], [$N_{8,8,8,H}$][DEHP], and [$P_{6,6,6,14}$][BTMPP] provide similar surface protection for both steel–steel and steel–iron contacts to ZDDP compound [47–49]. In choline based ILs, [choline][DEHP], [choline][DBDP], [$P_{6,6,6,14}$][BTMPP], [$P_{6,6,6,14}$][Tf_2N], [$P_{6,6,6,14}$][DMP], and [$P_{6,6,6,14}$][DEP]) have showed higher wear reduction to compared with the base oil, but only [choline][DEHP] and [$P_{6,6,6,14}$][Tf_2N] have only shown similar wear reduction property to ZDDP [50].

Phosphonium based ILs are used as additives in ester base oils and a VO, but only [$P_{2,4,4,4}$][DEP] and [$P_{6,6,6,14}$][FAP] have showed a stable >1% oil-solubility. At the same concentration, [$P_{2,4,4,4}$][DEP] and [$P_{6,6,6,14}$][FAP] ILs have showed comparable wear protection to ZDDP under low loads for a steel–steel ball-on-flat contact, [51–53].

Further, halogen-free ILs have been used for extended tribological properties of the steel-steel and DLC–DLC tribo-pairs. Lubricating and additive properties of bmimDCA and bmimTCM have been tested on the steel and DLC surfaces after the friction tests (**Figure 4**).

A chemical reaction film is observed on the sliding surface of the steel-steel tribo-pair. It is considered that a corrosive attack of ILs to the metal surface is also

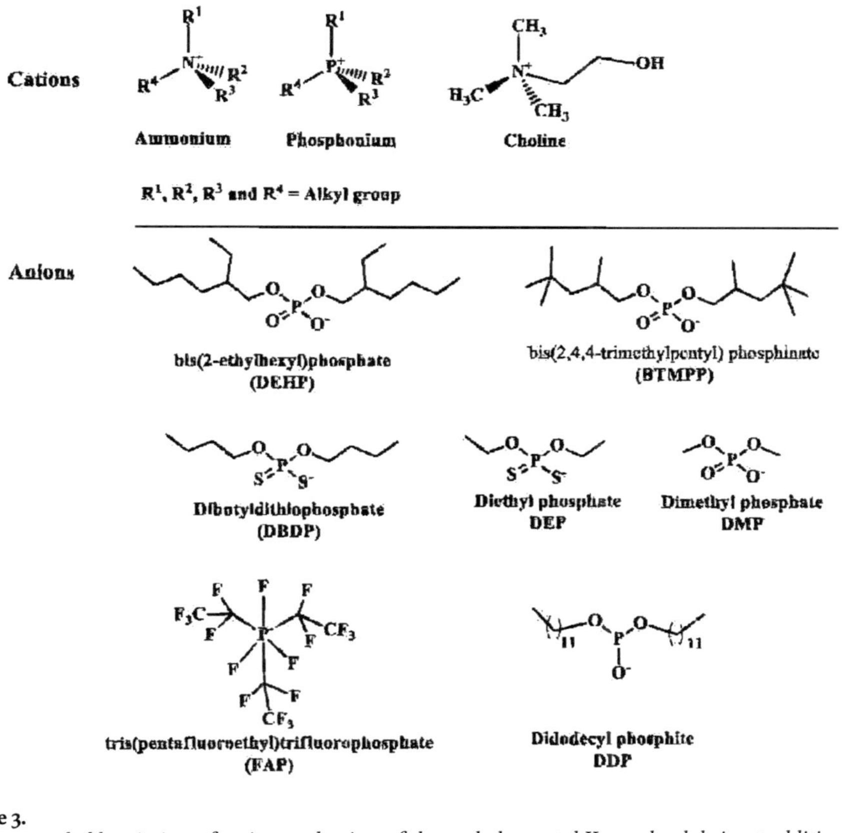

Figure 3.
Structures and abbreviations of cations and anions of the nonhalogenated ILs used as lubricant additives.

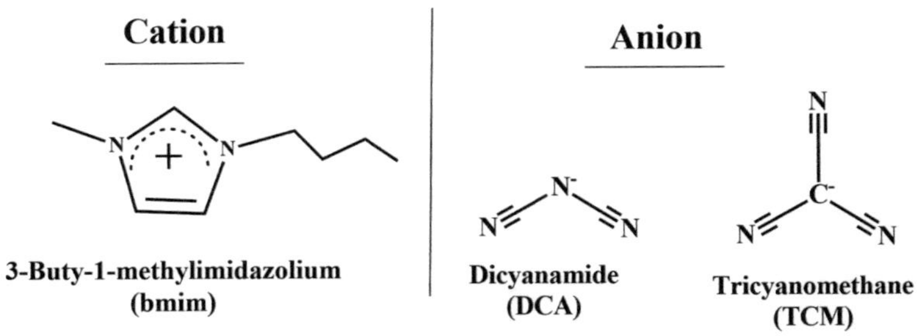

Figure 4.
Structures and abbreviations of cations and anions of the carbon-nitrogen atom based ILs used as lubricant additives.

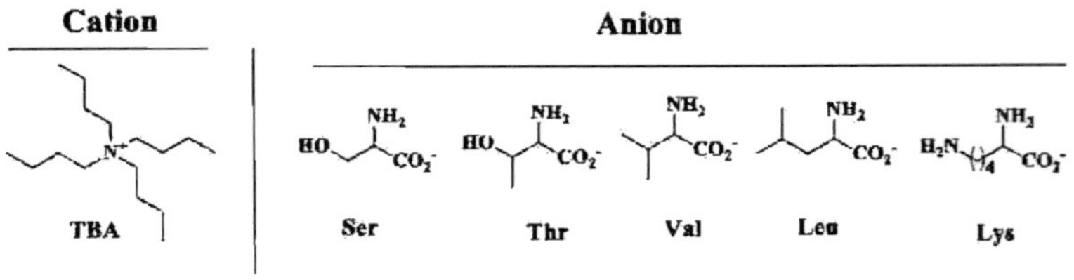

Figure 5.
Structures and abbreviations of cations and anions of amino acid ILs used as lubricant additives.

occurred because the chemical reaction film was mainly composed of the elements of the halogen-free ILs [54, 55]. The appearance of the chemical reaction film is similar to reported literature for tribo-films originating from zinc dialkyldithio-phosphates (ZDDP) [56–59]. Additional analysis of chemical reaction film is also needed to identify the species generated on the steel surface. On the other hand, the chemical reaction film formation is not observed on the DLC surfaces. As DLC films have high chemical stability, the inhibition of the chemical interaction between the DLC surfaces and the halogen-free ILs is observed. The bmimDCA has showed better reducing frictional properties than bmimTCM for the steel-steel tribo-pair, whereas bmimTCM has showed better reducing frictional properties than bmim-DCA for the DLC-DLC tribo-pairs. For explain the above phenomena, different lubrication mechanism is employed for DLC-DLC and steel–steel tribo-pairs [60].

A new family of green fluid lubricants (AAILs) have been designed for the lubrication of steel/steel, steel/copper and steel/aluminum contacts at room temperature (**Figure 5**). These AAILs can be obtained by simply neutralizing amino acids, which can be easily obtained in large quantities at low cost with the corresponding onium hydroxide. Use of natural amino acids as component ions makes the AAILs environmentally friendly with good biodegradability and reduced toxicity, making the AAILs as good potential green lubricants. The degree of hydrolysis of these AAILs are much higher than that of bmimBF_4 and the anti-corrosion properties of the AAILs are also far better than bmimBF_4 and hmimNTf_2, due to their halogen-free characters. The tribological properties of the AAILs (**Figure 5**) have been tested on steel-steel contacts as steel is the most widely used material in various machines in our everyday life. Generally, AAILs produce a lower friction coefficient value than hmimNTf_2 and prove the better friction-reducing performances where commercial oil PAO and a conventional IL hmimNTf_2 are chosen for comparison purposes.

From experimental results, the wear volume losses of the steel discs lubricated by all AAILs are lower than that of the hmimNTf_2 but higher than that of the PAO. The anti-wear properties of the AAILs should be improved compared with PAO. For improving anti-wear properties of the AAILs, [TBA][Ser] and [TBA][Thr] have been synthesized and exhibit the higher anti-wear properties due to attribution of their anionic moiety. Due to presence of hydroxyl groups in the anion structures of [Ser] and [Thr] effective protection films is formed on the metal surfaces. The effect of hydrogen bonding in [TBA][Ser] and [TBA][Thr] ILs provides effective separation of the steel surfaces, further reducing friction and wear. Besides, from an application point of view, [TBA][Ser] and [TBA][Thr] are more useful as lubricants than hmimNTf$_2$, because of the lower cost associated with their preparation, and their intrinsic environmentally friendly characters.

The AAILs are used to lubricate Cu alloys to change in friction coefficients and the wear volume losses of the copper discs under lubrication process [61, 62]. The evolution of friction coefficients shows that hmimNTf$_2$ and [TBA][Leu] start at a moderately high value and then tend to become lower and more stable. From experimental results, AAILs have good lubricating effects for steel/copper contact [62].

The lubrication of aluminum alloys has shown relatively poor wear-resistance, makes them especially difficult to be lubricated at a modest load [63]. It is also observed that the halogen-containing IL (like hmimNTf$_2$) is not an efficient lubricant for aluminum, and that severe wear may be caused by its tribo-corrosion during the sliding process [64]. On the contrary, the AAILs are effective lubricants for aluminum alloy, and their tribological properties are comparable to PAO.

The friction-reducing and anti-wear mechanism of the AAILs are explored by XPS analysis. However, characteristic peaks of N1s, which provide important information regarding the occurrence of a tribochemical reaction on a metal surface, are not detected. Besides, the lubricated metal surfaces by the AAILs and nonlubricated surface have shown similar binding energies of C1s, O1s, Fe2p, Cu2p and Al2p [65]. An AAIL adsorbed layer is formed via adsorption of cations and amino acid anions through an electrostatic attraction. The physical adsorption films by several AAIL adsorbed layer prevent close contact of metal–metal and further reduce the friction and wear on metal–metal surface [65]. The AAILs can substitute PAO and especially halogen containing ILs for use as neat lubricants for metal–metal contact. Additionally, the environmentally friendly and outstanding anti-corrosion properties of the AAILs also confirm that they are suitable for the lubrication of metal–metal surface contact.

Boron containing ILs are also class of non-halogenated ILs (**Figure 6**). Recently, boron containing ILs are reported as an efficient lubricant and additive [66, 67]. Development of halogen-free orthoborate anions based phosphonium ILs has been attracted research for tribological studies [68]. It is also reported that the boron constituted materials are well-known for exhibiting excellent friction-reducing

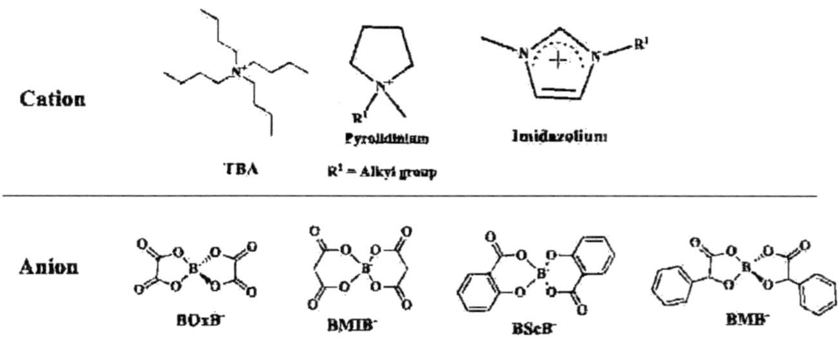

Figure 6.
Structures and abbreviations of cations and anions of boron based ILs used as lubricant additives.

and antiwear properties [69–72]. Boron-containing ILs have been attracted great interest in recent time. Chelated orthoborate anion $[BScB]^-$ with different cations provides large number of ILs [73]. With Cation $bmim^+$, $[BScB]^-$ has shown lower friction than $[FAP]^-$ or $[DBP]^-$, but $[DBP]^-$ has shown the most wear reduction [74]. Further, wear and friction are significantly reduced when $[BScB]^-$ anion paired with dicationic $[bis(imidazolium)]^{2+}$ and $[bis\text{-}(ammonium)]^{2+}$ [75]. For cation $[TBA]^+$, anions $[BScB]^-$, $[BMIB]^-$, and $[BOxB]^-$ show 50% or more in wear reduction under similar testing conditions [73–76].

The scope of chelated orthoborate anions based ILs are further extended with imidazolium, bis-imidazolium and pyrrolidinium cations for their application in tribological studies. The ILs with aromatic and aliphatic structures (**Figure 6**) which are reported recently with an aim to probe their structural effects on corrosion and tribo-physical properties compared with the halogenated analogue TBA-BF_4 [73]. It is also observed that TBA-BMdB, TBA-BOxB and TBA-BScB ILs exhibited higher thermal stability due to the presence of aromatic rings in their chelated structure and presence of various intermolecular interactions and rigidity to their anionic moieties.

Presence of halogen, phosphorus, and sulfur constituent components in the lubricant system facilitates the corrosion events and damages the engineering surfaces. Khatri and co-worker have investigated the corrosion property of boron based ILs (**Figure 6**) probed by copper strip test meth by optical and electron microscopic techniques [73]. It is also reported that the copper strip, exposed to TBA-BOxB, exhibited corrosion pits distributed throughout the substrate. The surface features of copper strips remain intact without any damage, exposed to TBA-BMdB, TBA-BScB and TBA-ILs. These experimental results suggested that TBA-BMdB, TBA-BScB and TBA-BMlB ILs (halogen-free), do not corrode the copper strips surface, whereas, presence of fluorine in TBA-BF_4, corrosive events on copper strips surface are facilitated. Furthermore, TBA-BOxB IL has poor thermal stability and its decomposed acidic (oxalic acid) product leads to corrosive events. As a result, TBA-BOxB showed higher friction and WSD compared to other chelated orthoborate ILs. Most of chelated orthoborate ILs has shown noncorrosive properties and can be tested for their lubrication properties.

Among all boron based ILs (**Figure 6**), maximum antiwear property is achieved by TBA-BMdB IL due to compact, rigid and stable structure of BMdB anion. To understand the effect of halogen, the friction and wear properties of fluorine constituted TBA-BF_4 ILs are examined under identical condition. It is observed that TBA-BF_4 has showed poorer tribo performance compared to the chelated orthoborate ILs. Poor tribo-performance and corrosion results suggest that corrosive products generation by BF_4 anion constituted ILs could be further facilitated by trapped water molecules in the lubricant [28].

The exact mechanism and role of boron based ILs in tribo-chemical thin film formation is believed to be complex because of their inherent polarity. Recently, Oganov et al. have revealed that boron containing ILs can generate partial negative charge and facilitate the interaction of chelated orthoborate anions with steel surfaces and forms the tribo-thin film under the high pressure [77]. Usually, under the tribo-stress, the positive charge is induced on metal surfaces. Chelated orthoborate anions are adsorbed on induced positive charge surface with counter cations. The layering structure on metal surface is formed through electrostatic attractions and generates the physico-chemically adsorbed tribo-thin film [78]. Furthermore, the very hard nature of boron is understood to provide durable tribo-thin film, which protects the steel interfaces and reduces wear significantly.

It has been suggested that the dangling bonds of carbon atoms on the metal surface are terminated by lubricant additives or the decomposition of lubricant additives and the formation of a monomolecular layer, which results in ultralow friction. These results suggest that an adsorbed film derived from the halogen-free ILs formed on the surfaces, which led to the ultralow friction. Moreover, a soft, thin layer on hard substrate materials is important for achieving an ultralow friction regime under boundary lubrication in accordance with the adhesion theory of friction [60]. The tribo-chemical thin film developed by chemical interaction of ILs and their decomposed products with steel interfaces could be an alternate to justify the tribo-mechanism [6, 10, 13]. Comparison of halogenated and non-halogenated ILs with conventional lubricants is listed in **Table 1** for better understanding the utility of ILs as lubricant.

Oil-soluble ILs, when used as lubricant additives, have repeatedly exhibited effective wear and friction reductions in tribological bench tests and demonstrated improved engine mechanical efficiency in engine dynamometer tests. The lubricating performance has shown a strong correlation with the ILs chemistry, concentration, compatibility with other oil additives, material compositions of the contact surfaces, and rubbing conditions. While some results simply showed improvement over the base oils, others have direct comparisons with commercial antiwear additives. Phosphonium based ILs with halogenated and non-halogenated anions are also used as additive for different contact surfaces [14]. Further, tribological study

Lubricants	COF	Wear	Contact	Reference
$emimBF_4$	0.56	$3.11x10^{-3}$ mm^3/m	Titanium-Steel	[79]
$bmimBF_4$	0.17	$0.02x10^{-3}$ mm^3/m		
bmimCl	0.17	$0.02x10^{-3}$ mm^3/m		
$hmimPF_6$	0.19	$0.08x10^{-3}$ mm^3/m		
$omimBF_4$	0.18	$0.1x10^{-3}$ mm^3/m		
Mineral Oil	0.45	$1.9x10^{-3}$ mm^3/m		
$hmimPF_6$	0.065	$9.3x10^{-3}$ mm^3/m	Steel-Steel	[80]
PAO	0.105	$9x10^{-3}$ mm^3/m		
$bmimBF_4$	0.045	$230x10^{-9}$ mm^3/Nm	Copper-Si_3N_4	[81]
Diesel oil	0.07	$210x10^{-9}$ mm^3/Nm		
$bmimBF_4$	0.041	$73.1x10^{-9}$ mm^3/Nm	Steel-Si_3N_4	
Diesel oil	0.105	$80.2x10^{-9}$ mm^3/Nm		
$bmimBF_4$	0.035	$75x10^{-9}$ mm^3/Nm	Crystalline Cr- Si_3N_4	
Diesel oil	0.075	$34x10^{-9}$ mm^3/Nm		
$(C_8H_{17})_3NHNTf_2$	0.05	$29.1x10^{-9}$ mm^3/Nm	Engine inner ring	[5]
$dmimNTf_2$	0.07	$24.5x10^{-9}$ mm^3/Nm		
Mineral Oil	0.11	$44.8x10^{-9}$ mm^3/Nm		
15w40 Engine oil	0.11	$36.9x10^{-9}$ mm^3/Nm		
$hmimPF_6$	0.085	$3x10^{-9}$ mm^3/Nm	Nickel-Steel	[82]
$omimPF_6$	0.1	$9x10^{-9}$ mm^3/Nm		
PFPE	0.145	$37x10^{-9}$ mm^3/Nm		
DSa	0.3	$0.26x10^{-9}$ mm^3/Nm	Copper-Copper	[83]
PAO	0.1	$4.54x10^{-9}$ mm^3/Nm		

Table 1.
Comparison of ionic liquids (ILs) and conventional lubricants.

Lubricants	COF	Wear	Contact	Reference
PAO	0.14	38.5×10^{-7} mm^3/Nm	Cast iron–steel	[46]
PAO@1.67% amine-phosphate	0.1	9×10^{-7} mm^3/Nm		
PAO@ 0.75% [P_{4444}][DEHP]	0.11	2.5×10^{-7} mm^3/Nm		
PAO@1.03% [P_{66614}][DEHP]	0.08	13×10^{-7} mm^3/Nm		
PAO@1.65% [P_{66614}][i-C_7H_{15}COO]	0.11	4×10^{-7} mm^3/Nm		
PAO@1.98% [P_{66614}][n-$C_{17}H_{35}$COO]	0.08	3×10^{-7} mm^3/Nm		
PAO@ 2.44% [P_{66614}][RSO_3]	0.11	7×10^{-7} mm^3/Nm		
PEG-200	0.12	730 mm (wear scar)	Steel–steel	[75]
PEG-200@ 1% MIm5-$(BScB)_2$	0.07	360 mm (wear scar)		
PEG-200@ 2% MIm5-$(BScB)_2$	0.07	330 mm (wear scar)		
PEG-200@ 3% MIm5-$(BScB)_2$	0.07	335 mm (wear scar)		
PAO	0.22	4.9×10^{-4} mm^3/Nm	Cast iron–steel	[11]
PAO@ 5% P_{66614} DEHP	0.1	5.6×10^{-7} mm^3/Nm		
5 W-30 engine oil	0.1	4.7×10^{-7} mm^3/Nm		
5 W-30 engine oil @ 5% P_{66614} DEHP	0.1	1.3×10^{-7} mm^3/Nm		
10 W base oil	>0.3	490×10^{-7} mm^3/Nm	Cast iron–steel	[12]
10 W C 5% PP-IL	0.09	4.7×10^{-7} mm^3/Nm		
10 W-30 engine oil	0.1	9×10^{-7} mm^3/Nm		
10 W-30 engine oil C 5% PP-IL	0.11	2.5×10^{-7} mm^3/Nm		

Table 2.
Tribological properties of ILs as lubricant additives.

of oil-miscible quaternary ammonium phosphites ILs as Lubricant additives in PAO is also investigated in different surface environment and shows efficient reduction of wear [53]. Biodegradable fatty-acid-constituted halogen-free ILs are efficient for renewable, environmentally friendly, and high-performance lubricant additives [76]. Halogen-free imidazolium/Ammonium-bis(salicylato)borate ILs act as high-performance lubricant additives and lower wear values on metal surfaces [74]. For better understanding the utility of ILs as lubricant additive in oils, COF and wear properties are for few ILs and listed in **Table 2**.

4. Conclusion

For ILs as lubrication, the major concerns included corrosion, thermal oxidation, oil-miscibility, toxicity, and cost. The recent successful development of noncorrosive, thermally stable, and oil-soluble ILs has largely been addressed and discussed in technical barriers and application point of views. The mainstream research of IL involved lubrication has been shifted from using ILs as neat or base lubricants to using them as lubricant additives. The development of ILs as new lubricating systems are encouraging and still challenging issues in present day. There must be considered the disintegration and corrosion problems of ILs related to their applications as lubricant. However, these fundamental issues can help us to the understanding of fundamental mechanisms of tribology. Now, the focus is to develop halogen and phosphorus-free ILs as energy efficient and environment-friendly lubricant additives for the steel-based engineering surfaces, and to establish the correlation between structure of anion and tribo-physical properties of ILs. Halogen free ILs (mainly borate based ILs) are more important for application as lubricant in near future.

Acknowledgements

SKP acknowledges Department of Chemistry, Uka Tarsadia University, Maliba Campus, Gopal Vidyanagar, Bardoli, Mahuva Road, Surat-394350, Gujrat, India. Conflict of Interest.

Conflict of interest

The author confirms that he has no conflict of interest to declare for this publication.

Acronyms and abbreviations

AAILs	amino acid ionic liquids
AW	anti-wear
BF_4	tetrafluoroborate
bmim	1-butyl-3-methylimidazolium
BMP	1-butyl-1-methylpyrrolidinium
BScB	bis(salicylato)borate
BTAG3	methoxy tris-ethoxy methylene benzotriazole
BTMPP	bis(2,4,4-trimethylpentyl) phosphinate
COF	coefficient of friction
DEHP	bis(2-ethylhexyl)phosphate
DLC	diamond Like Carbon
DOSS	dioctyl sulfosuccinate
DOP	dioctyl phosphite
DEHP	bis(2-ethylhexyl)phosphate
emim	1-ethyl-3-methylimidazolium
FAP	tris(pentafluoroethyl)trifluorophosphate
hmim	1-hexyl-3-methylimidazolium
ILs	ionic liquids
Leu	leucine
Lys	lysine
MO	mineral oil
MIm5	1,1′-(pentane-1,5-diyl)bis(3-methylimidazolium)
MMIm5	1,1′-(pentane-1,5-diyl)bis(2,3-dimethylimidazolium)
PAO	poly-α-olefin
PE	polyester
PEG	poly(ethylene glycol)
PF_6	hexafluorophosphate
POE	polyolester
PAO	poly-α-olefin
Ser	serine
Thr	threonine

Val	valine
$Tf_2N/NTf_2/TFSI$	bis(trifluoromethanesulfonyl)imide
[TBA][Ser]	tetrabutylammonium serine
[TBA][Thr]	tetrabutylammonium threonine
[TBA][Val]	tetrabutylammonium valine
[TBA][Leu]	tetrabutylammonium leucine
[TBA][Lys]	tetrabutylammonium lysine
[TBA][OH]	tetrabutylammonium hydroxide
$[P_{4,4,4,8}]$	tributyloctylphosphonium cation
$[P_{4,4,4,14}]$	tributyltetradecylphosphonium cation
$[P_{6,6,6,14}]$	trihexyltetradecylphosphonium cation
TMP	trimethylolpropane
VO	vegetable oil
XPS	x-ray photoelectron spectrometry
ZDDP	zinc dialkyldithiophosphate

Author details

Sumit Kumar Panja
Department of Chemistry, Uka Tarsadia University, Maliba Campus, Gopal Vidyanagar, Bardoli, Mahuva Road, Surat-394350, Gujrat, India

*Address all correspondence to: sumitpanjabhu@gmail.com; sumit.panja@utu.ac.in

References

[1] Carnes, K. The Ten Great Events in Tribology History. *Tribol. Lubr. Technol.* **2005**, 61, 38-39.

[2] Carpick, R. W.; Jackson, A.; Sawyer, W. G.; Argibay, N.; Lee, P.; Pachon, A.; Gresham, R. M. The Tribology Opportunities Study: Can Tribology Save a Quad? *Tribol. Lubr. Technol.* **2016**, 72, 44.

[3] Shah, F. U.; Glavatskih, S.; Antzutkin, O. N. Boron in Tribology: From Borates to Ionic Liquids. *Tribol. Lett.*, **2013**, 51, 281-301. [https://doi.org/10.1007/s11249-013-0181-3]

[4] Ye, C. F.; Liu, W. M.; Chen, Y. X.; Yu, L. G. Room-Temperature Ionic Liquids: A Novel Versatile Lubricant. *Chem. Commun.* **2001**, 2244-2245. [https://doi.org/10.1039/B106935G]

[5] Qu, J., Blau, P. J., Dai, S., Luo, H., and Meyer III, H. M. Ionic Liquids as Novel Lubricants and Additives for Diesel Engine Applications. *Tribol. Lett.*, **2009**, 35, 181-189. [https://doi.org/10.1007/s11249-009-9447-1]

[6] Minami, I. Ionic Liquids in Tribology, *Molecules*, **2009**, 14 (6), 2286-2305. [https://doi.org/10.3390/molecules14062286]

[7] Zhou, F.; Liang, Y.; Liu, W. Ionic liquid lubricants: designed chemistry for engineering applications. *Chem. Soc. Rev.*, **2009**, 38, 2590-2599. [https://doi.org/10.1039/B817899M].

[8] Palacio, M.; Bhushan, B. A Review of Ionic Liquids for Green Molecular Lubrication in Nanotechnology. *Tribol. Lett.* **2010**, 40, 247-268. [https://doi.org/10.1007/s11249-010-9671-8]

[9] Perkin, S. Ionic Liquids in Confined Geometries. *Phys. Chem. Chem. Phys.* **2012**, 14, 5052-5062. [https://doi.org/10.1039/C2CP23814D]

[10] Somers, A. E.; Howlett, P. C.; Macfarlane, D. R.; Forsyth, M. A. A Review of Ionic Liquid Lubricants, *Lubricants*, **2013**, 1, 3-21. [https://doi.org/10.3390/lubricants1010003].

[11] Qu, J.; Bansal, D. G.; Yu, B.; Howe, J. Y.; Luo, H. M.; Dai, S.; Li, H. Q.; Blau, P. J.; Bunting, B. G.; Mordukhovich, G.; Smolenski, D. J. Antiwear Performance and Mechanism of an Oil-Miscible Ionic Liquid as a Lubricant Additive. *ACS Appl. Mater. Interfaces* **2012**, 4 (2), 997-1002. [https://doi.org/10.1021/am201646k]

[12] Yu, B.; Bansal, D. G.; Qu, J.; Sun, X. Q.; Luo, H. M.; Dai, S.; Blau, P. J.; Bunting, B. G.; Mordukhovich, G.; Smolenski, D. J. Oil-Miscible and Non-Corrosive Phosphonium-Based Ionic Liquids as Candidate Lubricant Additives. *Wear* **2012**, 289, 58-64. [https://doi.org/10.1016/j.wear.2012.04.015]

[13] Bermudez, M. D.; Jimenez, A. E.; Sanes, J.; Carrion, F. J. Ionic Liquids as Advanced Lubricant Fluids. *Molecules*, **2009**, 14, 2888-2908. [https://doi.org/10.3390/molecules14082888]

[14] Zhou, Y.; Qu, J. Ionic Liquids as Lubricant Additives: A Review, *ACS Appl. Mater. Interfaces* **2017**, 9, 3209-3222. [https://doi.org/10.1021/acsami.6b12489]

[15] Welton, T. Room-Temperature Ionic Liquids. Solvents for Synthesis and Catalysis. *Chem. Rev.* 1999, 99 (8), 2071-2084. [https://doi.org/10.1021/cr980032t]

[16] Zhang, C. L.; Zhang, S. M.; Yu, L. G.; Zhang, P. Y.; Zhang, Z. J. Tribological Behavior of 1-Methyl-3-hexadecylimidazolium Tetrafluoroborate Ionic Liquid Crystal as a Neat Lubricant and as an Additive of Liquid Paraffin. *Tribol. Lett.*, **2012**,

46, 49-54. [https://doi.org/10.1007/s11249-012-9917-8]

[17] Qu, J.; Truhan, J. J.; Dai, S.; Luo, H.; Blau, P. J. Ionic liquids with ammonium cations as lubricants or additives. *Tribol. Lett.*, **2006**, 22, 207-214. [https://doi.org/10.1007/s11249-006-9081-0]

[18] Anand, M.; Hadfield, M.; Viesca, J. L.; Thomas, B.; Hernández Battez, A.; Austen, S. Ionic liquids as tribological performance improving additive for in-service and used fully-formulated diesel engine lubricants, *Wear* **2015**, 334-335, 67-74. [http://dx.doi.org/10.1016/j.wear.2015.01.055].

[19] Hernández Battez, A.; González, R.; Viesca, J. L.; Blanco, D.; Asedegbega, E.; Osorioa. A. *Wear* **2009**, 266, 1224-1228. [https://doi.org/10.1016/j.wear.2009.03.043]

[20] Yu G., Yan S., Zhou F., Liu X., Liu W., Liang Y. Synthesis of dicationic symmetrical and asymmetrical ionic liquids and their tribological properties as ultrathin films. *Tribol. Lett.* **2007**, 25, 197-205. [https://doi.org/10.1007/s11249-006-9167-8].

[21] Fox, M.F.; Priest M. Tribological properties of ionic liquids as lubricants and additives. Part 1: synergistic tribofilm formation between ionic liquids and tricresyl phosphate. *Proc. Inst. Mech. Eng., Part J: J. Eng. Tribol.* **2008**, 222, 291-303. [https://doi.org/10.1243%2F13506501JET387]

[22] Phillips, B. S.; John, G.; Zablinski, J. S. Surface chemistry of fluorine containing ionic liquids on steel substrates at elevated temperature using Mössbauer spectroscopy. *Tribol. Lett.*, **2007**, 26, 85-91. [https://doi.org/10.1007/s11249-006-9020-0].

[23] Uerdingen, M.; Treber, C.; Baser, M.; Schmitt, G.; Werner, C. Corrosion behaviour of ionic liquids. *Green Chem.*, **2005**, 7, 321-325. [https://doi.org/10.1039/B419320M]

[24] Pisarova, L.; Gabler, C.; Dorr, N.; Pittenauer, E.; Allmaier, G. Thermo-oxidative stability and corrosion properties of ammonium based ionic liquids. *Tribol. Int.*, **2012**, 46, 73-83. [https://doi.org/10.1016/j.triboint.2011.03.014]

[25] Jimenez, A. E.; Bermudez, M. D.; Iglesias, P.; Carrion, F. J.; Martınez-Nicolas, G. "1-N-alkyl-3-methylimidazolium Ionic Liquids as Neat Lubricants and Lubricant Additives in Steel–Aluminium Contacts. *Wear*, **2006**, 260, 766-782. [https://doi.org/10.1016/j.wear.2005.04.016]

[26] Kamimura, H.; Kubo, T.; Minami, I.; Mori, S. Effect and mechanism of additives for ionic liquids as new lubricants. *Tribol. Int.*, **2007**, 40, 620-625. [https://doi.org/10.1016/j.triboint.2005.11.009]

[27] Yao, M.; Liang, Y.; Xia, Y.; Zhou, F. Bisimidazolium Ionic Liquids as the High-Performance Antiwear Additives in Poly(ethylene glycol) for Steel-Steel Contacts. *ACS Appl. Mater. Interfaces.* **2009**, 1, 467-471. [https://doi.org/10.1021/am800132z]

[28] Bubalo, M. C.; Rado sevic, K.; Redovnikovic, I. R.; Halambek, J.; Gaurina Srcek, V. A Brief Overview of the Potential Environmental Hazards of Ionic Liquids. *Ecotoxicol. Environ. Saf.*, **2014**, 99, 1-12. [https://doi.org/10.1016/j.ecoenv.2013.10.019]

[29] Liu, W.; Ye, C.; Gong, Q.; Wang, H.; Wang, P. Tribological Performance of Room-Temperature Ionic Liquids as Lubricant. *Tribol. Lett.*, **2002**, 13, 81-85. [https://doi.org/10.1023/A:1020148514877]

[30] Zeng, Q.; Zhang, J.; Cheng, H.; Chena, L.; Qia, Z. Corrosion properties of steel in 1-butyl-3-methylimidazolium hydrogen sulfate ionic liquid systems for desulfurization application. *RSC*

Adv., **2017**, 7, 48526-34856. [https://doi.org/10.1039/C7RA09137K]

[31] Kumar, A., Srivastava, S., Gupta, G., Kumara, P., and Sarkar, J. Functional Ionic Liquid [Bmim][Sac] Mediated Synthesis of Ferrocenyl Thiopropanones via the 'Dual Activation of the Substrate by the Ionic Liquid. *RSC Adv.*, **2013**, 3, 3548-3552. [https://doi.org/10.1039/C3RA22543G]

[32] Hough-Troutman, W. L.; Smiglak, M.; Griffin, S.; Reichert, W. M.; Mirska, I.; Jodynis-Liebert, J.; Adamska, T.; Nawrot, J.; Stasiewicz, M.; Rogers, R. D.; Pernak, J. Ionic Liquids with Dual Biological Function: Sweet and Anti-Microbial, Hydrophobic Quaternary Ammonium–Based Salts, *New J. Chem.*, **2009**, 33, 26-33. [https://doi.org/10.1039/B813213P]

[33] Somers, A.; Howlett, P.; MacFarlane, D.; Forsyth, M.; A review of ionic liquid lubricants. *Lubricants*, **2013**, 1, 3-21. [http://dx.doi.org/10.3390/lubrica-nts1010003.

[34] Canter, N. Evaluating ionic liquids as potential lubricants. *Tribol. Lubr. Technol*. **2005**, 61, 15-17.

[35] He, L.; Tao, G. H.; Parrish, D. A.; Shreeve, J. M. Slightly Viscous Amino Acid Ionic Liquids: Synthesis, Properties, and Calculations. *J. Phys. Chem. B*, **2009**, 113, 15162-15169. [https://doi.org/10.1021/jp905079e]

[36] Viciosa, M. T.; Santos, G.; Costa, A.; Dan ede, F.; Branco, L. C.; Jordao, N.; Correia, N. T.; Dionisio, M. Dipolar Motions and Ionic Conduction in an Ibuprofen Derived Ionic Liquid. *Phys. Chem. Chem. Phys*. **2015**, 17, 24108-24120. [https://doi.org/10.1039/C5CP03715H]

[37] Sintra, T. E.; Shimizu, K.; Ventura, S. P. M.; Shimizu, S.; Canongia Lopes, J. N.; Coutinho, J. A. P. Enhanced Dissolution of Ibuprofen Using Ionic Liquids as Catanionic Hydrotropes. *Phys. Chem. Chem. Phys.* **2018**, 20, 2094-2103. [https://doi.org/10.1039/C5CP03715H]

[38] Hough, W. L.; Smiglak, M.; Rodrıguez, H.; Swatloski, R. P.; Spear, S. K.; Daly, D. T.; Pernak, J.; Grisel, J. E.; Carliss, R. D.; Soutullo, M. D.; Davis, J. H.; Rogers, R. D. The Third Evolution of Ionic Liquids: Active Pharmaceutical Ingredients. *New J. Chem.*, **2007**, 31, 1429-1436. [https://doi.org/10.1039/B706677P]

[39] Zhao, Y. S.; Zhao, J. H.; Huang, Y.; Zhou, Q.; Zhang, X. P. Toxicity of Ionic Liquids: Database and Prediction via Quantitative Structure–Activity Relationship Method. *J. Hazard. Mater.*, **2014**, 278, 320-329. [https://doi.org/10.1016/j.jhazmat.2014.06.018]

[40] Rogers, R. D. and Seddon. K. R. (Eds.) (2005), Ionic Liquids IIIB: Fundamentals, Progress, Challenges, and Opportunities—Transformations and Processes, ACS Symposium Series 902, American Chemical Society: Washington, DC.

[41] Fan, M. J.; Zhang, C. Y.; Guo, Y. N.; Zhang, R. R.; Lin, L. B.; Yang, D. S.; Zhou, F.; Liu, W. M. DOSS-Based QAILs: As Both Neat Lubricants and Lubricant Additives with Excellent Tribological Properties and Good Detergency. *Ind. Eng. Chem. Res.*, **2016**, 53, 17952-17960. [https://doi.org/10.1021/ie502849w]

[42] Battez, A. H.; Bartolome, M.; Blanco, D.; Viesca, J. L.; Fernandez-Gonzalez, A.; Gonzalez, R. Phosphonium Cation–Based Ionic Liquids as Neat Lubricants: Physicochemical and Tribological Performance. *Tribol. Int.*, **2016**, 95, 118-131. [https://doi.org/10.1016/j.triboint.2015.11.015]

[43] Shah, F. U.; Glavatskih, S.; MacFarlane, D. R.; Somers, A.; Forsyth, M. Novel Halogen-Free Chelated

Orthoborate–Phosphonium Ionic Liquids: Synthesis and Tribophysical Properties. *Phys. Chem. Chem. Phys.* **2011**, 13, 12865-12873. [https://doi.org/10.1039/C1CP21139K].

[44] Barnhill, W. C.; Luo, H.; Meyer, H. M., III; Ma, C.; Chi, M.; Papke, B. L.; Qu, J. Tertiary and Quaternary Ammonium-Phosphate Ionic Liquids as Lubricant Additives. *Tribol. Lett.* **2016**, 63, 22-33. [https://doi.org/10.1007/s11249-016-0707-6]

[45] Barnhill, W. C.; Qu, J.; Luo, H. M.; Meyer, H. M.; Ma, C.; Chi, M. F.; Papke, B. L. Phosphonium-Organophosphate Ionic Liquids as Lubricant Additives: Effects of Cation Structure on Physicochemical and Tribological Characteristics. *ACS Appl. Mater. Interfaces* **2014**, 6, 22585-22593. [https://doi.org/10.1021/am506702u]

[46] Zhou, Y.; Dyck, J.; Graham, T. W.; Luo, H. M.; Leonard, D. N.; Qu, J. Ionic Liquids Composed of Phosphonium Cations and Organophosphate, Carboxylate, and Sulfonate Anions as Lubricant Antiwear Additives. *Langmuir* **2014**, 30, 13301-13311. [https://doi.org/10.1021/la5032366]

[47] González, R.; Bartolomé, M.; Blanco, D.; Viesca, J. L.; Fernández-González, A.; Battez, A. H. Effectiveness of Phosphonium Cation-Based Ionic Liquids as Lubricant Additive. *Tribol. Int.* **2016**, 98, 82-93. [https://doi.org/10.1016/j.triboint.2016.02.016]

[48] Qu, J.; Barnhill, W. C.; Luo, H. M.; Meyer, H. M.; Leonard, D. N.; Landauer, A. K.; Kheireddin, B.; Gao, H.; Papke, B. L.; Dai, S. Synergistic Effects between Phosphonium-Alkylphosphate Ionic Liquids and Zinc Dialkyldithiophosphate (ZDDP) as Lubricant Additives. *Adv. Mater.* **2015**, 27, 4767-4774. [https://doi.org/10.1002/adma.201502037]

[49] Qu, J.; Luo, H. M.; Chi, M. F.; Ma, C.; Blau, P. J.; Dai, S.; Viola, M. B. Comparison of an Oil-Miscible Ionic Liquid and ZDDP as a Lubricant Anti-Wear Additive. *Tribol. Int.* **2014**, 71, 88-97. [https://doi.org/10.1016/j.triboint.2013.11.010]

[50] Sharma, V.; Doerr, N.; Aswath, P. B. Chemical-Mechanical Properties of Tribo-films and Their Relationship to Ionic Liquid Chemistry. *RSC Adv.* **2016**, 6, 22341-22356. [https://doi.org/10.1039/C6RA01915C]

[51] Otero, I.; Lopez, E. R.; Reichelt, M.; Villanueva, M.; Salgado, J.; Fernandez, J. Ionic Liquids Based on Phosphonium Cations as Neat Lubricants or Lubricant Additives for a Steel/Steel Contact. *ACS Appl. Mater. Interfaces* **2014**, 6, 13115-13128. [https://doi.org/10.1021/am502980m]

[52] Zhu, L.; Dong, J.; Ma, Y.; Jia, Y.; Peng, C.; Li, W.; Zhang, M.; Gong, K.; Wang, X. Synthesis and investigation of halogen-free phosphonium-based ionic liquids for lubrication applications. *Tribol. Trans.*, **2019**, 62, 943-954. [https://doi.org/10.1080/10402004.2019.1609638].

[53] Fu, X. S.; Sun, L. G.; Zhou, X. G.; Li, Z. P.; Ren, T. H. Tribological Study of Oil-Miscible Quaternary Ammonium Phosphites Ionic Liquids as Lubricant Additives in PAO. *Tribol. Lett.* 2015, 60, 23-35. [https://doi.org/10.1007/s11249-015-0596-0]

[54] Qu, J.; Truhan, J. J.; Dai, S.; Luo, H.; Blau, P. J. Ionic liquids with ammonium cations as lubricants or additives. *Tribol. Lett.* **2006**, 22, 207-214. [https://doi.org/10.1007/s11249-006-9081-0]

[55] Minami, I.; Inada, T.; Sasaki, R.; Nanao, H. Tribo-Chemistry of Phosphonium-Derived Ionic Liquids. *Tribol. Lett.*, **2010**, 40, 225-235. [https://doi.org/10.1007/s11249-010-9626-0]

[56] Spikes, H. The History and Mechanisms of ZDDP. *Tribol. Int.*, **2004**, 17, 469-489. [https://doi.org/10.1023/B:TRIL.0000044495.26882.b5]

[57] Taylor, L. J.; Spikes, H. A. Friction-Enhancing Properties of ZDDP Antiwear Additive: Part I-Influence of ZDDP Reaction Films on EHD Lubrications. *Tribol. Trans.*, **2003**, 46, 303-309. [https://doi.org/10.1080/10402000308982631]

[58] Taylor, L. J.; Spikes, H. A. Friction-Enhancing Properties of ZDDP Antiwear Additive: Part 2-Friction and Morphology of ZDDP Reaction Films. *Tribol. Trans.*, **2003**, 46, 310-314. [https://doi.org/10.1080/10402000308982630]

[59] Topolovec-Miklozic, K.; Forbus, T. R.; Spikes, H. A. *Tribol. Trans.*, **2007**, 50, 328-335. [https://doi.org/10.1080/10402000701413505]

[60] Glaeser, W. Materials for Tribology, (1992), Amsterdam, The Netherlands, Elsevier.

[61] Liu, X. Q.; Zhou, F.; Liang, Y. M.; Liu, W. M. Benzotriazole as the additive for ionic liquid lubricant: one pathway towards actual application of ionic liquids. *Tribol. Lett.*, **2006**, 23, 191-196. [https://doi.org/10.1007/s11249-006-9050-7]

[62] Song, Z.; Liang, Y.; Fan, M.; Zhou, F.; Liu, W. Ionic liquids from amino acids: fully green fluid lubricants for various surface contacts. *RSC Adv.*, **2014**, 4, 19396-19402. [https://doi.org/10.1039/C3RA47644H]

[63] Qu, J.; Blau, P. J.; Dai, S.; Luo, H. M.; Meyer, H. M.; Truhan, J. J. Tribological characteristics of aluminum alloys sliding against steel lubricated by ammonium and imidazolium ionic liquids *Wear*, **2009**, 267, 1226-1231. [https://doi.org/10.1016/j.wear.2008.12.038]

[64] Jimenez, A. E.; Bermudez, M. D.; Carrion, F. J.; Martinez-Nicolas, G. Room temperature ionic liquids as lubricant additives in steel-aluminium contacts: Influence of sliding velocity, normal load and temperature. *Wear*, **2006**, 261, 347-359. [https://doi.org/10.1016/j.wear.2005.11.004]

[65] Minami, I.; Mori, S. Concept of molecular design towards additive technology for advanced lubricants. *Lubr. Sci.*, **2007**, 19, 127-149. [https://doi.org/10.1002/ls.37]

[66] Choudhary, R. B., Pande, P. P. Lubrication potential of boron compounds: an overview. *Lubr. Sci*. **2002**, 14, 211-222. [https://doi.org/10.1002/ls.3010140208]

[67] Oganov, A. R.; Chen, J.; Gatti, C.; Ma, Y.; Ma, Y.; Glass, C. W.; Liu, Z.; Yu, T.; Kurrakevych, O. O.; Solozhenko, V. L. Ionic high-pressure form of elemental boron. *Nature*, **2009**, 457, 863-867. [https://doi.org/10.1038/nature08164]

[68] Taher, M.; Ullah Shah, F.; Filippov, A.; de Baets, P.; Glavatskih, S.; Antzutkin, O. N. Halogen-free pyrrolidinium bis(mandelato)borate ionic liquids: some physicochemical properties and lubrication performance as additives to polyethylene glycol. *RSC Adv.*, **2014**, 4, 30617-30623. [https://doi.org/10.1039/C4RA02551B].

[69] Minami, I.; Inada, T.; Okada, Y. Tribological properties of halogen-free ionic liquids. *Proc. Inst. Mech. Eng. J. J. Eng. Tribol*. **2012**, 226, 891-902. [https://doi.org/10.1177%2F1350650112446276]

[70] Schneider, A.; Brenner, J.; Tomastik, C.; Franek, F. Capacity of selected ionic liquids as alternative EP/AW additive. *Lubr. Sci.*, **2010**, 22, 215-223. [https://doi.org/10.1002/ls.120]

[71] Shah, F. U.; Glavtskih, S.; Hogloud, E.; Lindberg, M.; Antzutkin, O. N. Interfacial Antiwear and Physicochemical Properties of Alkylborate-dithiophosphates. *ACS Appl. Mater. Interfaces*, **2011**, 3 (4) 956-968. [https://doi.org/10.1021/am101203t]

[72] Reeves, C. J.; Menezes, P. L.; Lovell, M. R.; Jen, T. C. The Size Effect of Boron Nitride Particles on the Tribological Performance of Biolubricants for Energy Conservation and Sustainability. *Tribol. Lett.*, **2013**, 51, 437-452. [https://doi.org/10.1007/s11249-013-0182-2]

[73] Gusain, R.; Khatri, O. P. Halogen-Free Ionic Liquids: Effect of Chelated Orthoborate Anion Structure on Their Lubrication Properties. *RSC Adv*. **2015**, 5, 25287-25294. [https://doi.org/10.1039/C5RA03092G]

[74] Gusain, R.; Singh, R.; Sivakumar, K. L. N.; Khatri, O. P. Halogen-Free Imidazolium/Ammonium-Bis(Salicylato)Borate Ionic Liquids as High Performance Lubricant Additives. *RSC Adv*. **2014**, 4, 1293-1301. [https://doi.org/10.1039/C3RA43052A]

[75] Gusain, R.; Gupta, P.; Saran, S.; Khatri, O. P. Halogen-Free Bis(Imidazolium)/Bis(Ammonium) Di[Bis(Salicylato)Borate] ionic liquids as energy-efficient and environmentally friendly lubricant additives. *ACS Appl. Mater. Interfaces* **2014**, 6, 15318-15328. [https://doi.org/10.1021/am503811t]

[76] Gusain, R.; Dhingra, S.; Khatri, O. P. Fatty-Acid-Constituted Halogen-Free Ionic Liquids as Renewable, Environmentally Friendly, and High-Performance Lubricant Additives. *Ind. Eng. Chem. Res*. **2016**, 55, 856-865. [https://doi.org/10.1021/acs.iecr.5b03347]

[77] Deshmukh, P.; Lovell, M.; Sawyer, W. G.; Mobley, A. On the friction and wear performance of boric acid lubricant combinations in extended duration operations. *Wear* 2006, 260, 1295-1304. [https://doi.org/10.1016/j.wear.2005.08.012].

[78] Perkin, S.; Albrecht, T.; Kein, J. Layering and shear properties of an ionic liquid, 1-ethyl-3-methylimidazolium ethylsulfate, confined to nano-films between mica surfaces. *Phys. Chem. Chem. Phys*., **2010**, 12, 1243-1247. [https://doi.org/10.1039/B920571C]

[79] Jimenez, A. E.; Bermudez, M. -D. Ionic Liquids as Lubricants of Titanium–Steel Contact. *Tribol. Lett.*, **2009**, 33, 111-126. [https://doi.org/10.1007/s11249-010-9633-1]

[80] Xia, Y.; Wang, S.; Zhou, F.; Wang, H.; Lin, Y.; Xu, T. "Tribological Properties of Plasma Nitrided Stainless Steel against SAE52100 Steel under Ionic Liquid Lubrication Condition," *Tribol. Int.*, **2006**, 39, 635-640. [https://doi.org/10.1016/j.triboint.2005.04.030]

[81] Zeng, Z.; Chen, Y.; Wang, D.; Zhang, J. Tribological Behaviors of Amorphous Cr Coatings Electrodeposited from Cr (III) Baths under Ionic Liquid Lubrication. *Electrochem. Solid-State Lett.*, **2007**, 10, D85–D87. [https://doi.org/10.1149/1.2746129]

[82] Xia, Y.; Wang, L.; Liu, X.; Qiao, Y. A Comparative Study on the Tribological Behavior of Nanocrystalline Nickel and Coarse-Grained Nickel Coatings under Ionic Liquid Lubrication. *Tribol Lett.*, **2008**, 30, 151-157. [https://doi.org/10.1007/s11249-008-9322-5]

[83] Espinosa, T.; Sanes, J.; Jim enez; A. -E.; Bermudez, M.-D. Protic Ammonium Carboxylate Ionic Liquid Lubricants of OFHC Copper, *Wear*, **2013**, 303, 495-509. [https://doi.org/10.1016/j.wear.2013.03.041]

Chapter 2

Turbulent Flow Fluid in the Hydrodynamic Plain Bearing to a Non-Textured and Textured Surface

Bendaoud Nadia and Mehala Kadda

Abstract

Hydrodynamic bearing are components that provide the guiding in rotation of rotating machines, such as turbines, the reactors. This equipment works under very severe operating conditions: high rotational speed and high radial load. In order to improve the hydrodynamic performance of these rotating machines, the industrialists specialized in the manufacture of hydrodynamic journal bearings, have designed a bearing model with its textured interior surface. The present work is a numerical analysis, carried out to observe the effect of a turbulent fluid flow in a non-textured and textured plain bearing and to thus to see the improvement of the hydrodynamic and tribological performances to a non- textured and textured surface of the plain bearing, under severe operating parameters. The rotational velocity varies from 11,000 to 21,000 rpm and radial load ranging from 2000 N to 9000 N. The numerical analysis is performed by solving the continuity equation of Navier-Stokes, using the finite volume method. The numerical results show that the most important hydrodynamic characteristics such as pressure, flow velocity of the fluid, friction torque, are significant for the textured plain bearing under rotational velocity of 21,000 rpm and radial load 10,000 N compared to obtained for a non-textured plain bearing.

Keywords: plain bearing, turbulent flow fluid, textured surface, pressure, friction torque

1. Introduction

Tribology is the science that studies the interactions of two surfaces in motion with respect to each other. It encompasses the associated technique and all of the friction and wear sectors, including lubrication. She studies the interactions between contact surfaces, but also those of solids, liquids and gases present between these surfaces, such as hydrodynamic plain bearings.

The hydrodynamic bearings allow the various parts of the mechanical devices to move easily while ensuring reliability that eliminates any risk of rupture or premature wear. When the operating conditions are severe (high or rapidly changing loads, high frequency of rotation), working under a turbulent regime (like the turbojet), it becomes difficult to achieve this double objective without the help of powerful digital prediction models.

Friction is one of the most answered physical phenomena in hydrodynamic bearings. This is the reason why a new concept of bearings was invented, the aim of which is to minimize the losses of material and energy linked to wear and friction; it is therefore to manufacture mechanical systems with textured surfaces to improve the efficiency and life of the machines. The aim of this study is to better predict the effect of tribological behavior as well as the effect of turbulent flow behavior in the textured and non-textured hydrodynamic bearing.

2. Turbulent flow effect in plain bearing

Constantinescu has developed the phenomenon of turbulence in lubrication between years 1962 and 1965 [1, 2], Elrod and Ng in 1967 [3–5], are presented a linearized turbulent lubrication theory based on eddy-viscosity concept of Boussinesq and Reichardt's formulation, including the treatment of turbulent shear and pressure gradient flows in thin films. This theory can be applied to the journal bearings by assuming that the turbulent flow field in the clearance space can be represented by the small perturbations on the turbulent Couette flow. The first studies on determining the Reynolds number, which expresses the ratio, changed inertial forces and viscous forces in the field of bearings, were made by Fantinos and colleagues [6].

In 2005 Braunetiere [7], show that a number of theories for the turbulent lubrication film exist which are based on various well-established models of turbulent flow. Solghar and Nassab (2013) [8] carry out a study in to assess the turbulent thermohydrodynamic (THD) performance characteristics of an axially grooved finite journal bearing [8, 9]. They are mentioned in their research that the bearing of the operating characteristics are significantly changed by increasing the Reynolds number.

3. Texture effect in plain bearing

At the moment, little is known about the effect of variations in the profile of the bushing and on its performance. Surface texturing is expected to make a significant contribution to future bearing technologies.

In 2011, Ivan Krupka and al [10] presents an experimental and numerical study on the superficial textures effect of the lubricated contact, for the transitional phase. This study is done in order to observe the lubricant film behavior between two surfaces of a disc coated with chrome and a steel ball. According to their study, they showed that lubricant produced from the micro-dents helps to separate rubbing surfaces.

Tala-Ighil, Fillon and Maspeyrot in 2011 [11] indicated the effect of textured area on the performances of a hydrodynamic journal bearing. They examined the texture location effect on the hydrodynamic performance hydrodynamic of the journal bearing. Their results show that the most important characteristics can be improved through an appropriate arrangement of the textured area on the contact surface.

In 2014, Pratibha and Chandreshkumar [12], present an experimental study on the effect of the bearing surface texture and the profile pressure distribution in hydrodynamic performance of journal bearing. Their study shows that with the increase of the radial loads and at the constant velocity, the increase of maximum pressure is significant in textured journal bearing, in contrast, this pressure is less important for a non-textured journal bearing and with the increase of velocity and at constant radial load.

In 2015 Zhang and al [13], present a numerical study of surface texturing for improving tribological properties of ultra-high molecular weight polyethylene. Ultra-high molecular weight polyethylene (UHMWPE). Smooth UHMWPE surfaces are used for total joint replacements; however, smooth surface contacts have been shown to be inadequate in friction reduction and/or anti-wear.

Uddin and Liu present in 2016 [14], present design and optimization of a texture shape (star-like) for to improve the tribological performance. The triangle form of the texture tends to reducing the friction. A star-like texture consisting of a series of triangular pikes is positioned around the texture center's proposed. The increasing theses triangular shape, produce the increases the film pressure and on the other hand the reduction of the friction.

In 2016, Shahab Hamdavi, H. H. Ya and T. V. V. L. N. Rao [15], presented a research on the surface texturing effect on hydrodynamic performance of journal bearings. The authors study the effect of partially textured surface of long journal bearing on the pressure distribution. The results show that, applying partial surface texture has a positive and remarkable effect on operating characteristics of the bearings.

In 2017, Sedlaček and al [16], studied the geometry effect and the sequence of the surface texturing process in contact on the tribological characteristics. They tested the behavior of surfaces with and without hard coating for different textures shapes: pyramid, cone and concave. The authors have shown that pyramidal textures cause significant results for tribological behavior. Deposition of textured surface coating tends to reduce friction over that achieved for uncoated textured surface.

Wang et al. presented the study in 2018 [17] on lubrication performance of journal bearing with multiple texture distributions. They are able to compare two shaped concave textures and convex texture on a bearing lubrication performance. Their results show that the bearing load capacity is reduced by the concave spherical texture, but enhanced by the convex texture; both the concave and convex textures have a very slight influence on the friction coefficient. In the same year, Ji and Guan [18], analyses the effect of the micro-dimples on hydrodynamic lubrication of textured sinusoidal surfaces and rough surfaces. In order to characterize the non-textured surfaces, sinusoidal waves were used. Their results show that, the effect of roughness of the textured surface on the hydrodynamic pressure is significant and the load carrying capacity decreases with the increase of the roughness ratio because the roughness greatly suppresses the hydrodynamic effect of dimples.

In 2019, Manser et al. [19] studied the hydrodynamic journal bearing performance under the combined influence of textured surface and journal misalignment. This study is a numerical analysis is performed to test three texture shapes: square "SQ," cylindrical "CY," and triangular "TR," and shaft misalignment variation in angle and degree. The Reynolds equation of a thin viscous film is solved using the finite difference's method. Their results show that the micro-step bearing mechanism is a key parameter, where the micro-pressure recovery action present in dimples located at the second angular part of the bearing (from 180° to 360°) can compensate for the loss on performances caused by shaft misalignment, while the micro-pressure drop effect at the full film region causes poor performances.

4. Theoretical analysis

The pressure field is determined by the resolution of the generalized Navier-Stokes equation according to the classical assumptions in the $(O, \dot{\theta} \rightarrow, z)$ coordinate system. **Figure 1** illustrates the schematization of plain cylindrical journal bearing.

4.1 Equation fluid flow

4.1.1 Equation of continuity

The continuity equation can be expressed by the relationship (1) [20].

$$\nabla\left(\rho\,\vec{U}\right) = \vec{0} \tag{1}$$

where $\vec{U} = \vec{U}(u, v, w)$ is the velocity vector.
Eq. (1) can also be written as follows:

$$\frac{\partial u}{\partial x} + \frac{\partial v}{\partial y} + \frac{\partial w}{\partial z} = 0 \tag{2}$$

4.1.2 Navier-Stokes equations

The Navier-Stokes equation can be defined in the following form (2003):

$$\partial\nabla.\left(\vec{U}\otimes\vec{U}\right) = -\nabla p + \mu\nabla.\left(\nabla\vec{U} + \left(\nabla\vec{U}\right)^T\right) + B \tag{3}$$

With P static pressure (thermodynamic); U velocity; μ dynamic viscosity.
For fluids in a rotating frame with constant angular velocity ω source term B can be written as follows:

$$B = -\rho\left(2\,\vec{\omega}\times\vec{U} + \vec{\omega}\times\left(\vec{\omega}\times\vec{r}\right)\right) \tag{4}$$

Eq. (1) can also be expressed in the form:

$$\rho\left(u\frac{\partial u}{\partial x} + v\frac{\partial u}{\partial y} + w\frac{\partial u}{\partial z}\right) = -\frac{\partial p}{\partial x} + \mu\left(\frac{\partial^2 u}{\partial x^2} + \frac{\partial^2 u}{\partial y^2} + \frac{\partial^2 u}{\partial z^2}\right) + B_x$$

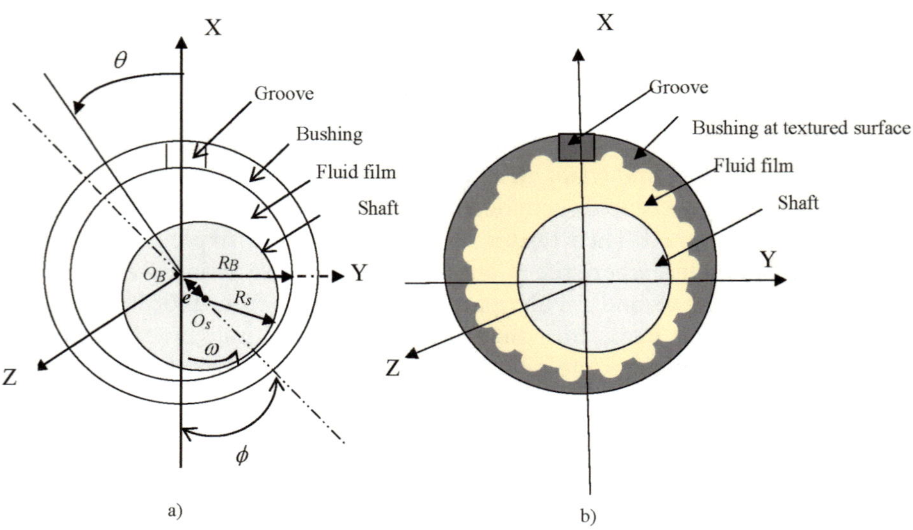

Figure 1.
Schematization of plain bearing. (a) Non-textured plain bearing. (b) Textured plain bearing.

$$\rho\left(u\frac{\partial v}{\partial x}+v\frac{\partial v}{\partial y}+w\frac{\partial v}{\partial z}\right)=-\frac{\partial p}{\partial y}+\mu\left(\frac{\partial^2 v}{\partial x^2}+\frac{\partial^2 v}{\partial y^2}+\frac{\partial^2 v}{\partial z^2}\right)+B_y$$
$$\rho\left(u\frac{\partial w}{\partial x}+v\frac{\partial w}{\partial y}+w\frac{\partial w}{\partial z}\right)=-\frac{\partial p}{\partial z}+\mu\left(\frac{\partial^2 w}{\partial x^2}+\frac{\partial^2 w}{\partial y^2}+\frac{\partial^2 w}{\partial z^2}\right)+B_z \tag{5}$$

ρ is fluid density.

Considering the Z axis as the axis of rotation, the components of B can be expressed as follows:

$$B_x=\left(\omega_Z^2\, r_x+2\omega_z v\right)$$
$$B_y=\left(\omega_Z^2\, r_y+2\omega_z u\right)$$
$$B_z=0$$

The finite volume method used to solve the continuity and Navier-Stokes equations consists in subdividing the physical domain of the flow into elements of more or less regular volumes; it converts the general differential equation into a system of Algebraic equations by relating the values of the variable under consideration to the adjacent nodal points of a typical control volume. This is achieved by integrating the governing differential equation into this control volume.

4.1.3 Discretization of governance equations

The main step of the finite volume method is the integration of governing equations for each control volume [20]. The algebraic equations deduced from this integration make the resolution of the transport equations simpler. Each node is surrounded by a set of surfaces that has a volume element. All the variables of the problem and the properties of the fluid are stored at the nodes of this element.

The equations governing the flow are presented in their averaged forms in a Cartesian coordinate system (x, y, z):

$$\frac{\partial}{\partial X_j}\left(\rho\, U_j\right)=0 \tag{6}$$

$$\frac{\partial}{\partial X_j}\left(\rho\, U_j U_i\right)=-\frac{\partial P}{\partial X_i}+\frac{\partial}{\partial X_j}\left(\mu\left(\frac{\partial U_i}{\partial X_i}+\frac{\partial U_j}{\partial X_i}\right)\right)+B_x \tag{7}$$

Eqs. (6) and (7) can be integrated into a control volume, using the Gaussian divergence theorem to convert volume integrals to surface integrals as follows:

$$\int_s \rho\, U_j dn_j=0 \tag{8}$$

$$\int_s \rho\, U_j U_i dn_j=-\int_s P\, dn_j+\int_s\left(\mu\left(\frac{\partial U_i}{\partial X_j}+\frac{\partial U_j}{\partial X_i}\right)dn_j+\int_V S_{ui}dv\right. \tag{9}$$

The next step is to discretize the known m's of the problem as well as the differential operators of this equation. All these mathematical operations will lead to obtaining, on each volume of control, a discretized equation that will link the variables of a cell to those of neighboring cells. All of these discretized equations will eventually form a matrix system. Considering an element of an isolated mesh, **Figure 2**.

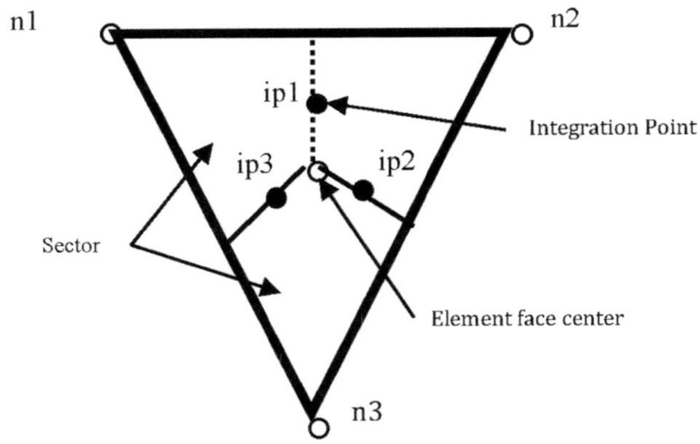

Figure 2.
Integration point in an element of a control volume control.

After the discretization and rearrangement of Eqs. (8) and (9) the following forms will be obtained:

$$\sum_{ip} (\rho\, U_j\, \Delta n_j)_{ip} = 0 \tag{10}$$

$$\sum_{ip} m_{ip}\,(U_i)_{ip} = \sum_{ip} (P\, \Delta n_j)_{ip} + \sum_{ip} \left(\mu \left(\frac{\partial U_i}{\partial X_j} + \frac{\partial U_j}{\partial X_i} \right) \Delta n_j \right) + \overline{S_{u_i} V} \tag{11}$$

$$N_j = \begin{Bmatrix} 1 & i = j \\ 0 & i \neq j \end{Bmatrix}$$

4.1.4 Coupling pressure-velocity

The method of pressure interpolation in pressure-velocity coupling is similar to that used by Rhie and Chow (1982). This method is among the methods that best save memory space and computation time. If the pressure is known, the discretized equations are easily solved [20]:

$$\left(\frac{\partial \mathrm{U}}{\partial \mathrm{x}} \right)_i + \frac{\Delta x^3 A}{4m} \left(\frac{\partial^4 P}{\partial x^4} \right) 0 \tag{12}$$

where:

$$m = \rho\, U_i\, \Delta n_j \tag{13}$$

4.1.5 Form functions

The physical quantity ϕ (p, u, v, w and p) of the flow in a volume element is a function of those in the nodes of the element is given by the following relation:

$$\phi = \sum_{i=1}^{\mathrm{Node}} N_i\, \phi_i \tag{14}$$

where Ni is the form function for node i and ϕi the value of the variable in the same node. A particularity of the form factors makes sure that:

$$\phi = \sum_{i=1}^{\text{Node}} N_i = 1 \tag{15}$$

These functions are also us5ed for the calculation of various geometric quantities, such as positions, coordinates of the integration point (ip), surfaces and different vectors. Form equations are also applicable for Cartesian coordinates, in which case they can be written in the following way:

$$x = \sum_{i=1}^{\text{Node}} N_i \, x_i \tag{16}$$

$$y = \sum_{i=1}^{\text{Node}} N_i \, y_i \tag{17}$$

$$z = \sum_{i=1}^{\text{Node}} N_i \, z_i \tag{18}$$

The shape functions are also used to evaluate the partial derivatives of the flow terms on the control surfaces and for each direction, the general formula of the different flows is as follows:

$$\left.\frac{\partial\phi}{\partial x}\right|_i = \sum_n \left.\frac{\partial N_n}{\partial x}\right|_{ip} \phi_n \tag{19}$$

4.1.6 Pressure gradients

The integration of the pressure gradient (P) on the control volume in the Navier-Stokes equations involves the evaluation of the following expression:

$$\left(P \, \Delta n_{ip}\right)_{ip} \tag{20}$$

where:

$$P_{\text{ip}} = \sum_n N_n \left(S_{\text{ip}}, t_{\text{ip}}, u_{\text{ip}}\right) P_n \tag{21}$$

For the improved treatment of fluctuations induced by turbulence in the motion of a particle of fluid, there are three methods of approach to address the notion turbulence. The first method is to decompose the field of velocity and temperature in a mean component and a turbulent fluctuation, to make a variety of models are now available, ranging from the simple model equation to zero to complex (model of the constraint equations Reynolds RMS).

The second is a method in which all the structures of turbulence (macro and micro-structures) are solved directly and models the effect of small structures by models more or less simple, so-called sub-grid models. This method is known as the large eddy simulation (Large Eddy Simulation, LES). The third method is a hybrid approach combines the advantages qm large eddy simulation (LES), with good results in highly separated zones, and model Reynolds-Averaged Navier-Stokes (RANS), which are most effective in areas close to the walls. The method is called (Detached Eddy Simulation, DES).

5. k-Epsilon model

One of the most prominent turbulence models, the (k-epsilon) model, has been implemented in most CFD codes [20]. It has proven to be stable and numerically robust and has a well-established regime of predictive capability; the model offers a good compromise in terms of accuracy and robustness. This turbulence model uses the scalable wall-function approach to improve robustness and accuracy when the near-wall mesh is very fine.

k is the turbulence kinetic energy and is defined as the variance of the fluctuations in velocity. It has dimensions of ($L^2\ T^{-2}$); for example, m^2/s^2. ε is the turbulence eddy dissipation (the rate at which the velocity fluctuations dissipate), and has dimensions of ***k*** per unit time ($L^2\ T^{-3}$); for example, m^2/s^3.

The k-ε model introduces two new variables into the system of equations. The continuity equation is following forms:

$$\frac{\partial \rho}{\partial t} + \frac{\partial}{\partial x_j}\left(\rho\, U_j\right) = 0 \tag{22}$$

and the momentum equation becomes:

$$\frac{\partial \rho\, U_i}{\partial t} + \frac{\partial}{\partial x_j}\left(\rho\, U_i U_j\right) = -\frac{\partial p'}{\partial x_i} + \frac{\partial}{\partial x_j}\left[\mu_{\text{eff}}\left(\frac{\partial U_i}{\partial x_j} + \frac{\partial U_j}{\partial x_i}\right)\right] + S_M \tag{23}$$

where $\boldsymbol{S_M}$ is the sum of body forces, $\boldsymbol{\mu_{eff}}$ is the effective viscosity accounting for turbulence, and ***p'*** is the modified pressure as defined in Eq. (22).

$$p' = p + \frac{2}{3}\rho\, k + \frac{2}{3}\mu_{\text{eff}}\frac{\partial U_k}{\partial x_k} \tag{24}$$

The *k-ε* model, like the zero equation model, is based on the eddy viscosity concept, so that:

$$\mu_{\text{eff}} = \mu + \mu_t \tag{25}$$

where $\boldsymbol{\mu_t}$ is the turbulence viscosity. The *k-ε* model assumes that the turbulence viscosity is linked to the turbulence kinetic energy and dissipation:

$$\mu_t = C_\mu \rho\, \frac{k^2}{\varepsilon} \tag{26}$$

where $C\mu$ is a constant.

With $C_\mu = 0.09\ \theta = k^{1/2}\ l = \frac{k^{3/2}}{\varepsilon}$.

The values of ***k*** and ***ε*** come directly from the differential transport equations for the turbulence kinetic energy and turbulence dissipation rate:

$$\frac{\partial(\rho\, k)}{\partial t} + \frac{\partial}{\partial x_j}\left(\rho\, U_j\, k\right) = \frac{\partial}{\partial x_j}\left[\left(\mu + \frac{\mu_t}{\sigma_k}\right)\frac{\partial k}{\partial x_j}\right] + P_k - \rho\,\varepsilon + P_{\text{kb}} \tag{27}$$

$$\frac{\partial(\rho\,\varepsilon)}{\partial t} + \frac{\partial}{\partial x_j}\left(\rho\, U_j\,\varepsilon\right) = \frac{\partial}{\partial x_j}\left[\left(\mu + \frac{\mu_t}{\sigma_\varepsilon}\right)\frac{\partial \varepsilon}{\partial x_j}\right] + \frac{\varepsilon}{k}\left(C_{\varepsilon 1}P_k - C_{\varepsilon 2}\,\rho\,\varepsilon + C_{\varepsilon 1}P_{\text{kb}}\right) \tag{28}$$

where $C_{\varepsilon 1}$, $C_{\varepsilon 2}$, σ_k and σ_ε are constants.

$\boldsymbol{P_{kb}}$ and $\boldsymbol{P_{\varepsilon b}}$ represent the influence of the buoyancy forces, which are described below. $\boldsymbol{P_k}$ is the turbulence production due to viscous forces, which is modeled using:

$$P_k = \mu_t \left(\frac{\partial U_i}{\partial x_j} + \frac{\partial U_j}{\partial x_i} \right) \frac{\partial U_i}{\partial x_j} - \frac{2}{3} \frac{\partial U_k}{\partial x_k} \left(3\mu_t \frac{\partial U_k}{\partial x_k} + \rho\, k \right) \tag{29}$$

The term $3\,\boldsymbol{\mu_t}$ in Eq. (37) is based on the "frozen stress" assumption. This prevents the values of $\boldsymbol{k}$ and $\boldsymbol{\varepsilon}$ becoming too large through shocks.

6. Numerical model

The purpose of this study is to highlight the behavior of the turbulent fluid flow fluid on the operating characteristics as well as the hydrodynamic behavior of a plain bearing This study is simulated by the CFD calculation code, which provides accuracy, reliability, speed and flexibility in potentially complex flow areas. Integrating the Reynolds equation on each control volume to derive an equation connecting the discrete variables of the elements that surround it, all of these equations eventually form a matrix system.

6.1 3D structure of the numerical model

Figure 3 illustrates the 3-D structure of the plain bearing with fluid and solid regions are shown. The supply holes are presented in a simplified manner without affecting the accuracy of the model. A tetrahedron element is adopted in the oil supply holes of the fluid region, and a hexahedral element is adopted in domain fluid. A hexahedral element is also applied to the solid region such as the bearing and the shaft (**Figure 4**).

The geometrical and operating parameters of the plain journal bearing is presented in the **Table 1**. As well as, parameters of the lubricant are showed in **Table 2**.

6.2 Boundary conditions of the numerical model

Boundary conditions of the numerical model of the plain bearing are shown in **Figure 5**, definite as follows: 1: the rotating speed is applied to the outer wall surface

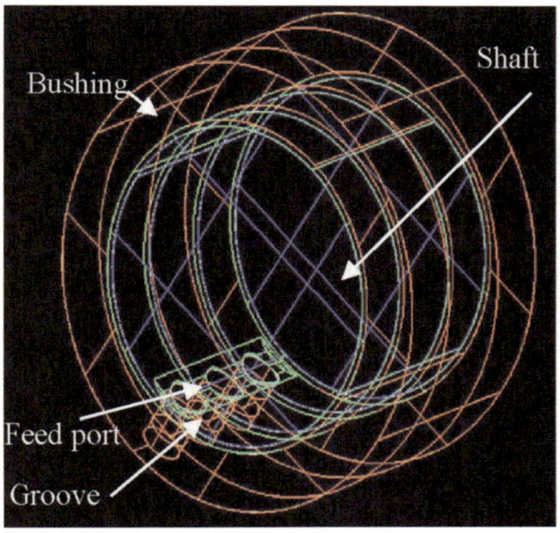

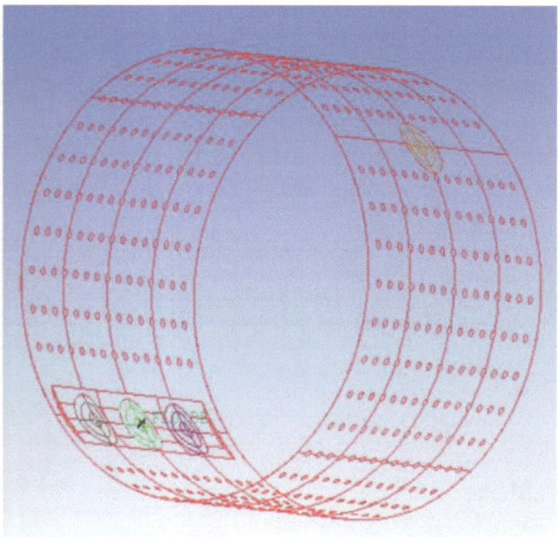

Figure 3.
3D structure of the non-textured plain bearing. (a) Non-textured bearing. (b) textured bearing.

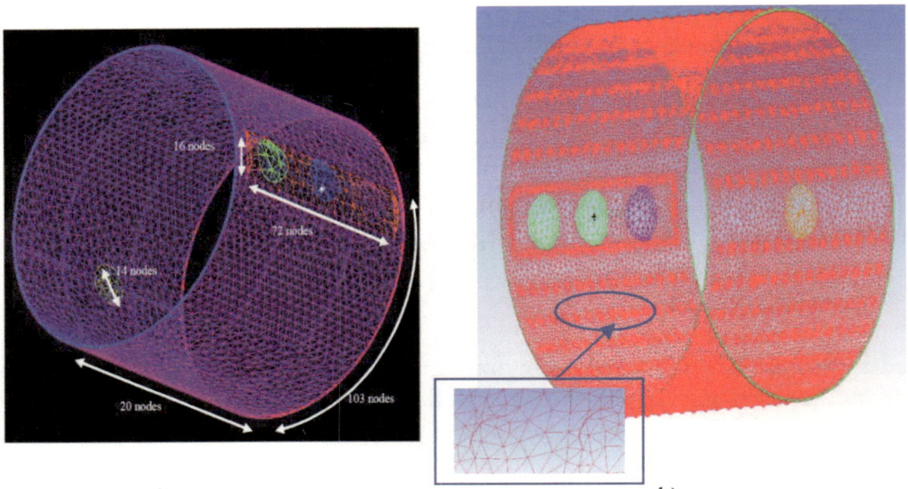

a) b)

Figure 4.
Mesh of the plain bearing. (a) Non-textured bearing. (b) textured bearing.

Item	Value
Bearing diameter (mm)	100
Shaft diameter (mm)	99.91
Bearing length (mm)	70
Radial clearance (mm)	0.09
Pad thickness (mm)	4
Feed port diameter (mm)	14
Feed groove length (mm)	70
Rotating velocity N (rpm)	11,000-- 21, 000
Radial load W (N)	2000–20- 10, 000
Supply temperature ambiaente Ta (°C)	40
Supply pressure Pa (MPa)	0.08

Table 1.
Geometrical and operating parameters of the plain bearing.

Item	Value
Lubricant type	PMA3
Density ρ (kg/m^3)	800
Specific heat capacity C (J/kg. K)	2000
Kinematic viscosity at 40 °C υ_1 (mm^2/s)	17.,49
Kinematic viscosity at 80 °C υ_2 (mm^2/s)	8,003

Table 2.
Parameters of the lubricant.

of the shaft; 2: the inner wall surface of the bushing is stationary; 3: the domain is simulated by the fluid region. The slip of the interface is ignored; 4: the oil supply pressure is 0.08 MPa and supply temperature is 40°C, are set in oil supply holes;

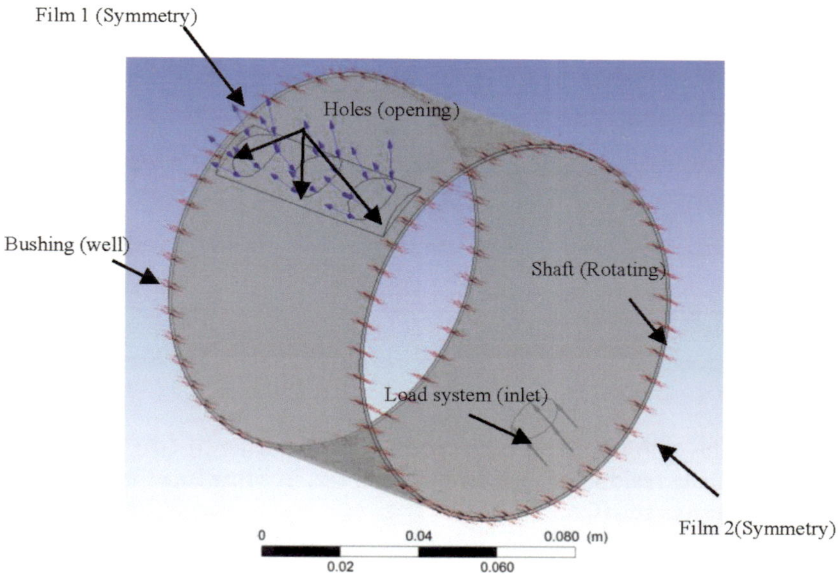

Figure 5.
Boundary conditions.

5: the two ends of the plain bearing domain, and the pressure is set to one bar; and is considered as symmetry.

6.3 Validation of the mesh independence of the numerical model

The setting is done by a graphical interface. The mesh used is a mixed mesh which understood elements of tetrahedral type with 6 nodes and hexahedral elements with 8 nodes. It's necessary to choose an appropriate mesh, consequently, a mesh independence study is carried out, and calculation results are shown in **Figure 6**. When the nodes number is greater than 4815, the evolution of the pressure stabilizes in the angular coordinate 205° of the plain bearing. Therefore, the number of nodes chosen for this numerical analysis corresponds to a number of nodes equal to 4815. The nodes number for textured bearing is 65,172. Convergence criterion of the numerical results is calculated for a maximum number of iterations

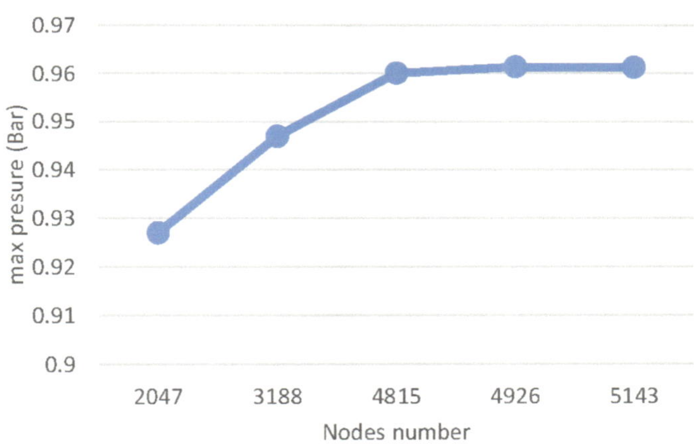

Figure 6.
Evolution max pressure according to the nodes number of the shaft mesh.

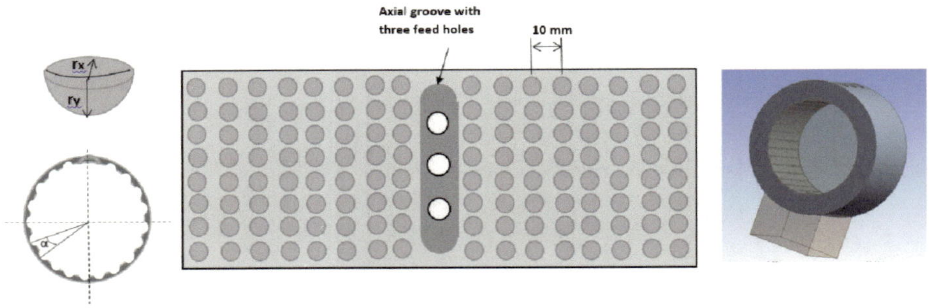

Figure 7.
Textured bushing parameters.

of 1000 iterations with a convergence criterion of the order of 10^{-4}. The solution converges when the residuals reach 10^{-4}. However, in some cases it is necessary to push the calculations to 10^{-6}.

6.4 Textures parameters

Surface texturing of the bushing is a technique used to improve the load capacity of various tribological conjunctions, as well as to reduce frictional losses. The texture spherical shape of diameter r_x = 3 mm and the depth of r_y = 0.5 mm, the axial distance between the textures d = 10 mm and their angular offsets α = 10°, (**Figure 7**).

7. Hydrodynamic and tribological performance of a hydrodynamic non-textured and textured plain bearing for the turbulent regime

7.1 Test of the different turbulent model

In this section, we will carry out a comparative study between two models of turbulence: k-ε model for turbulence in the vicinity of the walls and the RMS model (Reynolds shear stress) for turbulence in the vicinity and far from the walls. **Figure 8** illustrates the pressure distribution along the median plane of the plain bearing, for the k-ε model and the RMS model. Both models give the same pressure distribution. Since we are interested in examining the distribution of pressure, of

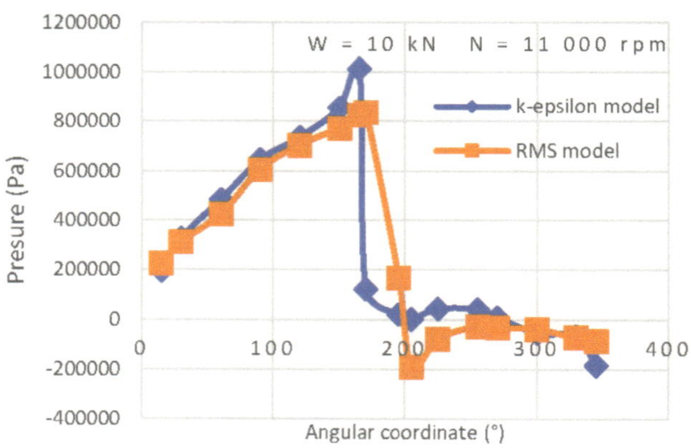

Figure 8.
Pressure evolution for k-epsilon model and Reynolds shear stress (RMS).

the friction torque between the fluid and the internal surface of the bearing, we used the k-ε model for the numerical analysis carried out in this study.

7.2 Radial load effect

To demonstrate the effect of the radial load on the operating performance of the non-textured and textured hydrodynamic plain bearing, such as pressure, fluid flow velocity and friction torque, the radial load is varied (W1 = 2000 N, W2 = 5000 N, W3 = 7000 N and W3 = 9000 N). The initial operating conditions of the bearing re a supply temperature Ta = 40° C, supply pressure Pa = 0.08 MPa and the rotational speed of the shaft equal to 11,000 rpm with a Reynolds number of Re = 3622.64 to ensure the turbulent regime.

7.2.1 Pressure

Figure 9 illustrates the distribution of the pressure along the median plane for non-textured and textured bearing, for different radial loads. The graph shows that increasing the load from 2000 N to 9000 N leads to an increase in pressure. Significant pressures are obtained for a bearing subjected to a radial load of 9000 N. This increase reaches 65 per cent for a textured bearing. Also for a no textured bearing, the increase in pressure will reach 81 per cent by varying the radial load from 2kN to 9kN. The curves also indicate that the maximum pressure is noted in the angular position from 160° to 175°, on the other hand, in the angular coordinates at 200°, the noted pressure is lower than the supply pressure, indicating the existence rupture zones of the oil film. The rupture zones of the oil film are observed in the angular positions between 190° and 335° and also between 300° and 350°. The values of circumferential pressure are significant for a textured bearing with respect to those recorded for a non-textured bearing (**Figure 10**).

7.2.2 Fluid flow velocity

The fluid flow velocity according to the angular position of the plain bearing, for different radial loads is presented in **Figure 11**. The maximum flow velocity is noted for a textured plain bearing working under a radial load of 9000 N and which is of the order of 61 m/s, on the other hand is of the order of 36 m/s for non-textured plain bearing. The increase in the radial load which reacts on the bearing causes the increase in the flow velocity. This increase is estimated at 21

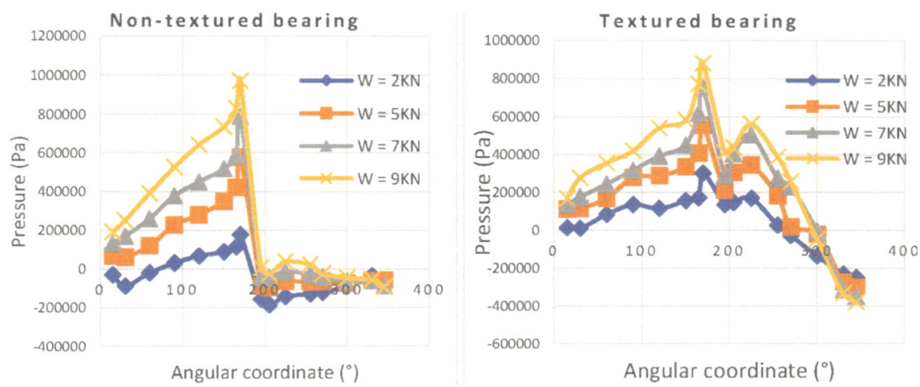

Figure 9.
Circumferential pressure for different radial load N = 11,000 rpm (Re = 3622.64 turbulent regime).

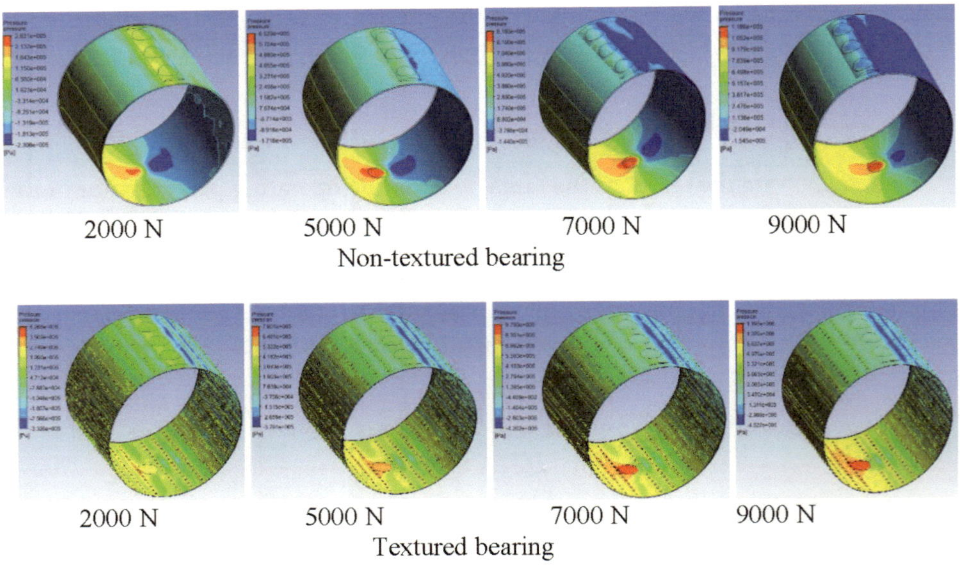

Figure 10.
Pressure evolution for different radial load N = 11,000 rpm.

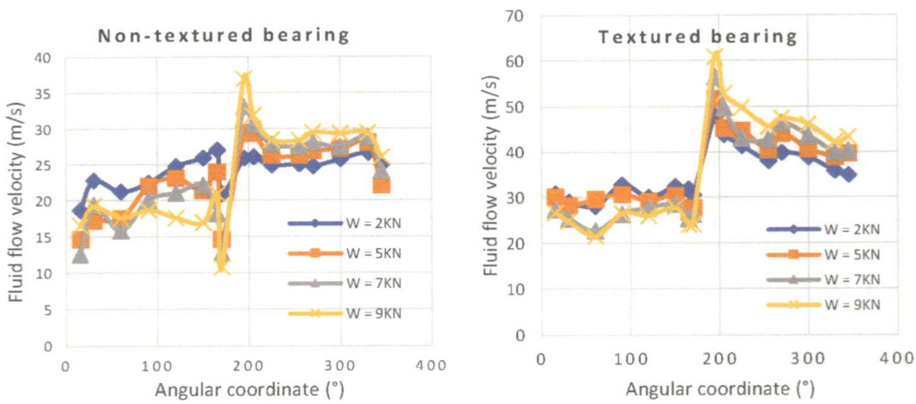

Figure 11.
Evolution of the fluid flow velocity according the angular position for different radial load N = 11,000 rpm (Re = 3622.64 turbulent regime).

per cent for textured bearing and estimated at 29 per cent for non-textured bearing (**Figure 12**).

7.2.3 Friction torque

The fluid friction torque or "viscous" friction is a particular friction force, which is associated with the movement of an object in a fluid (air, water, etc.). It is at the origin of energy losses by friction for the object moving in the fluid. The friction torque is calculated by integrating the shear stresses at the surface of the shaft or of the bushing, the shear stresses in the fluid are given by derivation the fluid velocity in the radial and tangential direction. Therefore, there is an empirical relationship between the flow velocity of the fluid and the friction torque, for this we obtain the same distribution for the fluid flow velocity and the friction torque along the median plane of the hydrodynamic bearing.

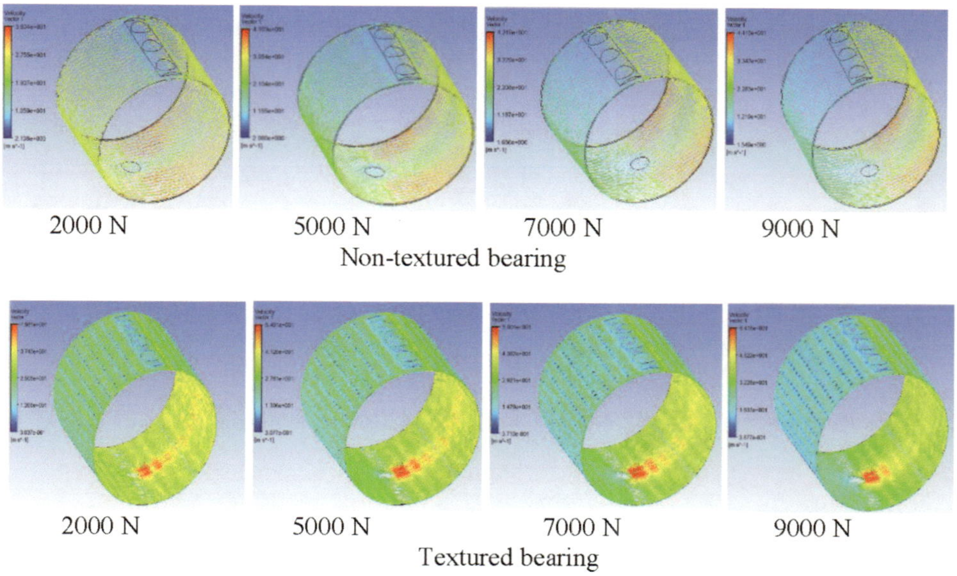

Figure 12.
Velocity evolution for different radial load N = 11,000 rpm.

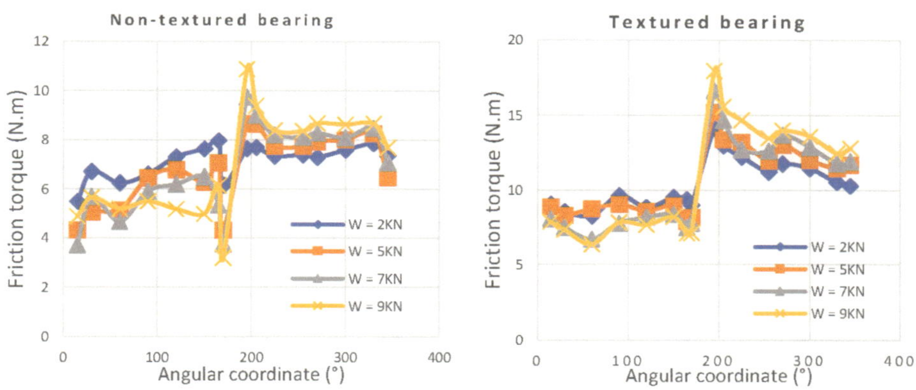

Figure 13.
Friction torque in the median plane for different radial load N = 11,000 rpm (Re = 3622.64 turbulent regime).

The friction torque along the circumference of the textured bearing is illustrated in **Figure 13**. The important values are noted for a radial load of 9000 N, the maximum value of the friction torque is of the order of 17.93 N.m for a textured bearing, and is the order of 10.83 N.m for non-textured bearing. These maximum values are noted in the angular positions at 180° and 195°. The increase in the radial load from 2000 N to 9000 N leads to an increase in the friction torque of 21 per cent and 29 per cent respectively for a textured and non-textured bearing.

7.3 Effect of the shaft rotation speed

7.3.1 Pressure

Figure 14 shows the pressure distribution along the bearing circumference, for four shaft rotation speeds (11,000 rpm, 14,000 rpm 17,000 rpm and 21,000 rpm). The supply conditions used for this numerical analysis are Ta = 40°C and Pa = 0.08 MPa. The radial load is 10,000 N. This rotational speed gives respectively

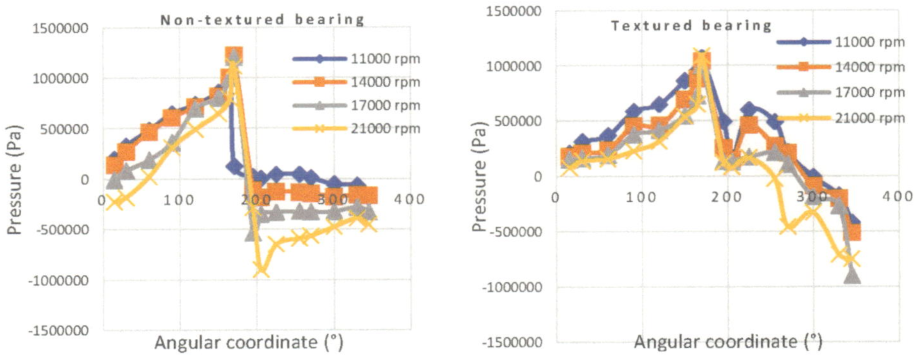

Figure 14.
Circumferential pressure for different rotational velocity W = 10 KN (turbulent regime).

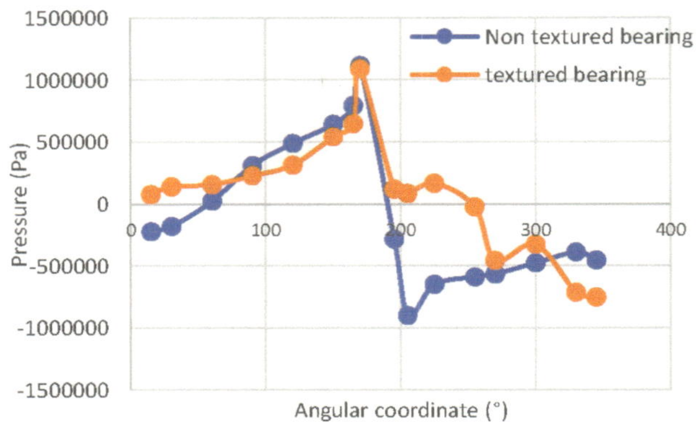

Figure 15.
Circumferential pressure according the angular coordinate of the non- textured and textured bearing W = 10 KN, N = 14,000 rpm (Re = 5187.6 turbulent regime).

a Reynolds number of Re = 3622.64, Re = 4687.53, Re = 5187.6 and Re = 6752.54, which indicates that the regime is turbulent.

The curve clearly shows that the maximum pressure is positioned at angular coordinates from 140° to 160°, while at angular positions between 170° and 200°, the pressure is lower than the supply pressure, which indicates the existence of the rupture zone of the oil film. It can also be said that increasing the rotational speed causes a slight decrease in pressure, this decrease being estimated at 24 per cent. The significant pressure is recorded for a very high rotation speed, which is of the order of 21,000 rpm.

Figure 15 shows the pressure distribution as a function of the angular position for a textured and non-textured bearing for a radial load of 10,000 N and a rotation speed of 14,000 rpm. The curve clearly shows that the pressure distribution along the median plane of the bearing is different in the case of a non-textured bearing and a bearing with a textured surface; the difference is estimated at 8.5 per cent (**Figure 16**).

7.3.2 Fluid flow velocity

Figure 17 illustrates the variation of flow velocity in the circumferential direction of the plain bearing, to a feed temperature of 40°C and feed pressure of 0.08 MPa. The shaft rotational speed varies from 11,000 rpm to 21,000 rpm (Turbulent regime) and a radial load of 10,000 N. The curve shows that the rotational speed leads to an increase in the fluid flow velocity. The increase reached 39 per cent. The flow velocity is significant for a bearing which rotates at a speed of

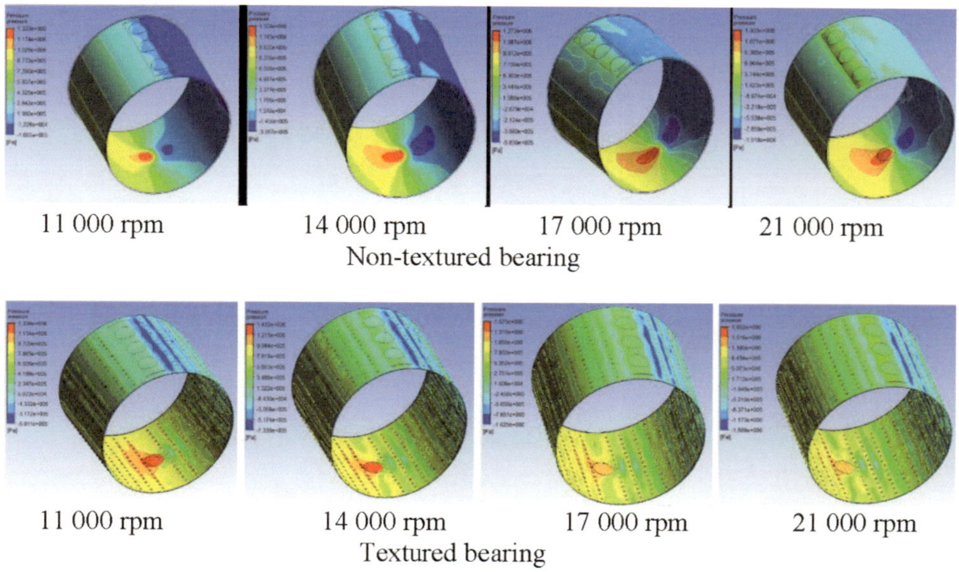

Figure 16.
Distribution circumferential of the pressure for differents rotational velocity.

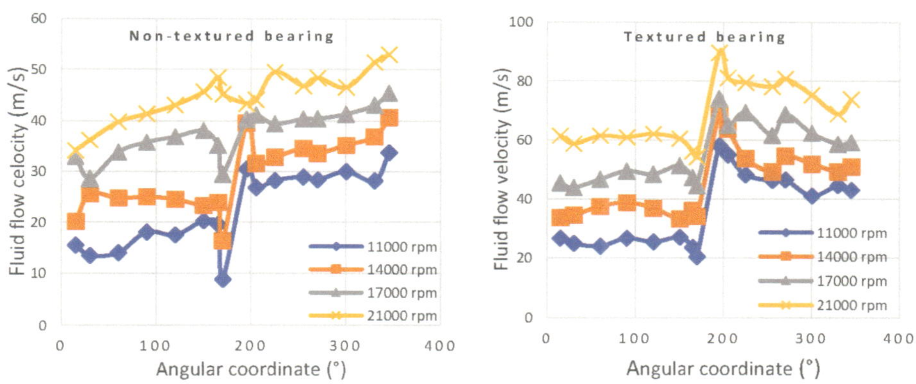

Figure 17.
Fluid flow velocity evolution according angular position angular for different rotational speed W = 10 KN (turbulent regime).

21,000 rpm (Re = 6752.54), on the other hand it is less important for a rotational speed of 11,000 rpm (Re = 3622.64). The significant value of the fluid flow velocity is noted for a textured plain bearing which is the order of 89.56 m/s. On the other hand, for a non-textured plain bearing, the maximum value of the fluid flow velocity is only of the order of 56.37 m/s.

For the different of the fluid flow velocity (**Figure 18**), has the same variation for the case of plain bearing without texture and a textured plain bearing. This speed takes a maximum value at the angular coordinate of 200° of the bearing. The difference between the fluid flow velocity for a non-textured and textured plain bearing is of the order of 38 per cent (**Figure 19**).

7.3.3 Friction torque

For the evolution of the friction torque as a function of the angular coordinates of the non-textured and textured plain bearing by varying the rotational speed of

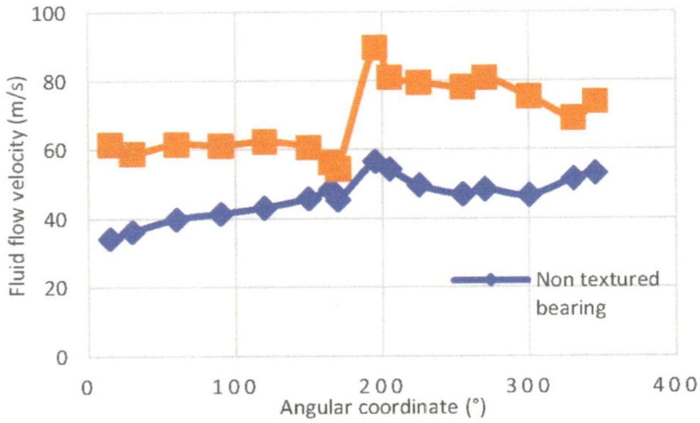

Figure 18.
Fluid flow velocity according angular position of the non-textured and textured bearing W = 10 KN, N = 14,000 rpm (Re = 5187.6 turbulent regime).

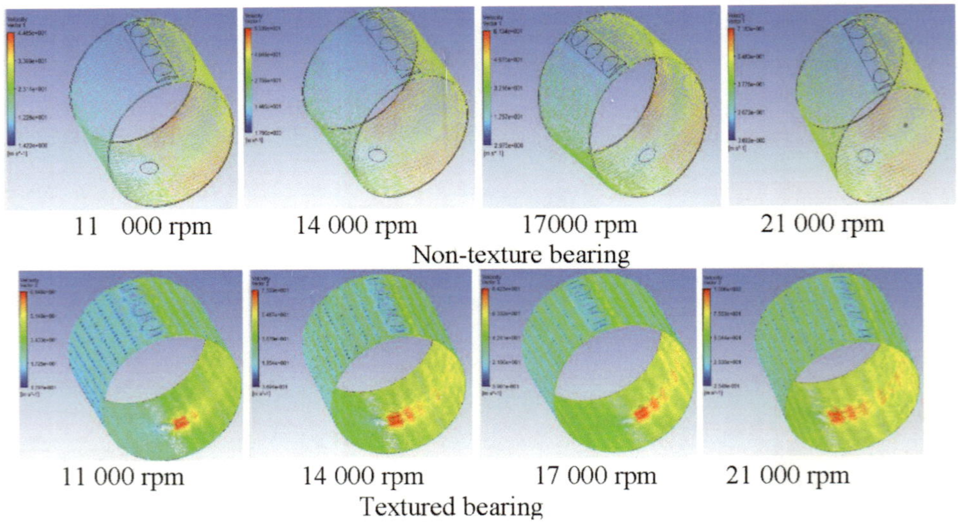

Figure 19.
Distribution circumferential of the fluid flow velocity for differents rotational velocity.

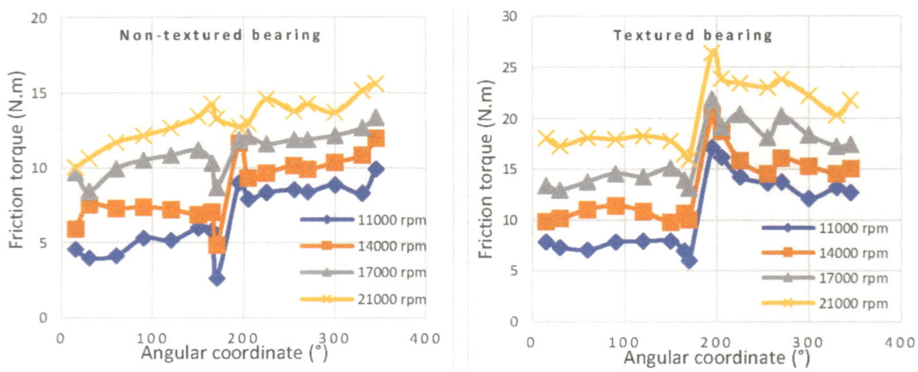

Figure 20.
Friction torque in median plane for different rotational speed W = 10 KN (turbulent regime).

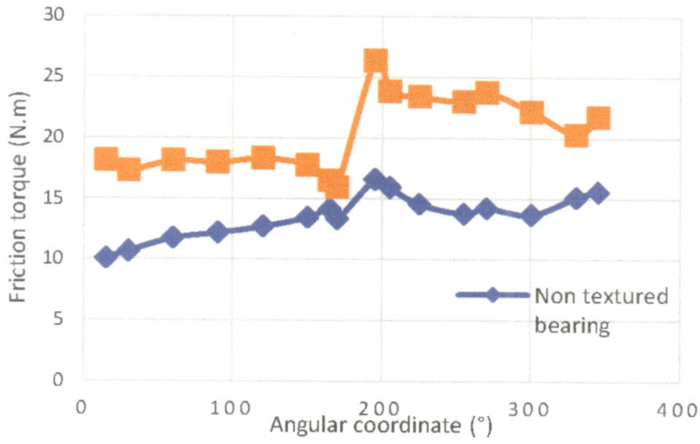

Figure 21.
Friction torque in the median plane of the non-textured and textured bearing W = 10 KN, N = 14,000 tr/min (Re = 5187.6 turbulent regime).

the shaft from 11,000 to 21,000 rpm and for a radial load of 10,000 N, is presented in **Figure 20**. The increasing the rotational speed causes a slight increase in the friction torque, this increase is of the order of 2 per cent. The important values are obtained for a rotational speed of 21,000 rpm; the maximum value of the friction torque is also positioned at the angular coordinate of 200°. The significant value of the friction torque for a non-textured plain bearing is of the order of 16 N.m, on the other hand for a textured plain bearing is 26 N.m.

Figure 21 illustrates the variation of friction torque along the circumferential non-textured and textured plain bearing. The evolution of the friction torque along the angular bearing position has the same shape for the two cases studied, the difference is estimated at 38 per cent at the 200° level.

8. Conclusions

This numerical study presents the evolution of the fluid flow for turbulent regime in hydrodynamic plain bearings with a non-textured and textured surface, in order to improve the hydrodynamic lubrication and tribological performance of plain bearing, using the finite volume method, such as pressure, friction torque and fluid flow velocity.

The results obtained for the textured plain bearing were compared to the non-textured plain bearing, the main conclusions drawn from this study are:

1. The pressure distribution according to the angular position for the textured and non-textured plain bearing for the radial load of 10,000 N and the speed of rotation of 14,000 rpm has the same appearance for the two cases studied; the difference is estimated at 8.5%.

2. The rupture zones of the oil film are observed in several angular positions at 190° and 300° for a plain bearing with textured surface, on the other hand for a plain bearing without texture, the rupture zone is positioned only in the angular position at 190°. This rupture of the oil film is due to the drop in pressure below the supply pressure.

3. The evolution of the friction torque, along the angular position, has the same distribution for the non-textured and the textured plain bearing, the difference is estimated at 38%.

4. The flow velocity of the fluid in the plain bearing takes a maximum value at the angular position of 165°. The difference between the flow velocity for a non-textured and textured plain bearing is estimated of 38%.

It should be emphasized, however, that the conclusions we give here are only valid for the cases we have studied, and that they are not independent of the characteristics of the plain bearing and of the lubricant.

The numerical results show that the most significant hydrodynamic characteristics such as pressure, flow velocity of the fluid and friction torque, are significant for the textured plain bearing under rotational velocity of 21,000 rpm and radial load 10,000 N compared to the results obtained for a non-textured plain bearing.

When one is interested in plain bearings operating under severe conditions, that is to say for the turbulent regime, the hydrodynamic pressures sometimes reach several hundred mega Pascal's.

Conflict of interest

The authors declare no conflict of interest.

Nomenclature

B	source term
C	radial clearance
e	eccentricity
L	bearing length
R_S	shaft radius
R_B	bush radius
N	rotational velocity [rpm]
P	pressure [Pa]
r_i	position vector [m]
U	peripheral speed [m/s]
U, v, w	velocity according x, y, z axis [m/s]
W	radial load [N]
Re	Reynolds number
ε	relative eccentricity
ϕ	Flow factor [°]
μ	dynamic viscosity [Pa.s]
μt	turbulent dynamic viscosity [Pa.s]
ω	shaft angular speed [rad/s]
K	turbulence kinetic energy
ρ	density [kg/m^3]

Indice

i	Interne
ip	Indice du point d'intégration
s	spécifique
t	théorique
u	utile

Author details

Bendaoud Nadia* and Mehala Kadda
Faculty of Mechanical Engineering, University of the Sciences and Technology of Oran Mohamed Boudiaf, Oran, Algeria

*Address all correspondence to: nadiamehala@yahoo.fr

References

[1] Constantinescu, V.N. (1959), "On turbulent lubrication", *Proceedings of the Institution of Mechanical Engineers*, Vol. 173 Nos. 3/8, pp. 881–900. doi.org/10.1243/PIME_PROC_1959_173_068_02

[2] Constantinescu, V.N. (1962), "Analysis of bearings operating in turbulent regime", *ASME Journal of Fluids Engineering*, Vol. 84 No. 1, pp. 139–151. DOI.org/10.1115/1.3657235

[3] Elrod, H.G. and Ng, C.W. (1967), "A theory for turbulent films and its application to bearings", *ASME Journal of Tribology*, Vol. 86, pp. 346–362. DOI.org/10.1115/1.3616989

[4] Ng, C.W. (1964), "Fluid dynamic foundation of turbulent lubrication theory", *ASLE Transactions*, Vol. 7 No. 4, pp. 311–321. DOI/abs/10.1080/05698196408972061

[5] Ng, C.W. and Pan, C.H.T. (1965), "A linearized turbulent lubrication theory", *ASME Transactions Journal of Fluids Engineering*, Vol. 87 No. 3, pp. 675–688. DOI.org/10.1115/1.3650640

[6] Fantinos B., Frêne J., Godet M., (1972), "Reynolds equation in viscous film theory", Trans, ASME, Journal of Lubrication Technology, vol. 94, n°3, pp: 287–288. DOI.org/10.1115/1.3451713

[7] Braunetiere, N. (2005), "A modified turbulence model for low Reynolds numbers: application to hydrostatic seals", *ASME Transactions Journal of Tribology*, Vol. 127 No. 1, pp. 130–140. DOI: 10.1115/1.1829721

[8] Solghar A.A and Nassab G., (2013), "Numerical analysis of turbulent lubrication in plain full journal bearings", Industrial Lubrication and Tribology, Vol.65 N°2, pp: 91–99. DOI/10.1108/00368791311303456

[9] Spalart, P.R., Jou, W.H., Strelets, M., and Allmaras, S.R., (1997), "Comments on the feasibility of LES for wings, aud on a hybrid RANS/LES approach." lst AFOSR Int. Conf. On DNS/LES, Aug.4–8. https://www.cobaltcfd.com/pdfs/DES97.pdf

[10] Krupka, I., Hartl, M., Zimmerman, M., Houska, P. and Jang, S., Effect of surface texturing on elastohydrodynamically lubricated contact under transient speed conditions, *Tribology International*, vol. **44**, no.10, pp. 1144–1150, 2011. DOI: 10.1016/j.triboint.2011.05.005

[11] Tala-Ighil, N., Fillon M. and Maspeyrot, P., Effect of textured area on the performances of a hydrodynamic journal bearing, *Tribology international*, vol. **44**, no. 3, pp. 211–219, 2011. DOI: 10.1016/j.triboint.2010.10.003

[12] Pratibha, S. and Chandreshkumar, R., Effect of Bearing Surface Texture on Journal Bearing Pressure, *International Journal of Science and Research (IJSR)*, vol. **3**, no. 6, pp. 2223–2226, 2014. paper_id=2014693

[13] Zhang, Y.L., Zhang, X.G. and Matsoukas, G. Numerical study of surface texturing for improving tribological properties of ultra-high molecular weight polyethylene, *Science direct Biosurface and Biotribology,* vol. **1**, pp.270–277, 2015. DOI.org/10.1016/j.bsbt.2015.11.003

[14] Uddin, M.S. and Liu, Y.W, Design and optimization of a new geometric texture shape for the enhancement of hydrodynamic lubrication performance of parallel slider surfaces, *Science direct, Biosurface and Biotribology* vol. **2**, pp. 59–69, 2016. DOI.org/10.1016/j.bsbt.2016.05.002

[15] Hamdavi, S., Ya, H.H. and Rao, T. V. V. L. N. Effect of surface texturing on hydrodynamic performance of journal bearings, ARPN Journal of Engineering

and Applied Sciences, vol. **11**, no.1, pp.172–176, 2016. https://www.researchgate.net/profile/T_V_V_L_N_Rao/publication/293820391

[16] Sedlaček, M., Podgornik, B., Ramalho, A. and Česnik, D. Influence of geometry and the sequence of surface texturing process on tribological properties, *Tribology international*, vol. **115**, pp. 268–273, 2017. DOI.org/10.1016/j.triboint.2017.06.001

[17] Wang, J., Zhang, J.,Lin, J. and Liang, M., Study on Lubrication Performance of Journal Bearing with Multiple Texture Distributions, *Applied Sciences*, vol. **224**, no.8, pp. 1–13, 2018. DOI.org/10.3390/app8020244

[18] Ji, J.-H., Guan, C.-W. Fu, Y.-H. (2018) Effect of Micro-Dimples on Hydrodynamic Lubrication of Textured Sinusoidal Roughness Surfaces, *Chinese Journal of Mechanical Engineering*, vol. **31**, art no. 67, 2018. DOI: 10.1186/s10033-018-0272-z

[19] Manser B., Belaidi I., Hamrani A., Khalladi S., Bakir F., Performance of hydrodynamic journal bearing under the combined influence of textured surface and journal misalignment: A numerical survey, *Comptes rendus Mécanique*, vol **347**, no. 2, pp.141–165, 2019. DOI : 10.1016/j.crme.2018.11.002

[20] ANSYS-CFX “Documentations Solver Theory”,(2009).

Chapter 3

Green Tribology

Nguyen Van Minh, Alexander Kuzharov, Le Hai Ninh, Nguyen Huynh and Andrey Kuzharov

Abstract

This chapter provides an overview of Green tribology, which is a new direction in the development of tribology, a new interesting area for scientific researches and a new way to turn tribology into a friend of ecological environment and saving energy. Green tribology is considered as well as close area with other "green" disciplines like green engineering and green chemistry. In the chapter, the various aspects of green tribology such as the concept, perspectives, role and goal, main principles, primary areas, challenges and directions of the future development have been discussed. It was clarified that green tribology can be defined as an interdisciplinary field attributed to the broad induction of various concepts such as energy, materials science, green lubrication, and environmental science. The most important role and goal of green tribology is improvement of efficiency by minimizing wear and friction in tribological processes to save energy, resources and protect environment, and consequently, improve the quality of human life. The twelve principles and three areas of green tribology were analyzed. Observation of these principles can greatly reduce the environmental impact of tribological processes, assist economic development and, as a result, improve the quality of life. The integration of these areas remains the major challenge of green tribology and defines the future directions of research in this field. This work also presents a rather detailed analysis of the most important effect in green tribology—the "zero-wear" effect (selective transfer effect). It was established that the "zero-wear" effect is due to self-organization in frictional interaction in tribological systems, which is the consequence of the complex tribo-chemical reactions and physico-chemical processes occurred in the area of frictional contact, that lead to the manifestation of unique tribological characteristics: super-antifrictional (friction coefficient ~ 10^{-3}) and without wear (intensity wear ~10^{-15}). This condition of tribo-system was provided by a protective nanocrystalline servovite film made of soft metal with unusual combination of mechanical properties.

Keywords: green tribology, friction, lubricants, wearlessness, zero-wear, selective transfer, biomimetics, self-lubrication, surface texturing, renewable energy

1. Introduction

Today, environmental and energy problems have become extremely serious and survival on a global scale. Scientists in all fields pay great attention to solving these problems. By this logic, the boom of recent decades, associated with the formation and development of nanotechnologies, including in relation to tribology and tribotechnics [1, 2].

It is noticeable that tribology has continuously developed into new phases. After opening in 1956, the effect of selective transfer (ST) ("zero-wear" effect) during friction, in tribology for the first time, was a basis to build the whole new paradigm of "friction without wear with minimal energy consumption", which was denied by all previous practical experience in operating movable coupling and by theoretical constructions of science of the friction and wear solids. At the same time, by the beginning of this century, the concept of "green tribology" was formed, which actually lists all the achievements in the study of the mechanisms of wearlessness (zero-wear) and super-anti-friction, as well as in the development of lubricants for their implementation in practice [3, 4]. In the 21st century and beyond, green tribology is expected to play an increasingly important role and become the key and strategies for solving a series of global problems in energy, environment and resources.

In recent years, there has been a rapid growth in research activities in green tribology field. A fairly large number of articles, world conference reports and academic books in related to this area have been published [2, 5–15]. However, there are still few publications that expounded the concepts, technological connotations, principles and disciplinary features of green tribology in precise, comprehensive definition and in an all-round way. The first scientific work completely devoted to green tribology, which emphasized the scientific rather than the economic and social aspects was published by M. Nosonovsky and B. Bhushan in 2010 [5].

Being a new field of tribology still in its infancy, an accurate understanding of the fundamentals of green tribology is important from both a scientific and practical point of view. In this regard, the aim of this work is to clarify the fundamental scientific and technological foundations of green tribology based on the analysis and generalization of the research achievements of green tribology.

2. Green tribology: concept, goal and role

It has been noted that the concepts in the main development direction of tribology are historically changed as follows (**Figure 1**).

Nowadays, the term "green tribology" has become part of the engineering dictionary. Green tribology is an emerging and actual area in tribological science with more focus on energy saving and environmental protection. Although green tribology is a fairly new concept; however, it already plays an important role in ensuring that all industrial systems can be able to function in an environmentally friendly manner. Green tribology is especially tuned to sustaining an ecological balance and biological effects on contact between surface systems from different materials. Green tribology ensures that any process of friction and wear is as environmentally friendly as possible. Thus, green tribology can be defined as an interdisciplinary field attributed to the broad induction of various concepts such as energy, materials science, green lubrication, and environmental science [8–11].

We have known the concept of green engineering for a long time. The United States Environmental Protection Agency (USEPA) defines green engineering as

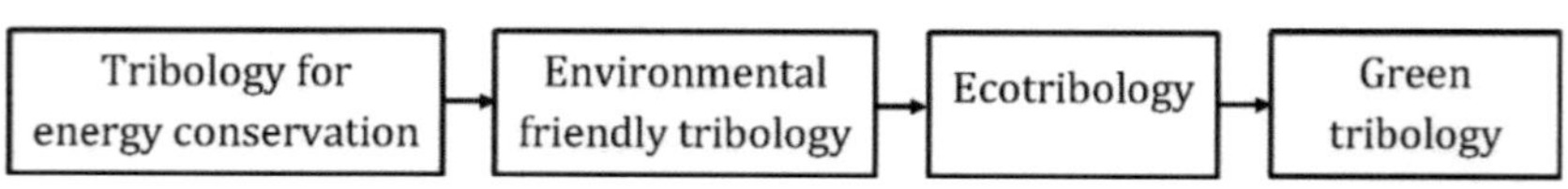

Figure 1.
Tribology concepts in the new main direction of development [1, 2].

"the design, commercialization and application of processes and products that technically and economically reduce sources of pollution and risks that adversely affect human health and the environment". Speaking of green tribology, one cannot ignore the terms "Green Technology", "Green Engineering", "Green Metalworking", etc., but the first association and historically, the first green science, as applied to science naturally, was "Green Chemistry". Green engineering and green chemistry are two closely related fields of green tribology that are actively engaged by researchers today [12, 13].

Specifically, green tribology has been identified as an area of engineering that could go beyond its original remit of improving efficiency by minimizing wear and friction in tribological processes to save energy and resources, minimize noise pollution, develop new bio-lubricants. In general, green tribology gives a positive contribution in reducing environmental harm. Inevitably, the term "green tribology" is spoken of in the context of quality of life. According to professor's Zhang opinion [2]: "Thus, the concepts and objectives of green tribology might be summarized into 3L + 1H, namely, low energy consumption, low discharge (CO_2), low environmental cost, and high quality of life. The mission of green tribology is researching and developing tribological technologies to reach the main objectives, thus making the sustained artificial ecosystems of the tribological parts and tribosystems in the course of a lifecycle".

Emilia Assenova and her colleagues in their work "Green tribology and quality of life" [14] reported: "Nowadays, losses resulting from ignorance of tribology amount to about 6% of the gross national product (GNP) in the United States alone. This figure is around USD 900 milliard annually. As far as China is concerned, they could save above USD 40 milliard per year by the application of green tribology or more than 1.5% of the GNP". It is clear that the basic goals of green tribology are "friction control, wear reduction and improved lubrication". Nevertheless, from a socio-economic point of view, it is possible to extend and confirm that the goal and essence of research works in the field of green tribology is to save material resources, improve energy efficiency, decrease emissions, shock absorption, investigate and apply novel natural bio- and eco-lubricants as well as to reduce the harmful effects of technical systems on the environment, and consequently, improve the quality and welfare of society. All advances in green tribology will lead to a high economic efficiency due to reduced waste and increased equipment service life, improved technological and environmental balance, decreased carbon footprint of mechanical systems, as a result, mitigate climate changes, and improve overall sustainability and safety in human life [15].

Green tribology will play an irreplaceable role in saving energy, material resources and environment. Trusted researches reveal that about 23% of energy consumption in the world today is the result of inefficient performance of tribological systems (**Figure 2**) [16]. In this case, approximately 18–20% of the energy is consumed to solve friction problems, and the remaining 3–5% is used to rebuild, repair and replace parts worn out due to wear and other failures associated with wear. The researchers estimated that by applying advances in green tribology in terms of new surfaces, materials and lubrication technologies, the total global energy loss in tribological systems could be decreased by 18% in the next 8 years and up to 40% in the next 15 years. An additional advantage of environmentally friendly green tribology is a significant reduction in carbon dioxide emissions and economic costs.

In works [17–23] the researchers applied green tribology concept using new class of eco-friendly lubricants and materials for manufacturing anti-friction contact surfaces, as a result of which their coefficient of friction is significantly reduced while the wear resistance and longevity are greatly increased.

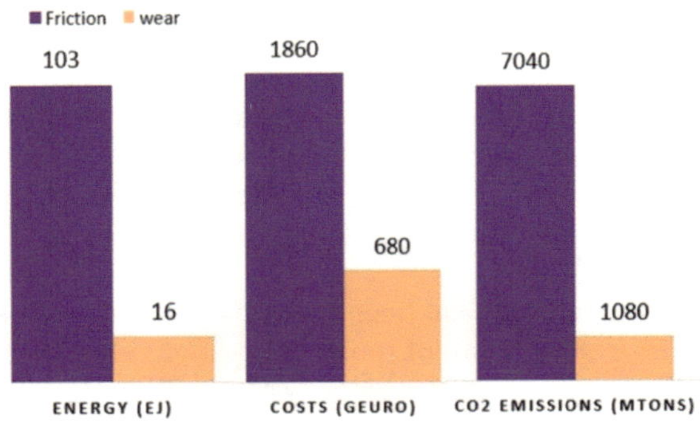

Figure 2.
Energy consumption, costs and CO_2 emissions due to inefficient performance of tribological systems globally [16].

A survey of gross energy consumption in the United States in four main areas: transportation, turbomachinery, power generation and industrial applications showed that savings of about 11% are achieved thanks to recent developments in lubrication and green tribology [24]. Chinese estimated that they could save more than $ 40 billion per year by applying advances in green tribology [25].

If we look at the share of wind energy in the total installed electricity capacity in Europe over the last decade, according to the European Wind Energy Association, it has increased more than quadrupled from 2.2% in 2000 to 10.5% in 2011 thanks to new developments in tribology, in particular as a result of the application of green tribology [26].

Many tribological problems can be put under the umbrella of "green tribology" and are mutually beneficial to each other. These problems are primary focus point of researchers and engineers, which include tribological technology that mimics living nature (biomimetic surfaces) and thus is expected to be environment-friendly, the friction and wear control that is important for energy conservation and conversion, environmental aspects of lubrication and surface modification techniques. These problems and aspects will be clarified in more detail in the next section.

3. Green tribology: principles, focus areas, and challenges

3.1 Principles of green tribology

As noted above, the interdisciplinary nature of green tribology often integrates aspects of chemical engineering and materials science in order to completely understand both chemistry and mechanics of surface. Since tribology is an interdisciplinary field, the principles of green engineering and green chemistry should also apply to green tribology. However, tribology includes not only chemistry of surfaces, but also other aspects related to the mechanics and physics of surfaces, there is a need to modify these principles.

Formulated by Paul Anastas in 1991, the 12 principles of green chemistry into a constant amount (12) upgraded to the 12 principles of green engineering [27, 28], and later, in the 12 principles of green tribology [1] mapped in **Table 1**.

These principles of green tribology can be assorted into 5 following groups: Friction, Wear, Lubrication, Material and surface production and treatment, and Tribology in the renewable energy sources.

Green chemistry	Green engineering	Green tribology
• Prevention. • Atom Economy. • Less Hazardous Chemical Syntheses. • Designing Safer Chemicals. • Safer Solvents and Auxiliaries. • Design for Energy Efficiency. • Reduce Derivatives. • Catalysis. • Design for Degradation. • Real-time analysis for Pollution. • Prevention. • Inherently Safer. • Chemistry for Accident Prevention.	• Inherent rather than circumstantial. • Prevention instead of treatment. • Design for separation. • Maximize mass, energy, space, and time efficiency. • Output-pulled versus input-pushed. • Conserve complexity. • Durability rather than immortality. • Meet need, minimize excess. • Minimize material diversity. • Integrate local material and energy flows. • Design for commercial "afterlife". • Renewable rather than depleting.	• Minimization of heat and energy dissipation. • Minimization of wear. • Reduction or complete elimination of lubrication and self-lubrication. • Natural lubrication. • Biodegradable lubrication. • Sustainable chemistry and green engineering principles. • Biomimetic approach. • Surface texturing. • Environmental implications of coatings. • Design for degradation. • Real-time monitoring. • Sustainable energy applications.

Table 1.
12 principles of green chemistry, green machine building and green tribology.

Friction *(minimization of heat and energy dissipation)*. Friction is the main source of energy dissipation, most of which is converted to heat. Controlling and minimizing friction, which results in both energy savings and the prevention of damage to the environment owing to heat pollution, is a top priority for green tribology. In addition, the friction in mechanical systems that operate on friction, such as clutches and brakes, also has to be well optimized.

Wear *(minimization of wear)*. This is the second most important task of green tribology. In most technological processes, wear is undesirable, it decreases the lifetime of elements/machine and creates the problems of their recycling/replacements which in turn leads to environmental damage by way of the emission. Wear can also lead to a large waste of material resources. In addition, due to wear, debris in the form of particles is generated, which pollutes the environment and in certain situations can be dangerous to humans.

Lubrication. *Reduction or complete elimination of lubrication and self-lubrication*. Lubrication is at the forefront of tribology as it reduces friction and wear. However, lubrication is also hazardous to the environment. It is desirable to reduce the use of lubricants or achieve a self-lubrication regime when no external lubrication is required. Tribological systems in living nature often operate in the self-lubricating mode. For example, the joints form a closed, self-sufficient system. Green tribology prompted researchers to think about self-lubricating materials, which also eliminated the external supply of lubricants.

Natural lubrication. In green tribology Natural lubricants such as vegetable oils should be used in cases when possible, since they are eco-friendly.

Biodegradable lubrication. Biodegradable lubricants should also be used when possible to avoid environmental pollution. In particular, water lubrication is an area that has attracted the attention of tribologists in recent years. Lubrication with natural oils is another good option.

Material and surface production and treatment. *Sustainable chemistry and green engineering principles*. These principles should be observed in the production of new materials, elements, parts, machines for tribological applications, coatings and lubricants.

Biomimetic approach. Wherever possible, biomimetic surfaces and materials, as well as other biomimetic and biological approaches, should be applied as they tend to be more environmentally friendly. Common engineered surfaces have occasional roughness, which makes friction and wear extremely difficult to overcome. On the other hand, many biological functional surfaces have complex structures with hierarchical roughness that determines their good properties for tribological systems.

Surface texturing. This technology should be used to provides a way to control many surface properties relevant to making tribo-systems more ecologically friendly.

Environmental implications of coatings. Environmental implications of coatings and other methods of surface modification (texturing, depositions, etc.) should be studied and taken into consideration.

Design for degradation. The ultimate degradation and utilization of contact surfaces, coatings, and tribological components should be considered during design.

Real-time monitoring. Tribological systems should be analyzed and monitored during operation to prevent the formation of hazardous substances.

Renewable energy sources (*Sustainable energy applications).* Sustainable energy applications should be a priority direction for tribological design, as well as engineering design in general.

Correct observation of discussed above principles of green tribology can greatly reduce the environmental impact of tribological process's products, assist economic development and, consequently, improve respectively the quality of life.

3.2 Focus areas of green tribology

Green tribology includes 3 main areas [1, 2, 5, 14], these are (1) Biomimetics (imitating living nature in order to solve complex human problems) and self-lubricating materials/surfaces; (2) Biodegradable and environmentally friendly lubrication and materials; and (3) Renewable and/or sustainable sources of energy. These 3 focus areas of green tribology aim to ensure a limited impact of tribological processes on the environment and human health. Below is a brief description and discussion about the features, contents, aspects of these areas and their relevance to green tribology.

Biomimetic and self-lubricating materials/surfaces. This is an important area of green tribology, the main task of which is the development and application of tribological technologies that mimic living nature (biomimetic surfaces). Many biological materials have amazing properties (superhydrophobicity, self-cleaning, self-healing, high adhesion, reversible adhesion, high mechanical strength, antireflection, etc.) that can hardly be achieved by conventional engineering methods. These properties of biological and biomimetic materials are reached due to their composite structure and hierarchical multiscale organization. It is noted that hierarchical organization and the ability of biological systems to grow and adapt also ensure a natural mechanism for the repair or healing of insignificant damage in the material. Biomimetic materials are also usually environmentally friendly in a natural way, since they are a natural part of the ecosystem. For this reason, the biomimetic approach in green tribology is especially promising.

In the field of biomimetic surfaces, a number of typical ideas have been proposed: (1) *The lotus effect based non-adhesive surfaces; (2) The Gecko effect based materials with the ability of specially structured hierarchical surfaces to exhibit controlled adhesion; (3) Fish-scale effect based micro-structured surfaces for underwater applications, including easy flow due to boundary slip, the suppression of turbulence and anti-biofouling; (4) Oleophobic surfaces capable of repelling organic liquids; (5) Microtextured surfaces for de-icing and anti-icing; (6) Various biomimetic microtextured surfaces to control friction, wear and lubrication; (7) Self-lubricating surfaces, using various principles, including the ability for friction-induced self-organization; (8) Self-repairing surfaces and materials, which are able to heal minor damage (cracks, voids); (9) The "sand fish" lizard effect, able to dive and "swim" in loose sand due to special electromechanical properties of its scale; (10) Nanocomposite materials tailored in such way that they can produce required surface properties, such as self-cleaning, self-lubrication, and self-healing.*

Figure 3 shows typical biological and biomimetic surfaces with hair or pillar like surface structures for various functions (**Figure 3**) [29]. Recently, the mechanisms of sand erosion resistance of the desert scorpion were studied to improve the erosion resistance of components in tribo-systems [30]. It was found that the biological surfaces used for sand erosion resistance of the desert scorpion were built by the special micro-textures such as bumps and grooves.

In works [31–33], the authors presented overview and studies of various biomimetic microtextured surfaces to control friction, wear and lubrication. Generally, biomimetic techniques have provided the different surface structures with strong adhesion, high hydrophobic properties, high coefficient of friction, self-lubrication, etc., which can be prospectively applied in green tribology field.

Biodegradable and environmentally friendly lubrication and materials. Advanced biomimetics is biomimicry used to identify best practices from nature on key tribological issues, such as finding improved lubrication solutions [1, 14, 34]. Natural lubrication is very effective at providing low coefficients of friction even at low speeds, and relies entirely on water as the base component, the effectiveness of which is ensured by the presence of many dissolved biomolecules.

Imitating such constructs of molecules, understanding their tribological performance is helpful. An example is the process of imitating natural lubricants, e.g. glycoproteins in synovial fluid [32]. By imitating this mechanism in the laboratory, molecules were synthesized that spontaneously produce polymer brushes on the surface. Brushes are formed on surfaces in an aqueous medium when end-grafted, water-soluble polymers are located at distance about one radius of gyration (Rg) from each other (**Figure 4**) [34] and stretch to maximize their interaction with water while reducing their interaction with each other.

The use of lubricants in machine components poses a serious threat to the environment, since they released into the environment not only contain harmful toxic waste but also contain the wear debris from machine parts. Development of environmentally acceptable lubricant products is one of priority direction in green tribology. Vegetable oils and animal fats have been used as lubricants for a very long time throughout human history. However, following the industrial revolution and the advent of lubricants made from mineral oils, bio-based lubricants have again come to be seen as an environmentally alternative for lubricant production and have only become effective in recent decades.

Researchers confirmed that properly formulated bio-lubricants are comparable with mineral based lubricants, so they could be used as an adequate substitution in appropriate cases. Vegetable-oil-based or animal-fat-based lubricants are potentially biodegradable that can be used for engines, hydraulic and metal-cutting applications. Vegetable oils i.e. corn, soybean and coconut oil, can have excellent lubricity, far superior than that of mineral oil [12, 14]. In general, the advantages of using bio-lubricants are non-toxic, biodegradable, renewable resources, good lubricity and high viscosity indices (**Table 2**) [35], while disadvantages are:

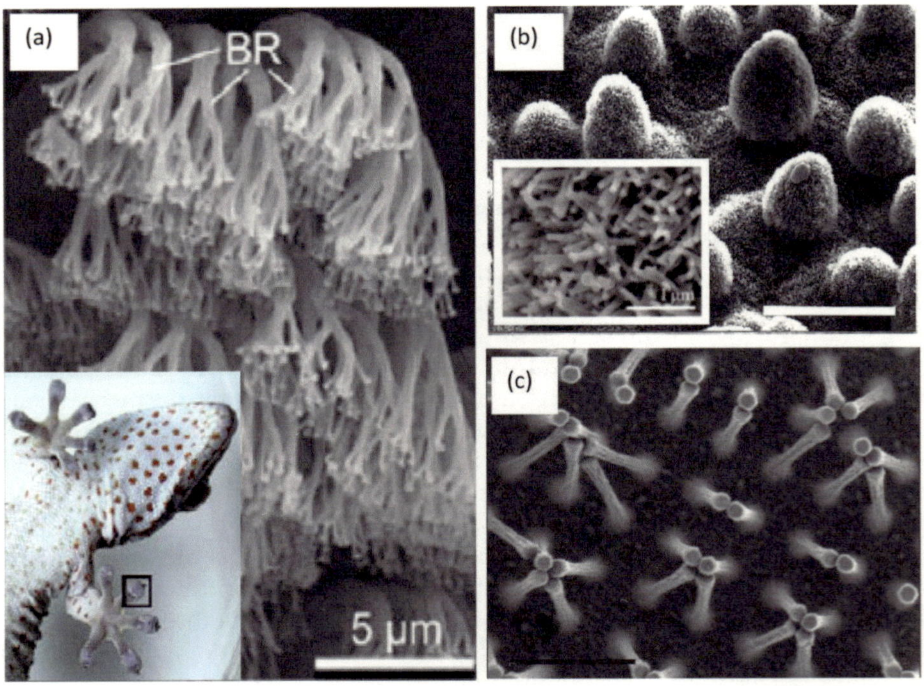

Figure 3.
Typical biological and biomimetic surfaces with hair or pillar like surface structures for various functions. (a) The nano to micro hierarchical hair-like surface structure of geckos' feet for strong adhesion; (b) The nano to micro hierarchical structure of plant leaves for superhydrophobic dewetting properties. (c) The microfabricated polyimide biomimetic hairs bunching together under the van der Waals interaction [29].

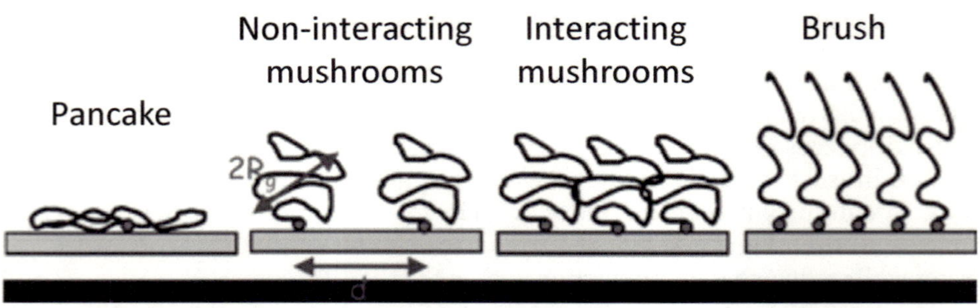

Figure 4.
The formation of polymer brushes on surfaces [34].

oxidative instability, poor low temperature properties, and hydrolytic instability. Applying chemical modification or additives can address these problems of bio-lubricants.

In the area of eco-friendly and biodegradable lubrication and materials we should also notice other following interesting ideas:

Hyrdo-lubrication. These are homogeneous lubricants containing water as a functional component. Tribological study and case analyses of the elastomeric bearings lubricated with seawater for marine propeller shaft systems were conducted [36].

Ionic liquids for green molecular lubrication. Ionic liquids (ILs) have been explored as lubricants for various device applications due to their excellent electrical conductivity as well as good thermal conductivity, where the latter allows frictional heating dissipation [37].

Oil type	Engine oil	Coconut oil	Palm oil
CO_2 (%)	4.5	2.9	3.4
CO (%)	0.92	0.67	0.73

Table 2.
The percentage content of CO and CO_2 in exhaust gas lubricated with regular mineral and vegetable oils [35].

Powder lubrication. Generally, these tend to be much more eco-friendly than the traditional liquid lubricants. Recent researches show that when using some nanoscale additives, such as boric acid and MoS_2 nanopowders to natural oils, their lubricity characteristics are significantly improved [38].

New eco-friendly coating materials for tribological applications. Recently, special attention has been paid to the development of "green" coatings in tribo-systems, which have improved tribological properties (low friction coefficient, high wear resistance), and therefore, not releasing a lot of worn-out waste into the environment, they are environmentally friendly [1, 2, 6].

Tribology in the Renewable Energy Sources (RES). Controlling and minimizing of friction and wear in tribology is important for energy and resources conservation. Sustainable energy applications have become priority of the tribological design, as well as an important area of green tribology. In contrast to the biomimetic approach and environmentally friendly lubrication, RES is not about manufacturing or operation, but about the application of the tribological system in production of renewable eco-friendly energies such as wind energy, marine energy, solar energy, geothermal energy, and so on.

In work [39] Wood et al. carried out the tribological studies on renewable sources of energy, namely three green energy systems: wind, tidal and wave machines. The authors also highlighted the role of design and durability for such large scale engineering systems from sustainability point of view. These systems are sensitive to operation and maintenance costs and thus depend on functioning tribological parts and lubrication. It was noted that weight reduction to reduce tribological and gravity loads would be beneficial for machines designs. Attention should also be paid to the knowing of dynamic loads to predict fatigue life and tribological loads on wind, tidal and wave machines. Structures and properties of tribological components must be considered for the inherent lack of stiffness of the turbines and wave devices.

Wind turbines have fairly many specific problems related to their tribology, which involve water contamination, electric arcing on generator bearings, wear of the main shaft and gearbox bearings and gears, the erosion of blades due to solid particles, cavitation, rain, hail stones, etc. The most commonly observed and discussed tribological problems in wind turbines are in the transmission system, in the gearbox. They are mainly the result of insufficient lubrication and/or lack of regular maintenance under extreme operating conditions. The solution to this problem is the use of lubricants and/or materials with improved tribological characteristics [40]. REWITEC nano-coatings is a metal treatment that can be applied to gearboxes and bearings during regular operation for restoration of its efficiency and economy. When examining certain micro-pitting areas on the metal surfaces of a wind turbine gear before applying REWITEC and after 6 months of treatment, it was found that the surface damage was filled and the asperities were smoothed out, and thus the surfaces became smoother with higher surface contact area (**Figure 5**) [2].

Tidal power turbines are another important way of producing renewable energy. Besides tidal, the ocean water flow and wave energy and river flow energy (without dams) can be used with the application of special turbines, which provides the same

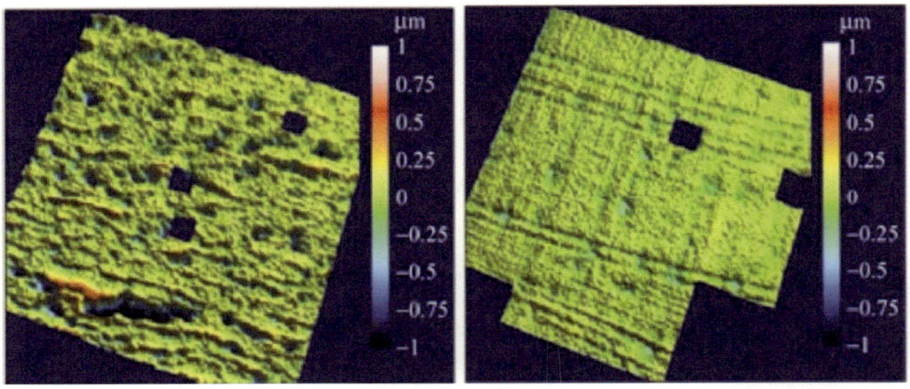

Figure 5.
3D-images of the metal surface before and after treatment with REWITEC 6 months [2].

direction of rotation independent of the direction of the current flow. Production processes of tidal, water flow and wave energy involve certain specific tribological problems such as lubrication of machine components (by seawater, oils, and greases), their erosion, corrosion, and biofouling, as well as the interaction between these modes of damage [1, 39].

Geothermal energy plants are widely used now, however, their application is limited to the geographical areas at the edges of tectonic plates. There are several specific tribological issues related to the geothermal energy sources which are discussed in the literature [1, 5, 15, 39].

3.3 Challenges of green tribology

Green tribology as a new area of tribology has a number of challenges. One obvious problem is integration, synergy of its above mentioned focus areas so that they can benefit from each other. Obviously, a lot of researches is needed to integrate the fields of green tribology. Some ideas can be borrowed from the related fields of green chemistry and green engineering, for example, the development of quantitative parameters for assessing the impact of tribological technologies on the environment. It is also important to develop quantitative measures and metrics that would allow us to compare which tribological material, technology, or application is "greener," i.e., produces lower carbon footprint, less waste from worn-out materials, and less chemical and heat pollution to the environment.

Green tribology should be integrated into world science and contribute to solving global problems such as resource depletion, environmental pollution and climate change. The application of principles of green tribology by itself, of course, will not solve world problems, and only major scientific achievements can become the key to their solution.

In the face of a large number of tribological problems requiring an early solution, which related to the environmental pollution, crisis of energy and resources on global scale, green tribology should be extended in the following directions [2].

- Large-scale deployment of existing knowledge, methods, and technologies of green tribology;

- Research and development of novel green tribological technologies;

- Research and development of tribo-techniques to support diversification and hybridization of renewable and clean energy;

- Making the traditional tribo-materials and lubricating materials "green" in the course of a lifecycle, namely, realizing cleaner production or eco-design of the these materials;

- Building up the theory and methodology of green tribology.

Consequently, tribologists should devote all their efforts to the investigation, application and development of green tribology, thereby making a valuable contribution to the existence and development of humanity.

4. "Zero-wear" effect: selective transfer

It seems expedient, at least briefly, to consider how the achievements that were obtained in the study of self-organizing tribo-systems, and in particular, the "zero-wear" (effect of wearlessness/effect of selective transfer—ST), play a role in the circle of tasks which green tribology is designed to solve.

The effect of ST in friction was registered as opening in 1966, with a priority in 1956. The authors of this discovery – D.N. Garkunov and I.V. Kragelsky – stated that the essence of the observed phenomenon as follows: "...that in the friction of couple copper alloys-steel under boundary lubrication, eliminating the oxidation of copper, there is a phenomenon of ST of a solid solution of copper from copper-alloy to steel and its transfer backwards from steel to copper alloy, with a reduction of the friction coefficient as liquid lubrication and leads to a significant reduction in wear of the friction pair..." [4].

In the closing years of the XX century the "zero-wear" effect is defined as one of the examples of self-organization in frictional interaction in tribological systems [41, 42], and since then, a synergistic approach at his description has become essential.

Classical tribo-system for realizing of ST is a system of "copper alloy (bronze or brass) – aqueous or alcoholic solution of glycerol – steel". The evolution of the tribological properties of this system visually demonstrated the self-organization in friction in ST mode, which is expressed in the ultra-low frequency vibrations of the friction coefficient and of the size of the rubbing bodies (**Figure 6** [42]).

Self-organization in the ST mode during friction is the consequence of the complex tribo-chemical reactions and physico-chemical processes occurred in the area of frictional contact, which lead to the manifestation of unique tribological characteristics: super-antifrictional (friction coefficient ~ 10^{-3}) and without wear (intensity wear ~10^{-15}). This condition of tribo-system was provided by a protective nanocrystalline servovite film made of soft metal with unusual combination of mechanical properties [43]. According to the results of nanoindentation, such a film has "super-hardness" at compression and "super-fluidity" at shear [44].

Within the framework of the I.V. Kragelsky's molecular-mechanical theory, the providing extremely low friction coefficients and practical absence of wear during friction of solids is possible either at spontaneous generation of wear auto-compensation systems or in the case of friction of perfectly smooth two-dimensional crystals, in which show up only molecular component of the friction force that occurs, such as, during friction of graphene [45].

In the engineering practice, the auto-compensation systems of wear during friction in the ST regime, usually are formed by selecting (a) the materials of tribo-coupling, (b) a composition of lubricants, and (c) a construction of the friction units. As a result of successful material science and engineering solutions, tribosystems are capable of self-organization, in which the process of frictional interaction

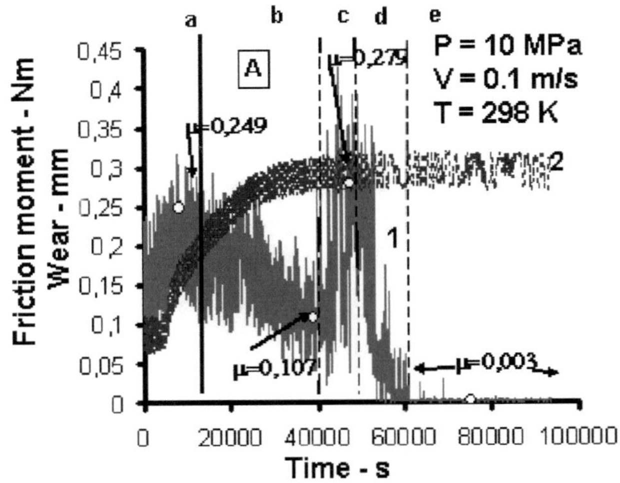

Figure 6.
The evolution of the tribological properties (1 – friction moment, 2 – the linear wear) in tribo-system brass-glycerol-steel. A (a, b) – running-in; c, d – transition mode; and e – the ST mode [42].

moved to the nanocrystalline quasi-liquid [3], and thus provides the friction coefficient, which is characterized for hydrodynamic friction, forming nanoclusters with almost perfect crystals, that leads to increases in load capacity and wear resistance of the friction surfaces.

In practice, "zero-wear" functioning of friction is achieved most often by application of metal-plating lubricants in the real friction units: oils, plastic lubricant, self-lubricating materials and coatings [3].

The mechanism of "zero-wear" effect during friction does not follow from the existing theoretical conclusions about the nature of the frictional interaction. Therefore, none of the attempts to propose developed in detail and experimentally substantiated scientific approach to explaining the "zero-wear" effect is currently generally recognized, although works aimed at clarifying the causes of friction without wear have been underway for more than half a century, during which reliable experimental facts have been accumulated and consistent approaches have been proposed that allowed to qualitatively explain the evolution of the tribotechnical characteristics of friction pairs during realization of the ST.

Currently, it has been reliably established that the composition, thickness and properties of servovite film during frictional interaction continuously change so that the extrapolated to the infinity friction surface is a pure copper (**Figure** 7), whose stability during friction is provided by the absorption of surfactants from the lubricating medium [42, 46].

Detailed studies on tribochemical reactions, as well as the evolution of the chemical composition of the servovite film on friction surfaces in the "zero-wear" regime, made it possible to characterize in detail the products arising during friction and to establish their role in the mechanisms of formation of boundary layers during self-organization of not only the classical tribosystem "copper-glycerin-steel", but also a number of more effective tribosystems using other lubricants, such as aqueous solutions of polyhydric alcohols, solutions of sucrose, glucose, galactose and other carbohydrates [3].

String of sequential and parallel chemical reactions: tribo-oxidation, tribo-coordination, tribo-restoration, tribo-reducing decay of coordination compounds, tribo-polymerization, tribo-clusterization and others, etc., accompanying, and/or generating vibrational tribo-chemical reaction, vibrational electrical and

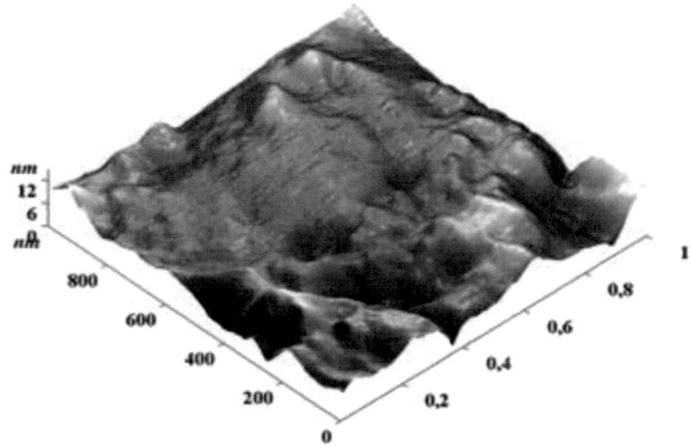

Figure 7.
AFM (3D-visualization) of the surface of the servovite transfer film (brass on steel in glycerol) [42].

electrochemical effects, vibration of the size of rubbing bodies and tribological characteristics of friction pairs - that's the one, is far from complete.

In the most general case, the evolution of the open tribo-system "copper alloy-glycerin-steel", classical for the realization of the "zero-wear" effect, from the thermodynamically equilibrium state of rest under constant external initial conditions (P, V, T) to the friction regime without wear always starts with high (more than 0.1) values of the friction coefficient and large running-in wear, which leads to an increase in the energy intensity of the frictional contact zone and triggers complex physicochemical transformations in the lubricating medium and on the contacting surfaces of copper alloy and steel. At the same time, in the initial period of time, the friction of the copper alloy against steel in glycerin does not differ in nature from the boundary friction, which manifests itself both in the tribo-technical, electrical and electrochemical contact characteristics. The products of wear accumulating at this time in glycerin have a very wide particle size distribution from 10^{-7} to 10^{-3} m and are almost exclusively particles softer from contact bodies of copper alloy. Wear, repeatedly increasing surfaces of copper alloy in the tribo-system, leads to the predominant role of topochemical and tribo-chemical effects, both in the lubricant composition and on the friction surfaces, which reflect in the tribo- and topochemical oxidation of glycerin with the accumulation of a wide range of oxygen-containing surfactants (aldehydes, ketones, carboxylic acids, ethers and esters, as well as oligomeric and polymeric products of their further transformations). Parallel to this, the formation of complex compounds (tribo-coordination) occurs both on the surface of wear particles and on the friction surfaces, and the soluble coordination metal compounds accumulate in the solution [3, 47–49].

Leading of tribo-chemical mechanisms on electrochemical reasons on the friction surface and on the surface of wear of particles is oxidation of copper $Cu^0 - 2e = Cu^{+2}$ (in the case of bronze) or zinc $Zn^0 - 2e = Zn^{+2}$ (in the case of brass) in result of its selective dissolution. Other metals that are part of the friction alloys such as Fe, Sn, Pb, etc., are also subjected to tribo-oxidation with the formation of metal-containing products, so that in the lubricating medium and on the frictional surface simultaneously there is a wide gamma of products, in which the explicitly pronounced tendency in the initial period is the accumulation of oxidized forms of different metals and oxidation products of the lubricant. An important event at this stage of evolution is the accumulation in the lubricating medium and reducing the size of the metallic wear particles with a simultaneous change in their composition due to above reasons. This leads ultimately to the accumulation of metal-containing products upto critical concentrations and thus, to the change of lubricant composition in becoming a metal-plating lubricant. There is a radical change of physico-chemical, electrochemical and tribological situation on the friction surfaces and in the zone of frictional contact. Tribo-system in the course of long enough evolution (in the laboratory it is about 10^3 m of sliding distance) reaches the bifurcation point with transition either to "zero-wear" friction regime or to the regime of the catastrophic wear.

In the transition and functioning of the tribo-system under ST, both contacting surfaces in friction as copper alloy and steel have the same composition and structure. This is another paradox of ST and an unusual combination of materials of the rubbing surfaces. It has been observed that during friction of the same materials (usually in the friction units dissimilar metals and alloys are combined) record-breaking parameters of frictional interaction can be achieved, wherein the self-organization of frictional systems was achieved by the special structure of surface layers.

In the transition from boundary friction to "zero-wear" friction, due to the non-equilibrium character of the processes occurring in the tribo-system (since the system is far enough from the position of thermodynamic equilibrium), and their description by systems of nonlinear differential equations, oscillatory mechanisms begin to appear, which associated with both tribo-chemical transformations in the contact zone, for example, with fluctuations in the concentration of copper-containing products in the lubricant, and with the electrical, electrochemical and tribological characteristics of the contact (**Figure 8**) [50]. Observing this type of oscillations, which always accompany friction in the "zero-wear" regime, prove the manifestation of the self-organization in friction, as well as the transition and functioning of the tribological system in one of the stationary states.

The transient regime from boundary to "zero-wear" friction lasts significantly less than the boundary friction regime, but at this time there are main events that lead to the unique tribological characteristics of tribo-system. It is in this transition that the ordering process occurs, which associated with the formation of a servovite film on the friction surface. The servovite film is formed under nonequilibrium, non-isothermal and topographically unequal conditions, which leads to inevitable differences in its composition and properties in different places of frictional contact. Nevertheless, formation of the film is always due to mutually complementary

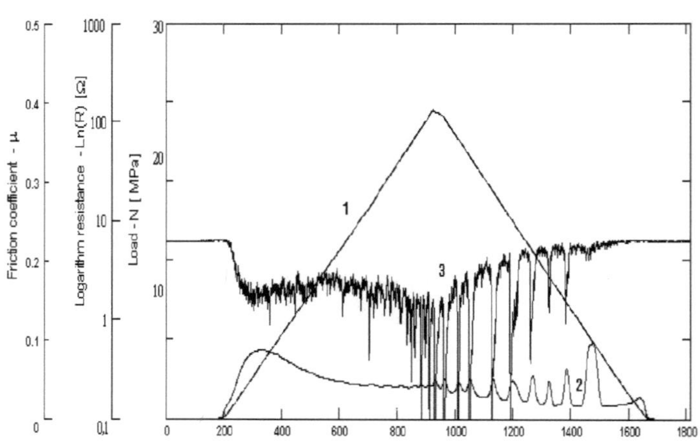

Figure 8.
Fluctuations in the transient regime of "boundary friction – ST" in the friction pair "AISI 1045 Steel–AISI 1045 Steel" in the lubricant of copper nanocluster in glycerol. 1 – load; 2 – friction coefficient; 3 – electrical resistance of contact [50].

processes of tribo- and electrochemical reduction of coordination compounds of soft metals on the friction surface, clustering their reduced forms, and optimization depending on the friction regimes (P, V, T), sizes (in the nanoscale) and the shape (triaxial ellipsoid) of the clusters in two ways "top-down" and "bottom-up" followed by the direct deposition of metal nanoclusters on the contact surfaces due to tribo-electrochemical effects. The formation of servovite film begins on the individual most active sections of the steel surface, which leads to reducing the friction coefficient and a decrease in the energy density the friction unit. Finally, it is accompanied by a decrease in wear and a transfer of the film formation process to less activated areas on the frictional contact surfaces.

Any system thermodynamically approaches to one of many possible stationary states, the choice of which is caused solely by the initial conditions. It should be noted that the trajectory of the tribo-system during evolution into the "zero-wear" regime is always strictly individual and can never be reproduced in detail. If the tribo-system self-organizes, which in the thermodynamic description is characterized by an increase in entropy and ordering, then its tribological and physicochemical characteristics in a stationary state become almost unchanged (**Figure 9**) [50].

This is due to the fact that a servovite film, formed from individual atoms and their small clusters, has a nanocrystalline structure and, on the one hand, is superstrong in compression, since its nanoparticles are fragments of almost ideal crystals, and on the other hand, the film is quasi-liquid and superplastic under tension and shear due to much weaker interactions between nanoparticles than between atoms in the metal crystal lattice [43].

In this regime, the system can function until continuously accumulating external disturbances or changing external conditions transfer it to a new stationary state, which may be characterized by other and not necessarily higher tribo-technical characteristics, which makes the practical implementation of "zero-wear" in real machines and mechanisms very complex and not always justified event.

At the same time, even with a partial realization of "zero-wear" friction, the effects can be impressive, since when functioning under self-organization conditions and with a slight change in external conditions, the transition to a nearby stationary state is accompanied, as a rule, by a slight change in the tribo-technical properties of the system.

Thus, the application of "zero-wear" effect in engineering practice opens a real opportunity for the design of friction units with significantly increased durability and ultra-high efficiency in terms of friction losses in moving machine interfaces. The "zero-wear" effect in the friction fully fits into the presentation and concepts of

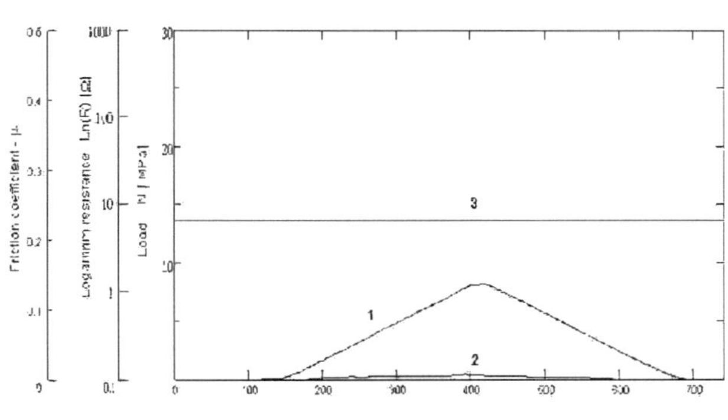

Figure 9.
Stationary regime in the realization of ST in friction pair "AISI 1045 Steel–AISI 1045 Steel" with lubricant of copper nanocluster in glycerol. 1 – load, 2 – friction coefficient, 3 – electrical resistance of contact [50].

green tribology and should be considered as the real embodiment in the theory and practice of modern engineering.

5. Conclusion

Green tribology is a novel area of science and technology. It is related to other areas of tribology as well as other "green" disciplines, namely, green engineering and green chemistry. In this chapter the main scientific and technological aspects of green tribology such as the concept, role and goal, principles, focus areas, challenges, "zero-wear" effect were considered in details.

The concept, role and goal of green tribology were clarified. Green tribology can be defined as an interdisciplinary field attributed to the broad induction of various concepts such as energy, materials science, green lubrication, and environmental science. The goal and essence of green tribology is to save material resources, improve energy efficiency, decrease emissions, shock absorption, investigate and apply novel natural bio- and eco-lubricants as well as to reduce the harmful effects of technical systems on the environment, and consequently, improve the quality of human life.

The twelve principles and three areas of green tribology were analyzed. Observation of these principles can greatly reduce the environmental impact of tribological processes, assist economic development and, consequently, improve the quality of life. The integration of these areas remains the major challenge of green tribology and defines the future directions of research in this field.

Within the framework of this work, one of the most important tribological effects, which is exclusively the basis for green tribology - the "zero-wear" effect in friction (selective transfer effect) was discussed.

As a result, this work allows us to conclude that green tribology is an environmentally friendly and energy-saving concept and, moreover, there are many opportunities for its inclusion in a sustainable society on a global scale. Furthermore, there is a need for tribologists to collaborate towards the development and application of Green tribology.

Acknowledgements

Our gratitude goes to the Institute of Technology and Don State Technical University for supporting this research. The authors also thank the reviewers for carefully reading the chapter and for their constructive comments and suggestions which have improved this paper.

Conflict of interest

The authors declare that there is no conflict of interest.

DOI: http://dx.doi.org/10.5772/intechopen.94510

Author details

Nguyen Van Minh[1*], Alexander Kuzharov[2†], Le Hai Ninh[1], Nguyen Huynh[1,2] and Andrey Kuzharov[2,3]

1 Institute of Technology, Hanoi 143315, Vietnam

2 Don State Technical University, Rostov-on-Don 344002, Russia

3 Southern Federal University, Rostov-on-Don 344006, Russia

*Address all correspondence to: chinhnhan88@gmail.com

† Deceased.

References

[1] Nosonovsky M, Bhushan B, editors. Green Tribology, Green Energy and Technology. Springer:Verlag Berlin Heidelberg; 2012. 632 p. DOI: 10.1007/978-3-642-23681-5.

[2] Zhang S. Green tribology: Fundamentals and future development. Friction. 2013;**1, 195**(2):186-194. DOI: 10.1007/s40544-013-0012-4

[3] Kuzharov S, The concept of wearlessness in modern tribology. Izvestiya vuzov: North Caucasian region. Series: Engineering Sciences 2014; 177: 23-31. [in Russian].

[4] Garkunov N, Kragelsky V, Selective Transfer Effect – Discovery N 41 with Priority, November 12th, 1956. [in Russian].

[5] Nasonovsky M, Bhushan B, Green tribology: principles, research areas and challenges. Phil Trans R Soc A 368. 2010; 4677-4694. DOI: 10.1098/rsta.2010.0202.

[6] Jost P. Development of Green Tribology - an Overview. Moscow: Seminar-New Direction in Tribotechnology; 2010

[7] Tzanakis I, Hadfield M, et al. Future perspectives on sustainable tribology. Renewable and Sustainable Energy Reviews. 2012;**16**:4126-4140. DOI: 10.1016/j.rser.2012.02.064

[8] Jost P, Tribology – from Basics to Productivity and Employment also commemorating the 40th Anniversary of the International Tribology Council. - 5th World Tribology Congress (WTC-2013); September 8-13, 2013; Torino; ISBN 978-88-908185-09.

[9] Jost P, The Presidential address, World Tribology Congress 2009; September 06-11, 2009; Kyoto, Japan.

[10] Bartz W. Ecotribology: Environmentally acceptable tribological practices. Tribology International. 2006;**39**(8):728-733. DOI: 10.1016/j.triboint.2005.07.002

[11] Kandeva M, Assenova E, Daneva M, Triboecology as a methodological center of modern science, in: Proceedings of the 2nd European Conference on Tribology ECOTRIB 2009; 07-10.06.2009; Pisa, Italy.

[12] Anand A et al. Role of green tribology in sustainability of mechanical systems: A state of the art survey. Materials Today: Proceedings. 2017; 3659-3665;**4**. DOI: 10.1016/j.matpr.2017.02.259.

[13] Zhang S. Green tribology - The way forward to a sustainable society. in Proceedings of the International Tribology Congress - ASIATRIB 2010; 5-9 December 2010; Perth, Western, Australia.

[14] Assenova E, Majstovovic V, Vencl A, Kandeva M. Green tribology and quality of life. International Journal of Advanced Quality. 2012;**40**:32-38

[15] Wood R, Green tribology, NCats newsletter, Ed. 6, October 2011.

[16] Holmberg K, Erdemir A. Influence of tribology on global energy consumption, costs and emissions. Friction. 2017;**5**(3):263-284. DOI: 10.1007/s40544-017-0183-5

[17] Nicolenco A, Tsyntsaru N, et al. Wear resistance of electrodeposited Fe-W alloy coatings under dry conditions and in the presence of rapeseed oil. Green tribology. 2018;**1**:16-23. DOI: 10.15544/greentribo.2018.04

[18] Jeong D, Erb U, Aust K, Palumbo G. The relationship between hardness

and abrasive wear resistance of electrodeposited nanocrystalline Ni-P coatings. Scripta Materialia. 2003;**48**:1067-1072. DOI: 10.1016/S1359-6462(02)00633-4

[19] Bochkov I, Varkale M. et al. Selected aspects of wear and surface properties of polypropylene based wood-polymer composites. Green tribology 1, Number 1. 2018; 5-8. DOI: 10.15544/greentribo.2018.02.

[20] Raspopov L, Matkovskii P. Wood–mineral–polymer composite materials. Polymers from Renewable Resources. 2011;**2**(3):117-130. DOI: 10.1177/204124791100200303

[21] Erhan S, Sharma B, Perez J. Oxidation and low temperature stability of vegetable oil-based lubricants. Industrial Crops and Products. 2006;**24**(3):292-299. DOI: 10.1016/j.indcrop.2006.06.008

[22] Bahadur S. The development of transfer layers and their role in polymer tribology. Wear. 2000;**245**(1-2):92-99. DOI: 10.1016/S0043-1648(00)00469-5

[23] Dong C, Zhang M, Xiang T, et al. Novel self-healing anticorrosion coating based on L-valine and MBT-loaded halloysite nanotubes. Journal of Materials Science. 2018;**53**(10):7793-7808. DOI: 10.1007/s10853-018-2046-5

[24] Levchenko V et al. Green tribology: Orientation properties of diamond-like carbon coatings of friction units in lubricating media. Russian Journal of Applied Chemistry. 2019;**92**(12):1603-1615. DOI: 10.1134/S1070427219120012

[25] Bronshteyn A, Kreiner H. Energy efficiency of industrial oils. Tribology Transactions. 1999;**42**(4):771-776. DOI: 10.1080/10402009908982281

[26] Jost P. 30th Anniversary and "Green Tribology" Report of a Chinese Mission to the United Kingdom, Tribology Network of the Institution of Engineering & Technology; 07-14.06.2009; London, UK; 2009.

[27] Anastas P, Warner J, Green Chemistry: Theory and Practice. Oxford University Press: New York; 1998. 135p.

[28] Anastas P, Zimmerman J, Design Through the 12 Principles Green Engineering. – Environmental Science & Technology. 2003; 94A – 101A. DOI: 10.1021/es032373g.

[29] Su Y, He S, Hwang C, Ji B, Why have not the hairs on the feet of gecko been smaller? Applied Physics Letters, 101. 2017; 173106. DOI:10.1063/1.4762822.

[30] Han Z, Zhang J, Ge C, Wen L, Lu R. Erosion resistance of bionic functional surfaces inspired from desert scorpions. Langmuir. 2012;**28**:2914-2921. DOI: 10.1021/la203942r

[31] Bhushan B. Biomimetics: Lessons from nature—An overview. Philosophical Transactions of the Royal Society A. 2009;**367**:1445-1486. DOI: 10.1098/rsta.2009.0011

[32] Nosonovsky M, Bhushan B. Thermodynamics of surface degradation, self-organization, and self-healing for biomimetic surfaces. Philosophical Transactions of the Royal Society A. 2009;**367**:1607-1627. DOI: 10.1098/rsta.2009.0009

[33] Varenberg M, Gorb S. Hexagonal surface micropattern for dry and wet friction. Advanced Materials. 2009;**21**:483-486. DOI: 10.1002/adma.200802734

[34] Spencer D, Understanding and Imitating Lubrication in Nature, in: Proceedings of the 2nd European Conference on Tribology ECOTRIB 2009, 07-10.06.2009; Pisa, Italy.

[35] Mannekote K, Kailas S. Performance Evaluation of Vegetable Oils as Lubricant in a Four Stroke Engine. Conference: World Tribology Congress. Vol. 2009. Kyoto, Japan: September; 2009

[36] Hirani H, Verma M. Tribological study of elastomeric bearings for marine propeller shaft system. Tribology International. 2009;**42**:378-390. DOI: 10.1016/j.triboint.2008.07.014

[37] Palacio P, Bhushan B. A review of ionic liquids for green molecular lubrication in nanotechnology. Tribology Letters. 2010;**40**:247-268. DOI: 10.1007/s11249-010-9671-8

[38] Lovell M, Kabir M, Menzes HC. Influence of boric acid additive size on green lubricant performance. Phil. Trans. Royal. Soc. A. 2010;**368**:4851-4868. DOI: 10.1098/rsta.2010.0183

[39] Wood K et al. Tribological design constraints of marine renewable energy systems. Philosophical Transactions of the Royal Society A: Mathematical, Physical and Engineering Sciences. 2010;**368**:4807-4827. DOI: 10.1098/rsta.2010.0192

[40] Kotzalas M, Lucas D. Comparison of Bearing Fatigue Life Predictions with Test Data, in Proceedings AWEA Wind Power. Vol. 3-6. Los Angeles [CD-ROM]: June; 2007

[41] Kuzharov S et al. Molecular mechanisms of self-Organization in Friction, P. 1, investigation of self-organization during hydrodynamic friction. Tr. i Iznos. 2001;**22**(1):84-91

[42] Kuzharov S, Marchak R, Features of evolutionary transition tribological system brass glycerol-steel mode "zero-wear" friction, Presentation RAS, Vol. 354.- No 5. 1997; 642-644. [in Russian].

[43] Kuzharov S, Specifics of deformation of copper during friction in condition of "zero-wear" effect, Vesnik DSTU, Vol. 5, No 1(23). 2005; 137-138. [in Russian].

[44] Kuzharov S. et al, Nanotribological "zero-wear" effect. - 5th world tribology congress (WTC-2013), September 8th-13th, 2013; Torino, Italy.

[45] Zhang Q et al, Tribological Properties and Mechanism of Graphene by Computational Study. 40th Leeds-Lyon Symposium on Tribology & Tribochemistry Forum 2013 September 4th–6th, 2013; Lyon, France.

[46] Kuzharov A, Tribological Properties of Nanoscale Copper Clusters, Abstract of Thesis, DSTU - Rostov on Don; 2004, [in Russian].

[47] Kuzharov S, Physical-chemical basics of lubricating action in regime of selective transfer, "zero-wear" effect and tribology, No 2. 1992; 3-14. [in Russian].

[48] Polyakova A, Role of surfing-film in selective transfer, Friction and wear, Vol. 12, No. 1. 1992; 108-112. [in Russian].

[49] Simakov S, Physical-chemical process during selective transfer.The selective transfer in heavily loaded friction units, pub. house: Engineering. 1982; 152 - 174. [in Russian].

[50] Kravchik K, Tribological identification of self-organization during friction with lubricant, Thesis of Doctor technical science, Rostov on Don; 2000. [in Russian].

Chapter 4

Tribological Behavior of Polymers and Polymer Composites

Lorena Deleanu, Mihail Botan and Constantin Georgescu

Abstract

This chapter means to explain the tribological behavior of polymer-based materials, to support a beneficial introducing of those materials in actual applications based on test campaigns and their results. Generally, the designers have to take into consideration a set of tribological parameters, not only one, including friction coefficient, wear, temperature in contact, contact durability related to application. Adding materials in polymers could improve especially wear with more than one order of magnitude, but when harder fillers are added (as glass beads, short fibers, minerals) the friction coefficient is slightly increased as compared to neat polymer. In this chapter, there are presented several research studies done by the authors, from which there is point out the importance of composite formulation based on experimental results. For instance, for PBT sliding on steel there was obtained a friction coefficient between 0.15 and 0.3, but for the composite with PBT + micro glass beads, the value of friction coefficient was greater. Adding a polymer playing the role of a solid lubricant (PTFE) in these composites and also only in PBT, decreased the friction coefficient till a maximum value of 0.25. The wear parameter, linear wear rate of the block (from block-on-ring tester) was reduced from 4.5 μm/(N·km) till bellow 1 μm/(N·km) for a dry sliding regime of 2.5 ... 5 N, for all tested sliding velocities, for the composite PBT + 10% glass beads +10% PTFE, the most promising composite from this family of materials. This study emphasis the importance of polymer composite recipe and the test parameters. Also there are presented failure mechanisms within the tribolayer of polymer-based materials and their counterparts.

Keywords: polymer, composites with polymer matrix, polymer blends, tribology, wear, friction, tribolayer, tribological behavior, wear mechanisms

1. Issues related to the use of polymeric materials in tribological applications

Plastics and materials based on plastic have become an acceptable replacement of metallic materials and, as a consequence, they have to face the challenge of having also a good tribological behavior, implying a set of characteristics favorable to a reliable functioning of the application.

Issues that an engineers (both designers and users) have to pay attention when using polymeric materials in tribological applications include dimensional stability. These materials have higher thermal expansion coefficients, shorter durability, sensitivity and particular behavior to high and low temperatures. As they are characterized by lower hardness, they are not prone to be introduced in rolling contacts, will few exceptions (here including car tires and gears), most applications being for sliding motion (belt, sliding bearing, seals, brakes etc.).

The advantages of using polymeric materials (polymers, blends and composites) [1–5] include self-lubricity, lower density as compared to metallic materials, resistance to tribocorrosion [6] or general oxidation, non-toxic nature and potential processing to final shape, usually, by injection molding. But their favorable properties come in a package with disadvantages. One is that a slight change in working conditions (load, velocity, temperature etc.) could substantially modify tribological characteristics [7], especially wear rate and low friction is not related to low wear rate. Also, negative temperatures have different influences on polymeric materials (some become brittle, some resist without problems and some are conditioned by the working conditions and environment).

Figure 1 presents materials based on polymers and elastomers that could be used in tribological applications.

When using polymeric materials, the designer should pay attention how the component will obey design requirements, if it has dimensional stability, mechanical characteristics with reliable values, if issues related to aging are acceptable for

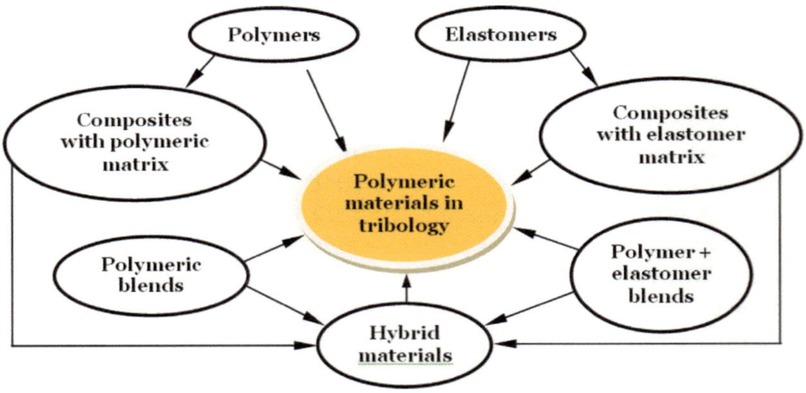

Figure 1.
Materials based on polymers and elastomers, involved in tribological applications [1–5].

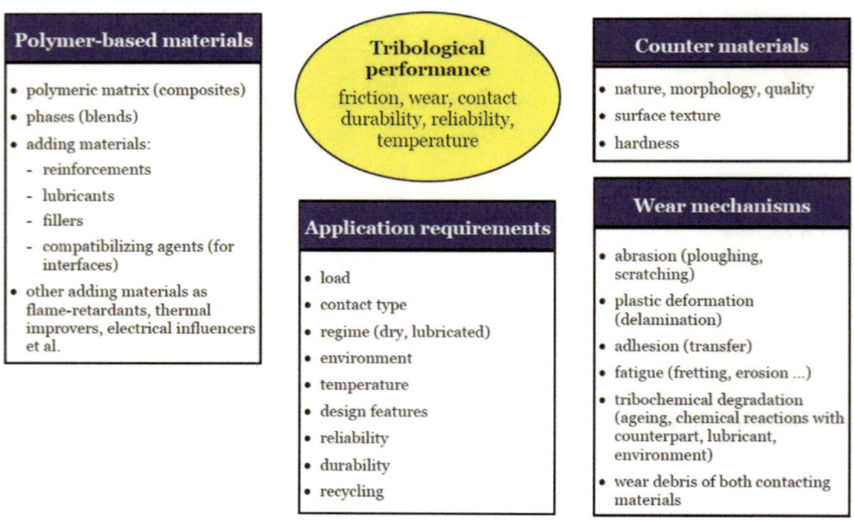

Figure 2.
A chart of significant aspects related to tribological performance implying polymeric materials.

the component durability (life time). The design should be done so that the working conditions will vary in narrow ranges (temperatures, load, velocity, material composition and morphology) [8].

The majority of tribological applications with polymeric materials are involving couples with one element made of metallic materials, the other being polymeric. Sometimes, the polymeric material is moving against a body made of the same materials, an example being gear transmissions.

Figure 2 summarizes the main aspects of tribological performance when using polymer-based materials.

2. Polymers, and materials with polymers for tribological applications

Table 1 presents the polymers used in tribological applications, several features and usual components made of them.

Polymer	Tribological characteristics
PTFE	Low friction, but high wear rate. Used both neat, as matrix and as solid lubricant. More recently, added in polymers, resulting polymer blends; in composite as solid lubricant or matrix in composite with reinforcements as glass fibers, carbon fibers, metallic powder as copper. High working temperature [9, 10]
PA	Moderate friction coefficient, low wear rate, but too sensitive to water and humidity. Working temperature quite low [11].
POM	Similar performance as PA. Good durability in rolling contacts.
PEEK	Polyetheretherketone, semicrystalline High working temperature and very good chemical resistance. Accept higher contact pressure but high friction coefficient as neat polymer [12–14]
UHMWPE	Very good wear resistance, especially against abrasion, even in water. Moderate friction coefficient. Modest working temperature.
PU	Good wear resistance in rolling contacts. Relatively high friction coefficient in sliding.
PI	High performance polymer with very good behavior in high contact pressure. Higher friction coefficient.
PBT	A reliable behavior in sliding contact, lower wear as PA, but more restrictive condition in molding. Usually with a solid lubricant or reinforcement [15, 16]
PEI	Amorphous thermal stability, very good mechanical and physical properties, easy processability, applicability and possibility of recycling and repair, thermosetting polyimides, blended with PEEK [13]
PES	Amorphous [17]
PPS	Semicrystalline, polyphenylenesulphide, water lubrication high glass transition and high melting temperature and high mechanical strength, high COF on steel in dry regime (0.4...0.5), PPS + SWCNT (0.5 wt.%) + WS_2 (1.5 wt.%) [18, 19]
PPP	Polyparaphenylene, semicrystalline, very high mechanical stability at room temperature, poor wear resistance [12]
PBI	Polybenzimidazole semicrystalline, high heat resistance and mechanical property retention, even under high temperatures [12]
Epoxy and phenolic polymers	Used especially as binder agents in composites, they induce high friction, but constant. Their brittleness induces wear by micro-detaching harder particles (as a dust) that could damage the smooth functioning of the tribosystem. The composites with these resins usually are designed for frictional applications (high and constant friction coefficient, with controlled wear evolution in time)

Table 1.
Tribological characteristics of thermoplastic polymers [5–8, 20].

Semi-crystalline polymers can be used even above their glass transition temperature (Tg), another added advantage against chemical constancy.

Various inorganic nanofillers, e.g., from metals (Cu, Fe), metallic and non-metallic oxides (CuO, ZnO, TiO_2, ZrO_2, SiO_2) and salts as silicon nitride (Si_3N_4), have been proved to not only enhancing mechanical properties, but also to lowering the friction coefficient and the rate of wear under various sliding circumstances. In particular, PEEK, PPS, and PTFE are the most widely studied polymers for different tribological applications and they are often blended with TiO_2, SiC, Si_3N_4, and carbon fiber fillers [19]. Nevertheless, it is also noted that there are no single or combined polymers or fillers that provide the best tribological performance in all conditions. Being a result of "system responses", friction and wear always depend on both the intrinsic material properties and the external environmental conditions. The beneficial effect of adding a certain material in a polymeric matrix is exemplified by tests did by Kurdi et al. [21] (**Figure 3**), 5–15% of TiO_2 reducing friction and wear at room temperature, but not at elevated temperature. Thus, functioning conditions are tremendously important when selecting a pair of materials for a good or at least acceptable tribological behavior.

Hanchi et al. [13] reported results on friction and wear under dry sliding of injection molded blends of PEEK and PEI, at temperatures from 20–232°C, on a pin-on-disk tribotester. It was found that tan δ peaks corresponding to α transitions occurring in the vicinity of the glass transition temperature (Tg) coincided with catastrophic tribological failure in the case of PEI and the amorphous PEEK/PEI blends. PEEK and the annealed 70% PEEK/30% PEI blend exhibited marked increases in friction and wear above the Tg. The absence of catastrophic tribological failure in PEEK and the annealed 70/30 blend in the vicinity of Tg corresponded to a transition of significantly lower strength those observed in PEI and the amorphous blends. Between 90°C and 105°C for PEI and 45°C and 70°C for the PEEK/PEI 50/50 blend, severe to mild friction and wear transitions were observed. It appeared that a substantial change in ductility associated with these β transitions resulted in the transitional tribological behavior.

Unal et al. reported the influence of test speed and load values on the friction and wear behavior of PTFE, POM and PEI, on a pin-on-disc tribotester. Tests were carried out at room temperature, under 5 N, 10 N and 15N and at 0,5 m/s, 0,75 m/s and 1m/s. The specific wear rates were deduced from mass loss. The results showed that, for all tested polymers, the coefficient of friction increases linearly with the increase in load. For the load and speed range of this investigation, the wear rate showed very low sensitivity to the applied load and large sensitivity to speed, particularly at high load values [22].

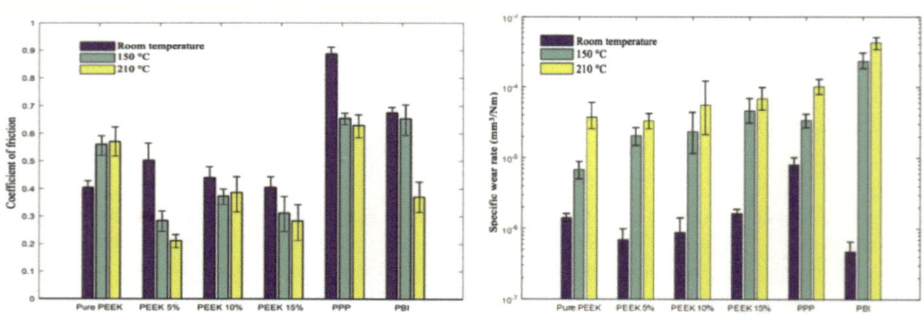

Figure 3.
Influence of percentage of TiO_2 on (a) friction coefficient and (b) specific wear rate, for a pin-on-disk configuration, in sliding at v = 0.1 m/s, average pressure p = 1 MPa, for 2 hours [21].

What do the engineers want from polymeric materials when introducing in tribological applications? A set of characteristics including thermal, mechanical and tribological ones:

- higher softening temperatures, sometimes obtained by adding short glass fibers;
- higher toughness; reinforcement could rise the flexural modulus till 11,000 MPa, a value that is overpass only by PPS in the thermoplastic polymers;
- low or acceptable friction and high wear resistance;
- good strength al negative temperature, including impact resistance;
- no or very less liquid absorption (including water)
- chemical resistance at fluids circulated in application (as lubricant or/and environment);
- good dimensional stability; low thermal expansion;
- good ability for compounding (mixing), when adding materials for reinforcement, solid lubricants, anti-ignition agents etc.,
- good processing capability (uniform flow, fast solidification and acceptably low cost and improvement by treatment).

Based on important works on tribology of polymer-based materials [3, 20, 23–25]. **Figure 4** presents a classification of adding materials taking into account the function of these materials in polymers. Generally, reinforcements [24–27] and solid lubricants in polymer-based materials improve their tribological behavior, but it is not a rule and the new recipes should be tested at laboratory scale and then the designed components at actual scale and under functioning conditions. Some solid lubricants, especially with sheet-like aspects (graphite, graphite, sulphides etc.) weaken the bulk materials as they reduce the superficial energy, but the mechanical properties are diminishing. Reinforcements in polymers make their resistance greater, but generate a more intense abrasive wear on the counterpart surface and

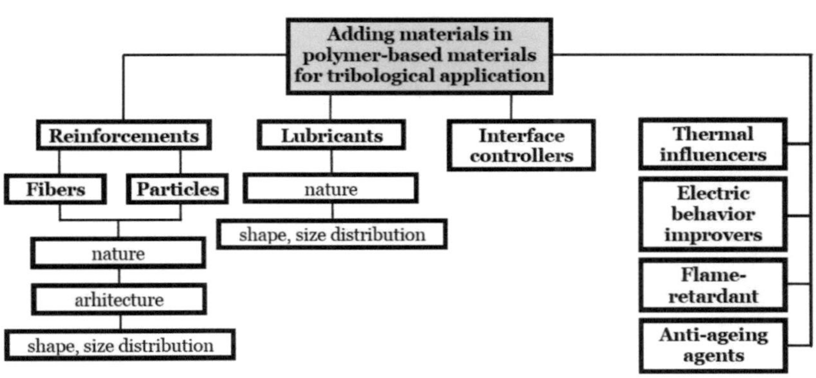

Figure 4.
A classification of materials in polymeric-based materials.

the friction coefficient is higher and the surface quality of both rubbing surfaces becomes worse. Reinforcements reduce or even damage the protective transfer films [28]. They could generate a sliding regime characterized by severe and third body wear [29]. For instance, the composite PA + 50% glass beads [11] exhibit a third body friction and wear, especially at low velocity (see **Figures 35b** and **c**).

Figure 4 points out that adding materials in polymers have different roles (sometimes, they could act in two or more directions) and the influence of the set added in the basic polymer could be synergic [30], difficult to enclose in formula, thus, testing is a necessity. **Figure 5** presents several reinforcements: a) micros glass beads with large dispersion of the bead radius (this is a favorable aspect because this large distribution allow for the small beads to fix the matrix next the bigger ones and wear is considerably reduced), b) short glass fibers with diameters of 8 ... 20 μm and length of hundred microns [31] (similarly, carbon fibers are added in 10 ... 20% wt), c) aramid fibers [16, 32] (more flexible and with nail-shape ends that help them to fix the polymer matrix).

Harder polymers and polymer composites with hard components are helped to reduce friction by adding solid lubricants with plaquette-like shape (graphite [33], graphene, disulphides [33], several examples being given in **Figure 6**) or polymers as PTFE, with more uniform transfer and having very low friction coefficient.

Tribologists considers that short fibers are more beneficial for tribological application, but recently, the polymers with long fibers were also introduced as materials for moving parts due to the advances in fibers and polymer technology. There is a short discussion about fiber architecture. Usually, short and tangle fibers are randomly organized in the material, they rarely could be oriented, but the cost will increase. Long fibers could be organized in woven, unidirectional, multi-axial,

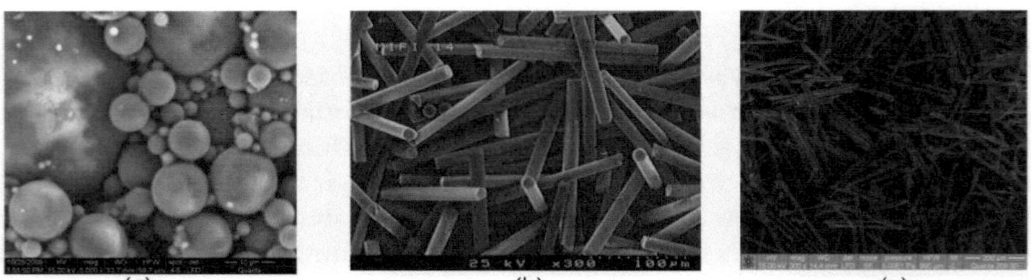

(a) (b) (c)

Figure 5.
Aspect of reinforcements in polymers. (a) Glass beads used in [11, 15]. (b) Short glass fibers from LANXESS [31]. (c) Short aramid fibers Twaron [16].

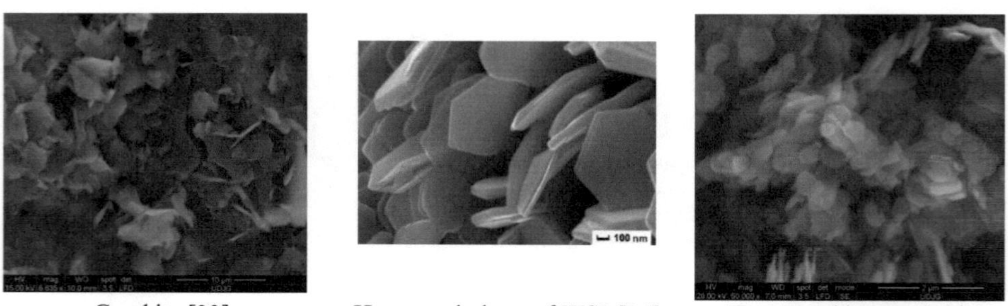

Graphite [33] | Hexagonal plates of WS2 [34] | Hexagonal boron nitride (h-BN) [33]

Figure 6.
Aspect of several solid lubricants introduced in polymers. Graphite [33]. Hexagonal plates of WS2 [34]. Hexagonal boron nitride (h-BN) [33].

depending on the other requirements besides the tribological one. Being organized, the wear of materials of long fibers is usually in steps, characterizing the damage of each layer of fibers. As fibers could have 5 to 50 microns, the wear of the first layers or two ones will end the life of the triboelement. The nature of the fibers is natural, synthetic or combination. For tribological application, there are used carbon fibers, carbon nanotubes, glass fibers (if short, from tens microns to hundreds of millimeters but more efficient being those of several hundred microns to several millimeters), polymer fibers, more recently, aramid fibers [16]. Particles as reinforcement could have different shapes, from almost spherical (as for glass beads) to sheet-like or plaquettes (one dimension being very small as compared to the other two). A particular aspect of wearing polymeric composites or blends is the initially preferential wear of the softer material, the result being an increase concentration of harder particles or fibers; then the counterpart body will "attack" these harder materials; they could be fragmented and embedded into the soft matrix or they are torn off becoming wear debris, "traveling" in contact and induces oscillations of friction coefficient, but when their concentration increases, the component of abrasive wear becomes dominant and wear is greater; when the tribolayer loses its hard particles, the cycle is repeating. Thus, wear is a dynamic process, in steps, depending on local concentration of material constituents [9].

Figure 7 presents a process of consolidation of the tribolayers by embedding the fragments broken from short glass fibers a) PTFE +25% glass fibers, water lubrication, partial bearing (Ø60 mm, 30 mm width) and steel shaft: some glass fibers within the superficial layer cannot bear the local load and were broken; the fragments are embedded into the PTFE matrix [9].

Sometimes, adding materials in polymers could worsen the tribological behavior. For instance, too much concentration of glass fibers increases both and friction coefficient and wear (especially abrasive wear on both surfaces in contact).
A relation between mechanical characteristics in tensile tests and tribological one could be triky. Tensile strength could be improved by adding reinforcements, but strain at break is usually decreased. In sliding contact, a deformability ensures the contact conformability and in fluid lubrication helps generating the fluid film. But, even from 1979, Evans and Lancaster [35] reported that fibers in polymers have beneficial effects on wear and only rarely worsen this parameter. Some adding materials could have the role of a reinforcement but also could help for heat evacuating. A greater interest in using polymer composites and blends pointed out that the designer of the material has to do compromises that have to be accepted only by experimental results, models for predicting tribological behavior being difficult to establish in quantities [36].

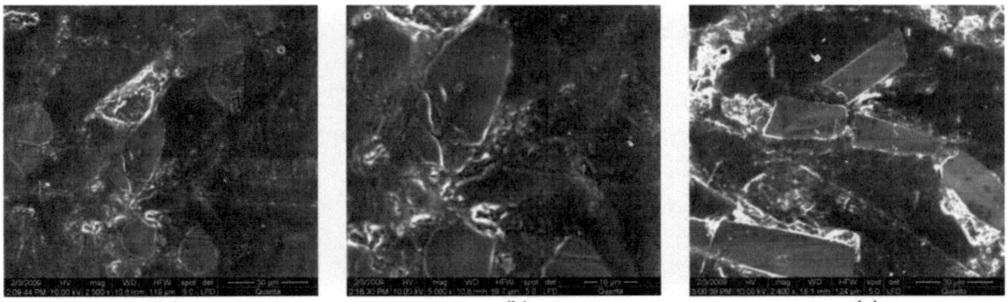

(a) (b) (c)

Figure 7.
Consolidation of the soft polymer matrix by glass fiber fragments, water lubrication, composites PTFE+glass fibers, large contact, partial bearing (120°) (Ø 60 mm x 30 mm width) [9]. (a) PTFE+15% glass fibers. (b) Detail of (a). (c) PTFE+25% glass fibers.

The addition of short carbon fibers (SCF) in a concentration from 5% to 20 vol% can improve the wear resistance of neat PEI remarkably, especially at high temperature and under high working pv-factor. The increased test temperature from room temperature to 150°C leads to a seven times increase in the wear rate of neat PEI and five times for the composites. SCF/PEI can withstand much higher pv-factor than that of neat PEI. When the pv-factor increased from 1 to 9 MPa m/s, the time-related wear rate of SCF/PEI almost linearly enhanced from 1.5×10^{-3} to 7×10^{-2} m/h. However, the wear rate of neat PEI increased from 0.214 to 3.42 m/h when the pv-factor was only increased from 0.25 to 3 MPa m/s. The micrographs of the worn counterface and specimens indicated that the sliding of neat PEI against metal counterface did not form a transfer film, and wear mechanisms varied from fatigue wear to plastic plowing at the increased temperatures. The presence of short carbon fibers helped generating transfer films both on the counterface and worn surface of specimens. The transfer film became more continuous with the increased test temperature. The composite wear was mainly undertaken by fibers [37].

Even if the process of wearing the polymeric composites comprises same stages, the aspect, dimensions and the concentration of added materials make the aspects of worn tribolayers very different. When sliding two bodies one against the other, the matrix is more deformable and the adding materials are like pebbles in the bottom of a shallow river. A partial detaching between matrix and particles/fibers could happen, the fibers change their position and the particles could roll or be dragged on the surface. The space left behind the hard element accumulate fine wear debris from both bodies or even from lubricant (when lubricating), stiffening the tribolayer. The random position of the hard materials and their agglomeration by wearing the soft material increase the probability of detaching conglomerates. This is why an optimum concentration of hard reinforcement in polymer-based material is around 15 ... 25% and depends on the nature of reinforcement. For instance, 20 ... 25% wt is an optimum in PTFE [9, 38, 39], but short aramid fibers are usually added at 10% wt due to the difficulty of injection molding as they block the injection nozzle [16]. As for particles with similar dimensions in all directions Georgescu [15] and Maftei [11] proved that 20% is the optimum concentration for glass beads of micron size.

If one analyzes the soft phases introduces in polymer-based materials, usually a solid lubricant, and with particular reference to PTFE [40] as solid lubricant, this concentration varies from 5 to 15% wt depending on the nature of the involved material. In PBT, the best concentration of PTFE was 5 ... 10% the preferred criterion being the wear rate of the polymeric blend on steel [15].

3. Testing rigs, standard and non-standard testing methodologies

Laboratory tests, on simplified specimens, are useful for ranking materials, but these results could not be extrapolated to actual component, especially for polymeric materials.

Test campaign has to answer how the material pair behaves in a series of parameters

- lubrication regime,
- environment
- working regime (load, speed etc.),

- family of tested pairs of materials

In the ISO standard collection, the word wear is mentioned in 118 items, the test methology being adapted to the application, as, for instance, road and tire wear, implants, but for testing plastics there is.

ISO 7148-2:2012 Plain bearings — Testing of the tribological behavior of bearing materials — Part 2: Testing of polymer-based bearing materials.

ISO 6601:2002. Plastics — Friction and wear by sliding — Identification of test parameters.

ISO 20329 Plastics — Determination of abrasive wear by reciprocating linear sliding motion.

ISO 9352:2012 Plastics — Determination of resistance to wear by abrasive wheels.

ISO/DIS 7148–2 Plain bearings — Testing of the tribological behavior of bearing materials — Part 2: Testing of polymer-based bearing materials.

ISO/TR 11811:2012 Nanotechnologies — Guidance on methods for nano- and microtribology measurements.

The selection of tests necessary for assessing tribological behavior of a material pair including polymer-based materials depends on

- the research level (laboratory, application under development, design of new materials, failure investigation),

- the characteristics of the tribosystems, distinct regimes of sliding wear are "severe" and "mild".

- the actual working conditions

Many different approaches could be seen in literature for assessing the tribological behavior of a system, differentiate in scale and complexity of the tested system. A logical order will be.

Laboratory tests → Model tests→ Component bench tests → System bench tests→ Machine bench test→ Machine field test [41, 42].

In the same direction there are increasing complexity and costs, but first types of tests have increasing control and scale investigation and flexibility.

Depending of the novelty degree of the solution, one or more of the stages mentioned above could be omitted. New materials and original design solutions ask for all, but they have to be solved quite rapidly in order to gain the market.

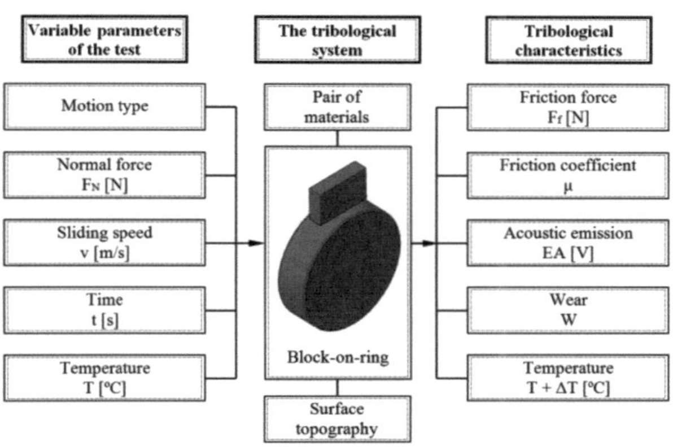

Figure 8.
Characteristics and relevant parameters for the tribotester block-on-ring [15, 16].

A testing campaign is suggestively given in **Figure 8**. This plan was elaborated by Georgescu [15], but also used by Botan [16]. It was the result of consulting adapted from [42]. Polymeric blocks have the dimensions (10 mm x 16.5 mm x 4 mm). The values are quite small and it is very probably not to have actual component of such dimensions, but such a test campaign is very useful for ranking the materials and to investigate modifications in the tribolayers by the help of electron scanning microscopy, AFM, Raman microscopy as the test specimens are small.

4. Tribological parameters and evaluation of experimental results

The set of tribological parameters are characterizing the materials Laboratory tests, on simplified specimens, are useful for ranking materials, but these results could not be extrapolated to actual component, especially for polymeric materials.

When designing a test campaign, for assessing the tribological behavior of a material pair, the tribosystem has to be identified as one in **Figure 9** [42], this simplified initial system being tested at laboratory level with as many as possible parameters closer to those from actual application.

The coefficient of friction is a convenient method for reporting friction force, since in many cases F_f is approximately linearly proportional to F over quite large ranges of N. The equation, known as Amonton's law is

$$F_f = \mu F \tag{1}$$

where the value of μ depends significantly on working regime (lubricated or not), the composition, topography and history of the tribolayers, the environment in which they are working and the loading conditions. Ashby [43] gave a suggestive diagram, positioning the polymeric materials with lowest wear rate, but wear rate values could scan o two-order of magnitude. He also suggests by this diagram that wear rate field could be extended, especially towards low values by filling the polymers. A special position is noticed for PTFE (**Figure 10**), unique polymer as tribological behavior (the lowest friction coefficient, high wear rate, high working temperature and very resistant in aggressive media).

Usually, when a component if made of polymeric material, the other is harder, made of steel, but recently contact could be between the same polymeric materials of different. Thus, friction has to be treated for these cases.

In the case of harder counterpart, the friction polymer-metal has the following components: plowing as a form of abrasion with larger elasto-plastic deformation and micro-cracks and adhesion [3, 8]. These processes are severely depending on many factors including the hardness and asperity shape of the counterpart, contact

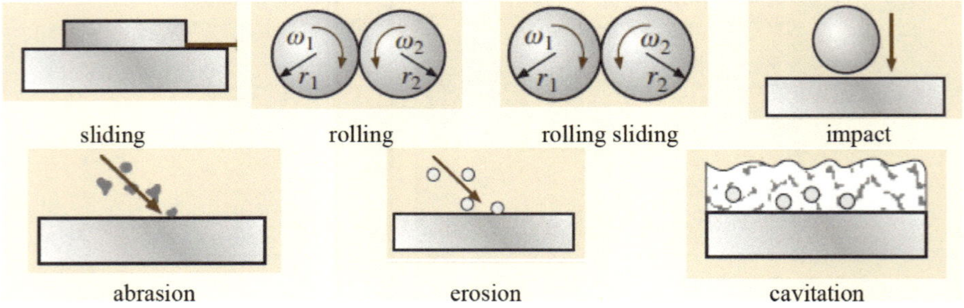

Figure 9.
Testers for assessment of tribological behavior of polymers and polymer composites [42].

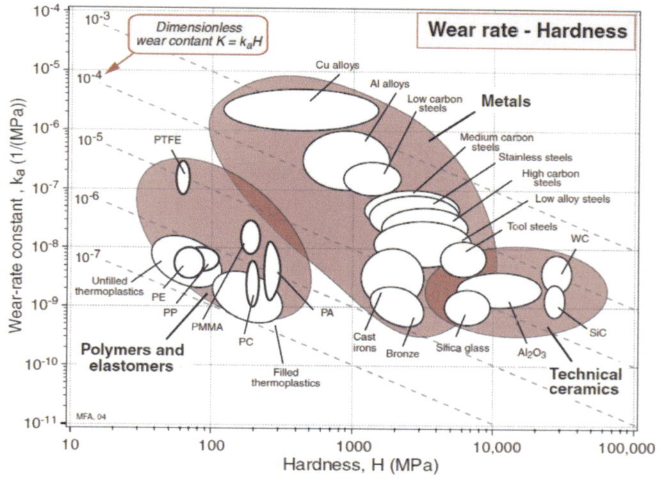

Figure 10.
Positioning of polymers and polymer composites in a space hardness-wear rate constant [http://www.mie.uth.gr/ekp_yliko/2_materials-charts-2009.pdf] [43].

load, speed, temperature. This component of friction could be reduced by introducing a lubricant in contact or/and by re-design the system to have rolling or rolling-sliding motion and by an adequate cutting (usually grinding, honing) of the metallic counterpart.

The adhesion is present both in static friction and dynamic friction of polymeric materials: at the interface motion generates shear and deformation of a very thin layer of the polymeric material, directly in contact with the counterpart. As adhesion and transfer on the counter part are developed in steps, the friction loss, and consequently, the friction coefficient will vary in time, especially for sliding contacts.

Values of friction coefficient are given by producers, researchers but they are depending on test conditions. Thus, they could give a ranking of the tested materials under the same conditions, but they could not be the same with actual components. Sometimes, especially under low load, negative values of μ may be noticed: they are rather artificial, due to contact separation and inertia of the tester components; values of μ greater than 1 are physically logical, especially in material processing, in the interaction between a car tire and a dry road. Sampling could vary depending on the gauge measuring the resistance force. Researchers usually use a moving average to draw the curve of friction force or coefficient in time. For instance, the curve in **Figure 11** was done by moving average of 200 values with sampling 2 values per second. But extreme values are also important as they limit a range that could explain failure mechanisms as adhesion or local melting, especially for polymeric materials.

In most cases, a single value of coefficient of friction is not adequate. This can be seen from the examples in **Figure 11**, depicting the evolution of friction coefficient for three sliding distance. The aspect of evolution is kept for PBT, but these three tests gave values between 0.16 ... 0.19, with stable evolution, a characteristic of polymer sliding as compared to metal–metal contacts.

The evolution of COF in **Figure 11** points out that, for polymer on steel in dry regime, it is less sensitive to time, but these conclusion has to mention the time range for which the researchers had obtained this results, here for 2500 ... 7000 m.

Czichos [41] modeled the evolution of COF for a dry regime in four stages: 1-increasing trend as the surfaces accommodate by wear, 2 - shorter stage of maximum values of COF, 3 - decrease of COF by the generation of a tribolayer favorable

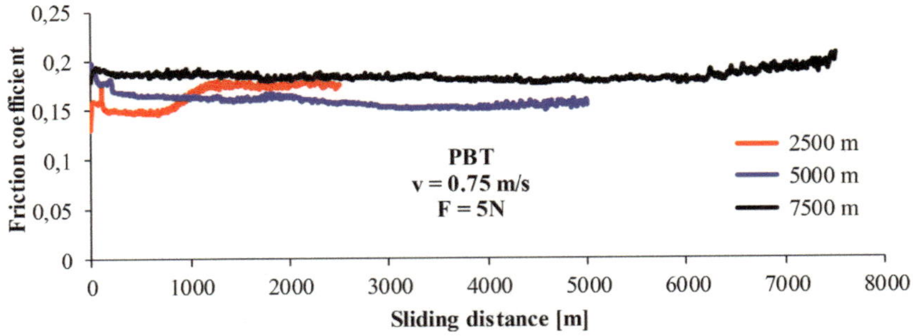

Figure 11.
Friction coefficient for three tests block-on-ring, with different sliding distances [15].

to reduce friction, for instance, a soften or molten layer of polymer, transfer films on harder surface etc. and the abrasive wear and deformation intensities decrease, 4 - stable evolution of friction. For polymer on steel or even on themselves, the authors will add a stage, 5 - slowly or sudden increase of COF meaning worsening the surface in contact due to severe wear, fatigue etc., in many times this increase announcing the life end of at least one triboelement (**Figures 12** and **13**).

Too low load makes the friction coefficient to have higher oscillations as superficial layer of the polymer is not compresses and hard asperities will easier tear up micro-sheets or plaquettes. As the load increases, the tribolayer is compacting and the energy loss by tearing decreases. This phenomenon of oscillating the friction coefficient in dry contact of polymers have been notice also by Jones in 1971 [44]. Higher concentration of reinforcement increases the friction coefficient and makes its evolution wavy (high amplitude could mean an increase of the glass bead concentration in the tribolayer and low values could happen when the tribolayer is richer in polymer.

Convergence of the curves for higher velocity (in **Figure 13**, for sliding speed of 0.5 m/s and 0.75 m/s) means that friction process is similar, very possible involving

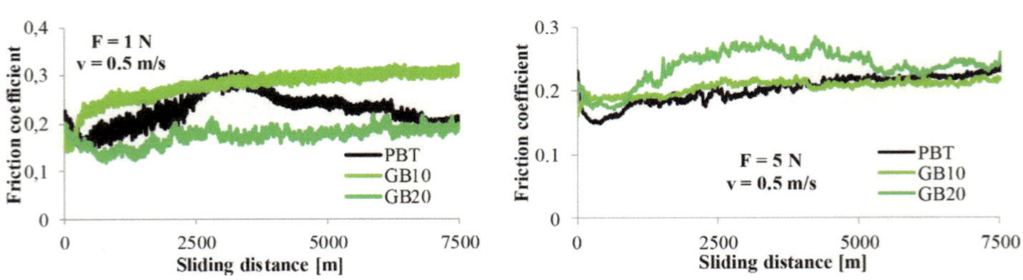

Figure 12.
Influence of load at the same sliding velocity (GB10 - PBT +10% glass beads, GB20 - PBT +20% glass beads) [15].

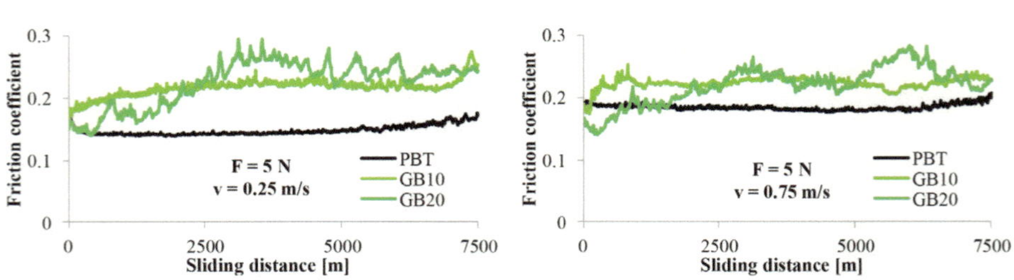

Figure 13.
Influence of sliding velocity under the same load [15].

a very thin soften/melted layer of polymer. This is obvious in another study [11], using pin-on-disk tribotester.

This example point out the influence of the nature of polymeric materials: the composite (the composites with hard micro-particles in a PBT matrix have higher and rough aspect of the curve, the blends PBT+ PTFE having lower values even the polymer PBT, considered a polymeric blends with soft drops of PTFE in PBT matrix). **Figure 14** presents the influence of sliding velocity on the friction coefficient, and the curves in **Figure 15** show the friction coefficient evolution in time depending on the highest load and velocity. The last plot is given only once as it could be related both to load and velocity dependence. The abbreviations for the materials are: PF5 - PBT + 5% PTFE, PF10 - PBT + 10% PTFE, PF15 - PBT + 15% PTFE). The composition of the hybrid composite GB10 + PF10 (having 10% glass beads and 10% PTFE) makes the friction coefficient to be higher at low velocity (0.25 m/s), but for the other two tested velocity, this tribological parameter evolves in a similar manner, but with higher oscillations, probably because of hard glass beads in the tribolayer (**Figure 16**).

Wear is not only a process of material removal in moving contacts, but a more complex one, defined recently as damage of the solid bodies caused by working or testing conditions, generally involving progressive loss of material, elasto-plastic deformations, tribo-chemical reactions caused by local pressure and heat generation in friction and their synergic interactions [8, 20]. In majority cases, the relative motion is intentional: for example, in plain bearings, pistons in cylinders, automotive brake disks interacting with brake pads, or in material processing (cutting, injection, rolling or extrusion). But in some cases, there are also undesired motion(s), resulted because of particular working conditions, as in the small cyclic displacements, known as fretting, produced by vibrations, elasto-plastic and tribological behavior of components in contact. If solid particles are passing through the contact, as contaminants in lubricant or, intentionally, as abrasive material for processing, then they will have a tremendous influence on wear process and, thus, on system durability.

Wear is a complex process, quantified by the volume or mass of removed material, from each body in contact, the change in some linear dimension after a time period of functioning. Thus, wear is obviously a function of material pair, working time and conditions and it is related to a particular tribosystem (materials, dimensions, shapes and working conditions).

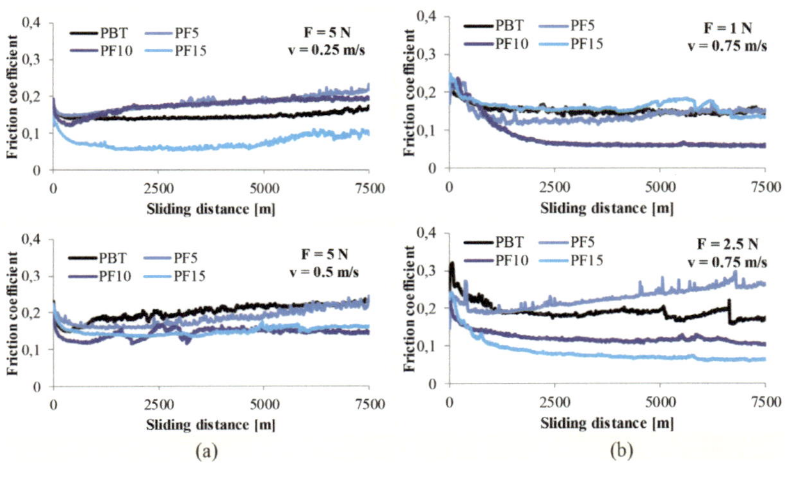

Figure 14.
Influence of sliding velocity, at F = 5 N, for PBT, PF5, PF10, PF15 [15].

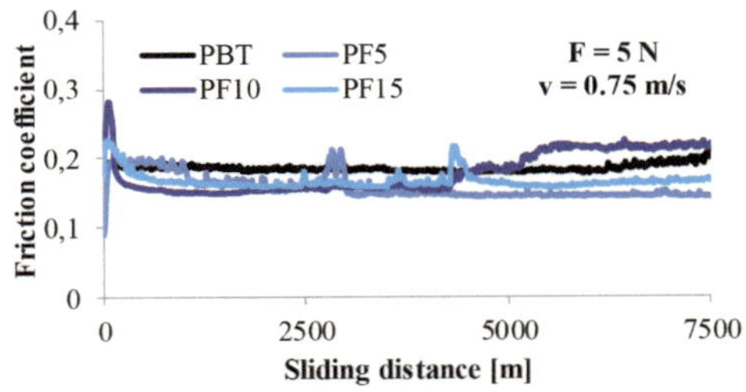

Figure 15.
Influence of load, at v = 0,75 m/s, for PBT, PF5, PF10, PF15 [15].

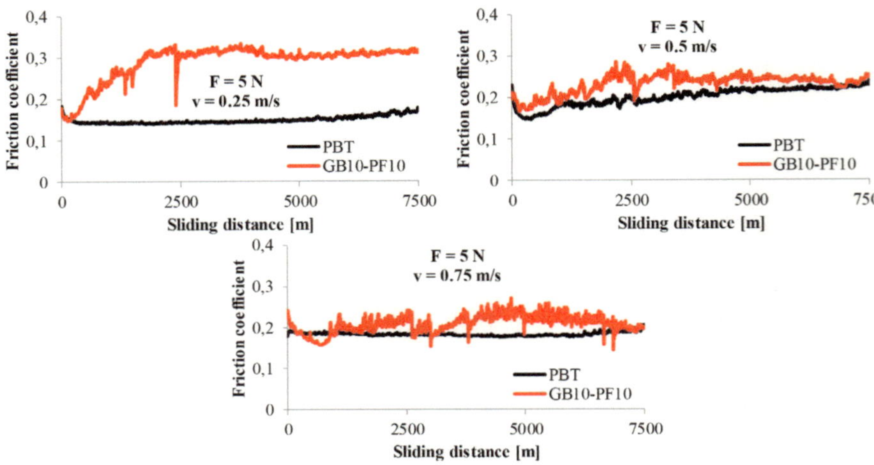

Figure 16.
Evolution of the friction coefficient for PBT and a hybrid composite (PBT + 10% glass beads+10% PTFE) [15].

In some cases, material may be lost from both triboelements, or significant transfer of material may occur between the triboelements, and particular care is needed in both measuring the magnitude of wear and describing the damage it generates (material removals, abrasion, adhesion, transfer, plastic deformation, fragmentation and mixing the constituents of the tribolayer changes in the topography, the last one being investigating by the help of advanced non-contact profilometers [45].

The wear of polymeric material implies an aspect that is of interest only in pairs with a polymeric material: melting wear. A part of heat generate by friction is transferred to the polymeric materials and as thermal conductivity of polymers is low, a very thin layer could soften or even melt, the material latent heat of melting imposing a temperature limit in dry contacts. Stachowiack and Batchelor [46] described the scenario of temperature evolution in contact with a polymeric material (**Figure 17**). Similar observations are done by Briscoe and Sinha in [8], relating the polymer softening and its nature to transfer process on the harder counterface.

Experiment work validated this process of keeping constant the temperature in contact when a triboelement is made based on polymers. In order to support this conclusion, two studies are presented. First one is shortly presented in **Figure 18**. A cylindrical pin made of bearing steel is sliding against a disk made of composites PA + 10% wt glass beads +1% black carbon [11]. The thermo-image in left side presents the positions and their codes where the temperature was recorded with a thermo-camera. The temperature evolutions in time for these three points re given in the right. It is obvious the tendency of maintaining the temperature almost

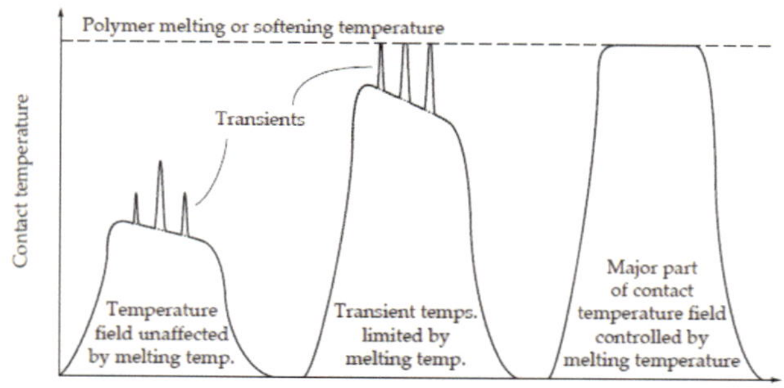

Figure 17.
Evolution in time of temperature of polymeric surface in sliding contact [46].

constant for v = 0.5 m/s and v = 1 m/s. As for the highest tested velocity, the plateau is zigzagged at almost regular time period. This could be explained by the polymer softening or even melting, followed by easier removal from the tribolayer, enrichment in glass beads of the tribolayer, with higher friction and thus, generating heat and rising the temperature. When the glass beads are embedded in the remaining matrix or removed, the temperature would reach a minimum.

Another study [16] for emphasizing the importance of testing composites with polymer matrix has the results obtained on block-on-disk tribotester (**Figure 19**). The block is made of composite with 10% short aramid fibers (Twaron, grade, 225 μm in length see **Figure 5c**), with two different matrices: PA and PBT and the ring is the outer ring of a taped rolling bearing (the quality of rolling bearing ring keeps contact the influence of the counterbody in sliding). Analyzing **Figure 20**, the friction coefficient for PAX on steel has a steady evolution, in narrow ranges, for low loads (F = 5 N and F = 15 N) but for F = 30 N, for higher velocities, it increases and becomes steady at higher values, around 0.3. Temperature is steady for the same low loads, but it increases with different slopes for highest load. A too low load on polymer-based material - steel could rise COF and temperature in contact because the hard body does not contact the polymeric tribolayer enough and thus, the wear has a more intense abrasive component, tearing-off easier the polymer.

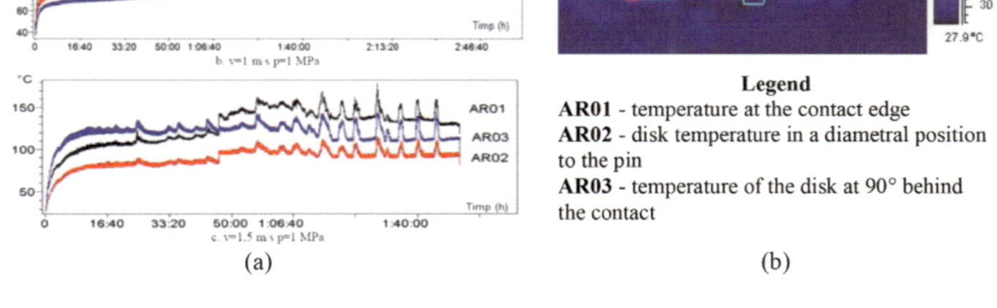

Figure 18.
Temperature evolution in time (a) for pin-on-disk tester, pin made of hard steel and disk made of PA + 10% grass beads + 1% black carbon, dry sliding for 10000 m and a thermal image during the test (b) (the rotation of the disk is clockwise) [11].

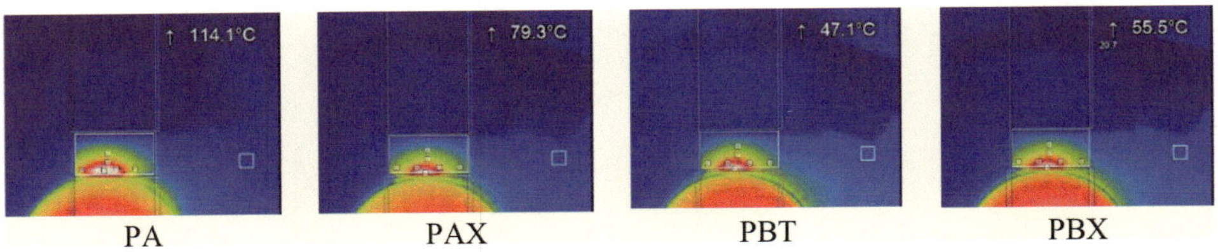

Figure 19.
Images of thermal recordings of the temperature at the end of the test, for temperature at the contact edge, F = 30 v = 0,75 m/s (block made of PA- polyamide, PAX - polyamide +10% aramid fibers +1% black carbon, PBT - polybythylene therephtalate, PBX - PBT +10% aramid fibers +1% black carbon) [16].

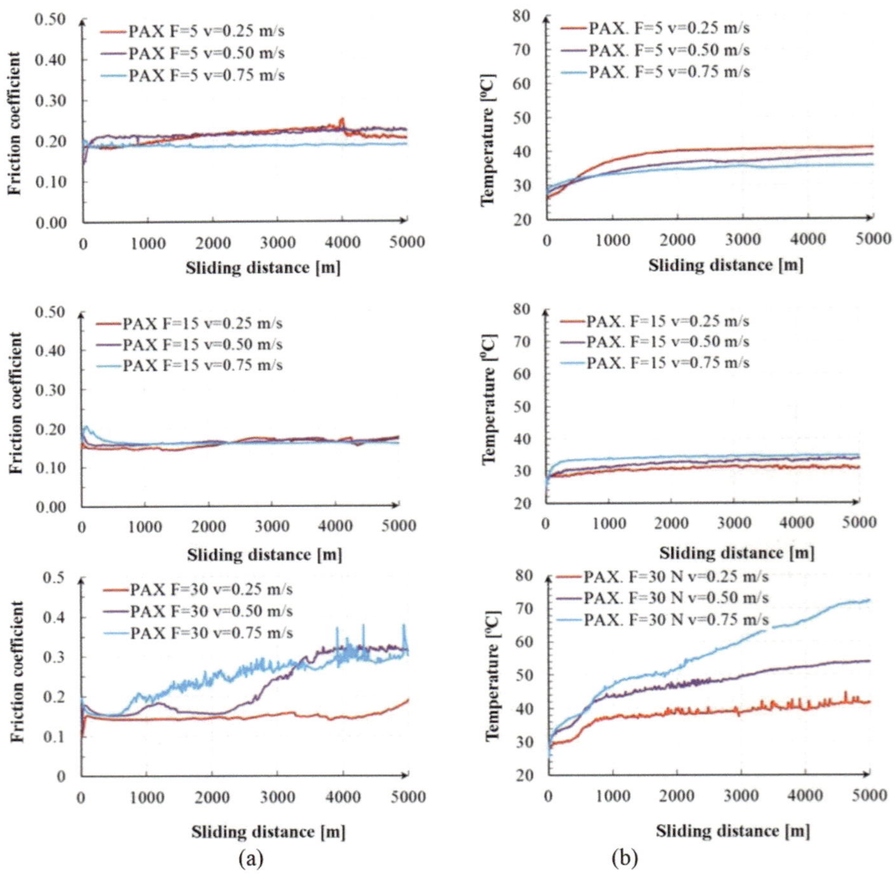

Figure 20.
Evolution of friction coefficient and temperature at the contact edge in time, depending on load, sliding velocity, for a sliding distance of L = 5000 m, block made of PBT +10% aramid fibers, L = 5000 m, block made of PAX (PA6 + 10% aramid fibers) [16].

The combined analysis of two tribological parameters could reveal a qualitative change of the working regime. For instance, analyzing COF and temperature at the contact edge (**Figure 20**),

- a too low load and sliding velocity make the temperature rising due to abrasive wear (more intense under low load)

- a higher speed makes the temperature curve higher for v = 0.75 m/s, but the COF is kept low meaning a softening process happened,

- a too high load makes the temperature to have a slope, greater as velocity increases; a mild regime (thus, a favorable regime) will keep the temperature constant in contact as for tests under F = 15 N. The severe regime is marked by high oscillation of friction coefficient or even a constantly increased value and also by the same shape of the temperature curves.

Comparing curves in **Figure 20**, regimes with F = 30 N and high sliding velocity (v = 0.5 ... 0.75 m/s) could be considered as severe because they do not make tribological parameters as friction coefficient and temperature in contact, stable.

The composite with PBT matrix with the same adding materials (10% short aramid fibers and 1% black carbon) has a similar evolution of COF, but temperature increases only for the extreme tested regime (F = 30 N, v = 0.75 m/s).

The applications involving the friction couple polymeric material - metallic counterpart are preferred by mechanical requirements of the design solution and the better tribological behavior by monitoring and measuring a set of tribological characteristics (wear, friction, temperature in contact, changes in materials' structures etc.) as compared to sliding polymers against themselves (**Figure 21**).

Wear process of polymeric materials are characterized by a transfer film, generated when sliding against a harder surface, strongly influencing on the tribological behavior of the system [8].

A favorable transfer film should be continuous, very thin and regenerating without inducing troubles in the working systems. This is the ideal transfer film of a

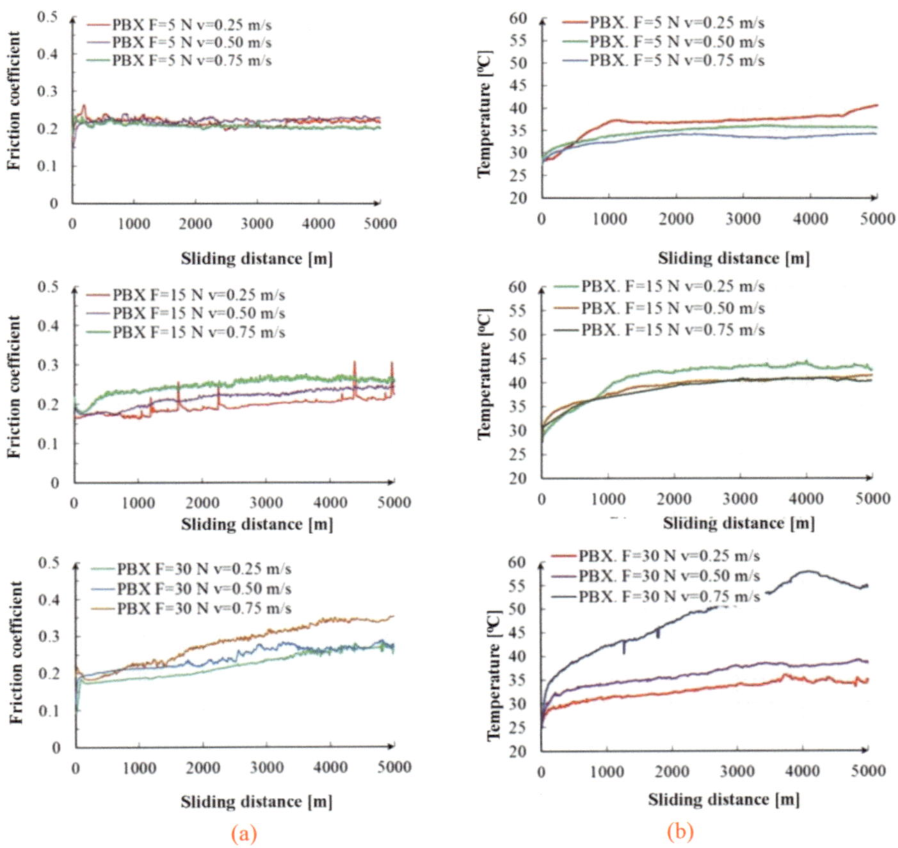

Figure 21.
Evolution of friction coefficient and temperature at the contact edge in time, depending on load, sliding velocity, for a sliding distance of L = 5000 m, block made of PBX (PBT +10% aramid fibers) [16].

polymeric material but, actually, there are two types of polymers, those generating an almost continuously transfer film as high density polyethylene (HDPE) and ultra-high-molecular weight polyethylene (UHMWPE), and those that form lumps or islands, more or less regular. Transfer process is influenced by contact temperature and texture of the counterpart. Only few polymers have only a mechanical component of the transfer film (again, PTFE and UHMWPE have to be given as examples) and polymers that could chemically interact with the metallic surface.

Myshkin et al. [7] pointed out that the dependence of friction coefficient with velocity has different shapes depending on the polymer sliding on steel or on itself, and even for the same polymer, the curve depends on temperature of the environment. At low velocity (10^{-3} ... 10^{-2} m/s), friction coefficient has an almost constant evolution, but at higher speed, its evolution could be with velocity could be parabolic, with minimum when the material is softening or has a thin melt layer, than it could increase. The conclusion of this work is that tests in the same conditions as the application are tremendously necessary for a reliable working of the tribosystem involving polymer-based materials in order to correct assess the power loss by friction and to prevent component failure by frictional heat.

The wear rate can then be defined as the rate of material removal or dimensional change per unit time, or per unit sliding distance. Because of the possibility of confusion, the term "wear rate" must always be defined, and its units stated. It is usually the mass or volume loss per unit time.

The Archard model of sliding wear [47] leads to the equation:

$$w = KW/H, \tag{2}$$

where w is the volume of material removed from the surface by wear per unit sliding distance, W is the normal load applied between the surfaces, and H is the indentation hardness of the softer surface. Many sliding systems do show a dependence of wear on sliding distance which is close to linear, and under some conditions also show wear rates which are roughly proportional to normal load. The constant K, usually termed the Archard wear coefficient, is dimensionless and always less than unity. The value of K provides a means of comparing the severities of different wear processes.

For the tribotester block-on-ring the wear parameter that reflects well the behavior of the materials could be the linear wear rate

$$Wl = \Delta Z/(F \cdot L)\ [\mu m/(N \cdot km)], \tag{3}$$

where ΔZ is the change in distance between ring and block at the end of the test, F is the normally applied load and L is the sliding distance. **Figure 22** presents test parameters, as recorded by the tribometer UMT-2, including friction coefficient (COF), wear depth (Z).

For pointing out wear parameters in a tribosystem with polymer-based material (s), the same two cases are analyzed (**Figure 23**).

A study has another objective [16]: to assess the tribological behavior of two polymer matrices, PA and PBT, with the same concentration of reinforcement, 10% wt short aramid fibers (Twaron, 225 microns as average length). There were measured several tribological parameters, average values of friction coefficient (COF, **Figure 24**), wear rate (**Figure 25**) and maximum value of the temperature at the contact edge (**Figure 26**). Wear rate in **Figure 25** was calculated as

$$W = \Delta m/(F \cdot L)\ [mg/(N \cdot km)], \tag{4}$$

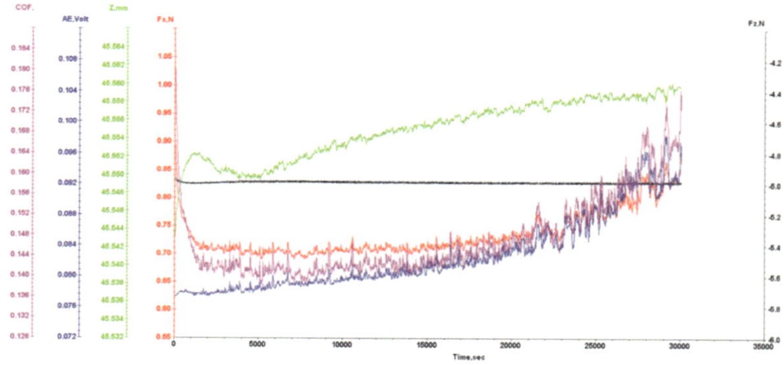

Figure 22.
Example of parameters monitored in actual time real on the tribotester UMT-2, block-on-ring test, block made of PBT, ring made of steel (100Cr6), F = 5 N (= Fz), v = 0,25 m/s, L = 7500 m, COF –friction coefficient, Fx – Resistant force (friction), AE – Acoustic emission, Z – Wear depth (linear wear) (linear change between ring and block), Fz – Normal load [15].

where Δm is the mass loss of the block, L is the sliding distance and F is the applied load in contact.

Temperature in contact is very important in tribosystem with one or both elements made of polymeric materials as a jump in contact temperature of less amount as for metals (even 10°C) could change their mechanical and thermal properties, could even change the chemical organization of the molecular chains; the power dissipated in the contact is given by ($\mu \cdot F \cdot v$) where μ is the friction coefficient, F is the normal load and v is the sliding velocity. The local temperatures in the contact areas can therefore become much higher than the bulk temperatures. This factor needs to be considered when designing wear tests or interpreting test results.

In Botan's study [16], neat PBT had a very good tribological behavior (being analyzed, average values of COF during 5000 m of sliding on steel, low wear as compared to PA) but adding 10%wt short aramid fibers in PBT substantially improves wear resistance. Thermal monitoring of the contact edge allows for ranking the tested materials having the temperature as criterion (**Figure 26**).

In study from 2012, Pei et al. [12] present the tribology of three polymers, considered as high-performance materials, introducing for evaluating the product pv (p being the average pressure in contact and v the sliding velocity). This parameter has to be used with precaution. Comparison should be done for the same tribosystem (dimensions and shapes) and under the same testing conditions. It is not recommended to extrapolate the results outside the investigated parameters. From **Figure 27**, one may notice that PPP grades exhibited low wear resistance as compared to PEEK and PBI had the lowest wear rate, due to its high value for heat resistance and very low decrease in mechanical characteristics under higher temperatures.

Obviously, in dry regime friction coefficient of a polymer on steel is lower than that for steel-on-steel and long and aligned carbon chain (as in PTFE and PE, even PA) will give lower dynamic friction coefficient, around 0.2 … 0.3, lower for PTFE, but polymers with higher mechanical characteristics as PPS and PEEK will have this parameter higher 0.3 … 0.5. Wear rate exhibits values that could not be deduced from the mechanical and structural characteristics. For instance, in **Figure 28**, the lowest wear rate among tested polymers under the same conditions was obtained for PA6, and wear rate increases from this to PI, PPS, PE-UHMW till PEEK, but high values were obtained for POM and PTFE.

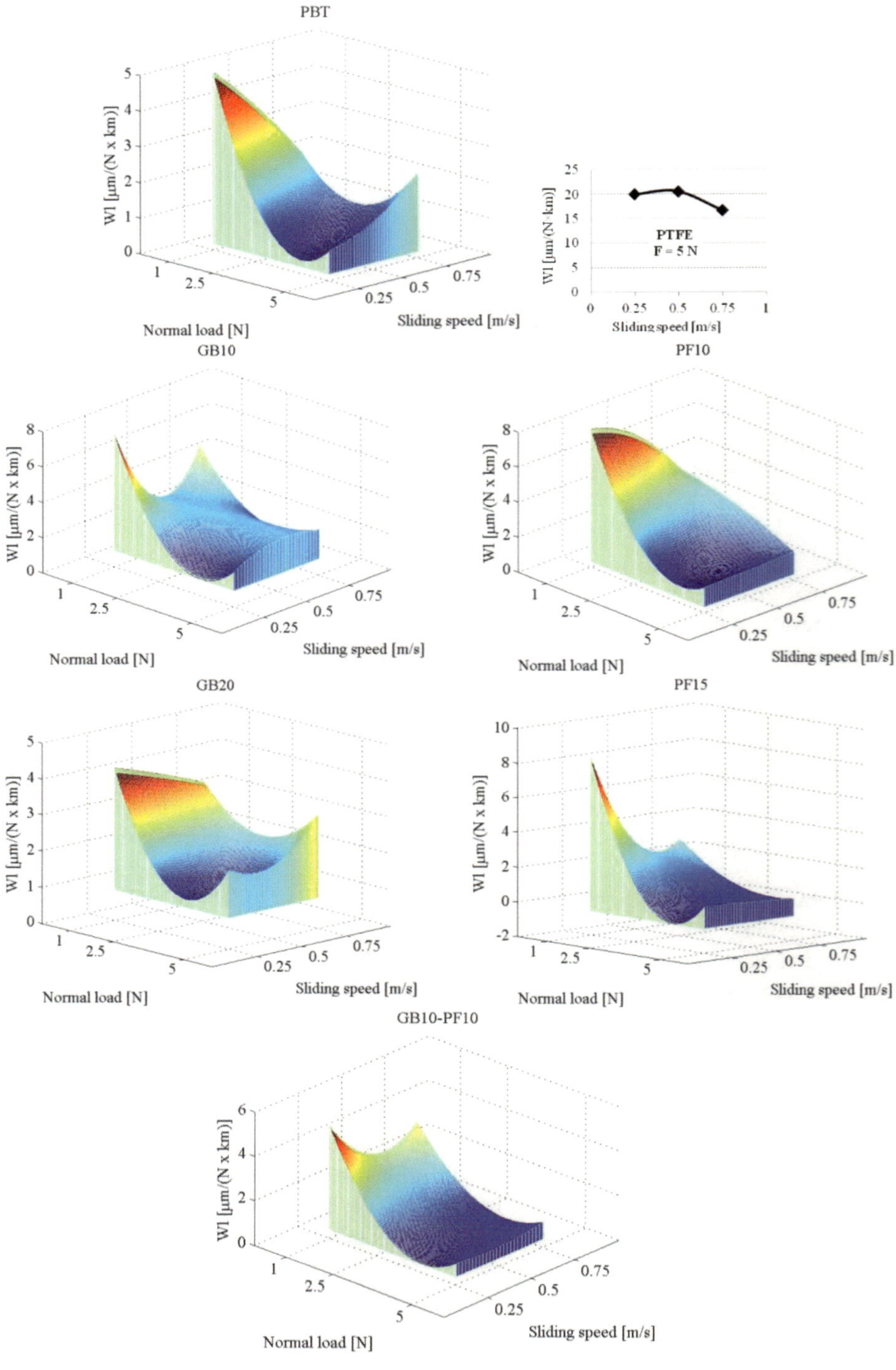

Figure 23.
Linear wear rate of the blocks made of polymer-based materials.

Worn surfaces and the debris resulting from wear, may be examined for several reasons:

- to study the evolution of wear during an experiment, or during the life of a component in a practical application,

- to compare features produced in a laboratory test with those observed in a practical application,
- to identify mechanisms of wear,
- (by studying debris) to identify the source of debris in a real-life application.

Figure 29 presents two virtual images, reconstructed with SPIP The Scanning Probe Image Processor SPIPTM, Version 5.1.11/2012, from a study done by

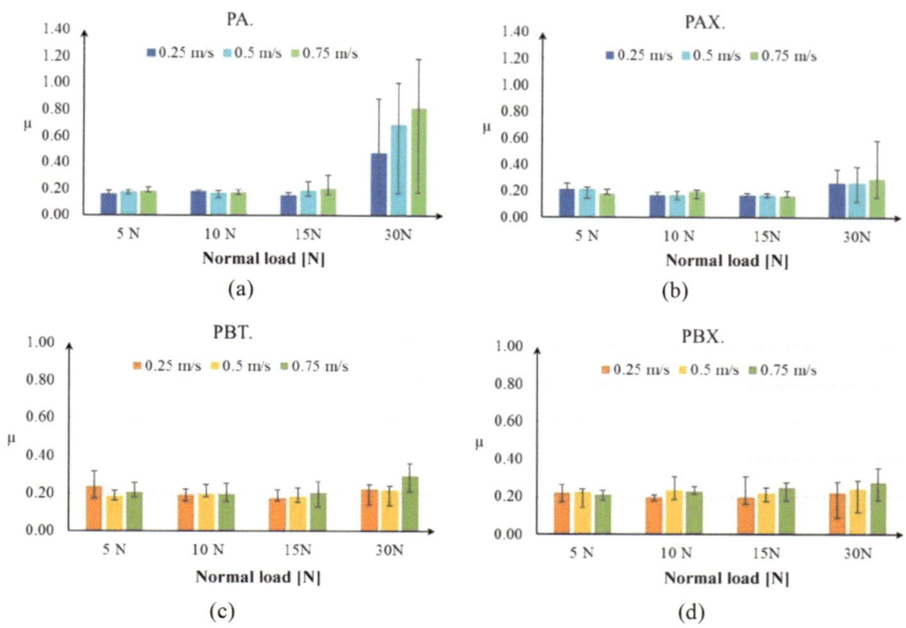

Figure 24.
Average values for COF for 5000 m of dry sliding on steel (same scale for PA and PAX and PBT and PBX, respectively) [16].

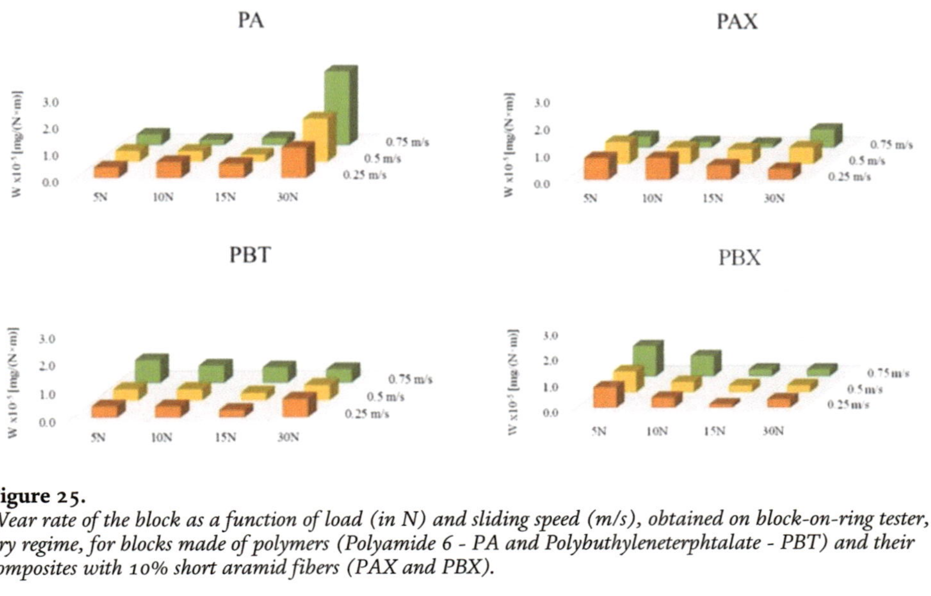

Figure 25.
Wear rate of the block as a function of load (in N) and sliding speed (m/s), obtained on block-on-ring tester, dry regime, for blocks made of polymers (Polyamide 6 - PA and Polybuthyleneterphtalate - PBT) and their composites with 10% short aramid fibers (PAX and PBX).

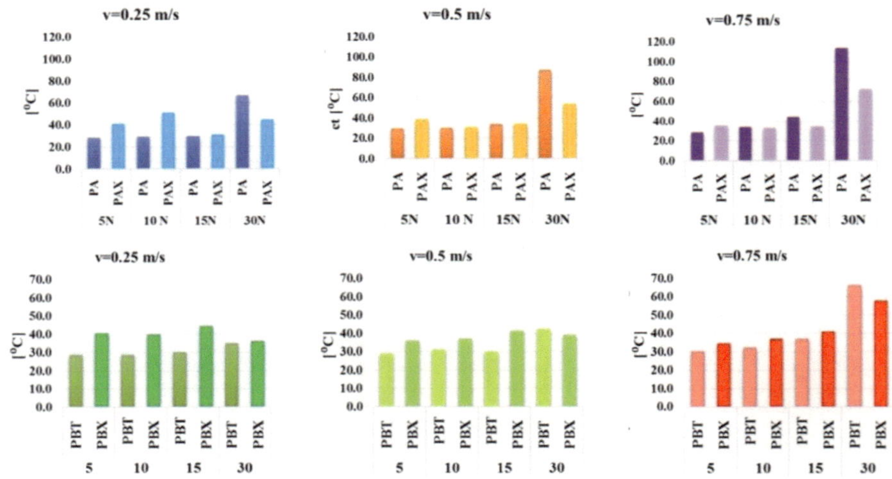

Figure 26.
Maximum value of temperature at the contact edge, for all four tested materials in [16] (material codes as in previous figure).

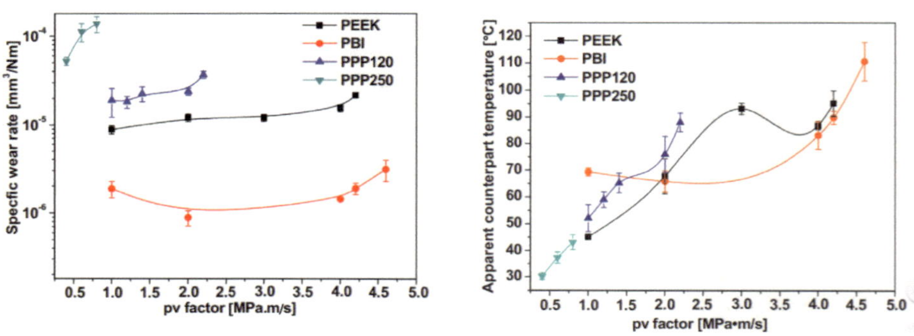

Figure 27.
Specific wear rate of polymer sliding on steel and counterpart temperature for [12].

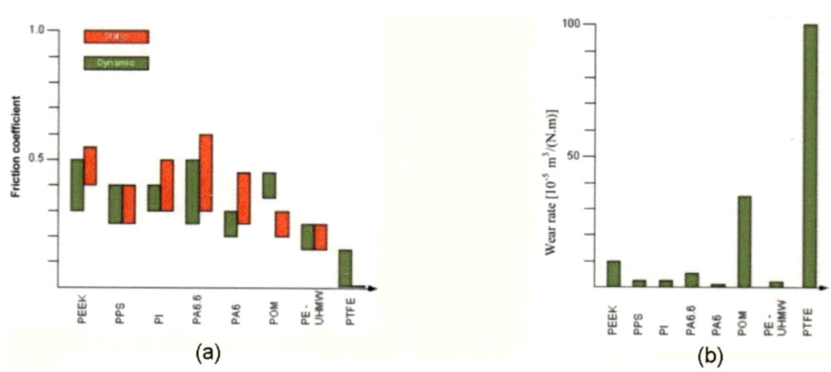

Figure 28.
Two tribological parameters for polymer in dry sliding on steel [http://www.appstate.edu/~clementsjs/polymerproperties/$p$lastics_$f$riction$5f$w$ear.pdf]. (a) Friction coefficient. (b) Wear rate [48].

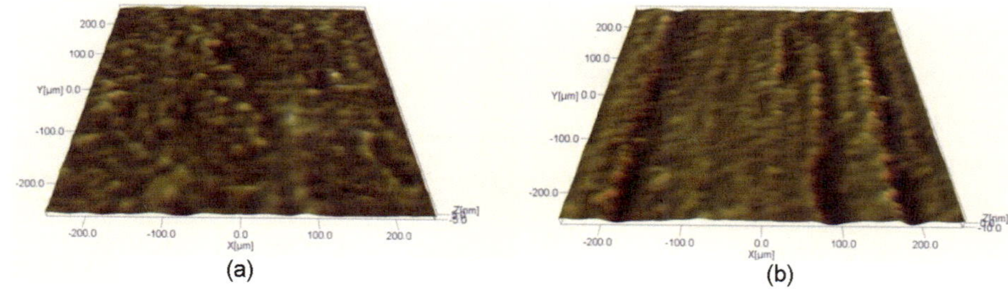

Figure 29.
Virtual images of block surfaces made of PBT + 20% glass beads. (a) Initial surface. (b) Used surface (F = 5 N, v = 0,75 m/s, L = 7500 m) [15].

Georgescu [15], pointing out initial surface (a) and traces as result of abrasive wear on the composite.

After testing, the worn surface quality of the composite with only 10% glass beads was better, meaning a lower value for Sa, Sz (**Figure 30**). In tribological evaluation a ratio Sz/Sa, bringing together an averaging parameter with an extreme one (Sz) is important because singular or rare high peaks have a great influence on the tribological behavior, especially for composites with hard fillers. Adding micro glass beads in PBT increases the amplitude parameters (these are plotted for v = 0 m/s, in **Figure 30**). Ssk has high positive values for 20% glass beads in PBT, but the polymer and the composite with only 10% glass beads have lower values, oscillating between 1 and − 1. If Ssk < 0, it can be a bearing surface with holes and if Ssk > 0 it can be a flat surface with peaks. Values numerically greater than 1.0 may

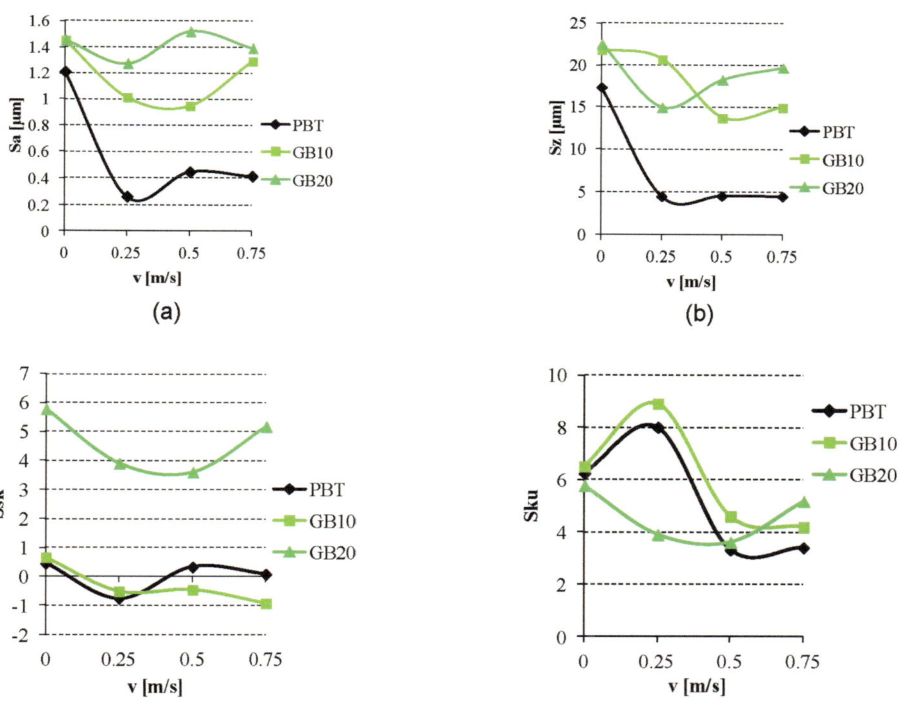

Figure 30.
Roughness for worn surfaces of the block made of PBT, PBT + 10% glass beads (GB10) and PBT + 20% glass beads (GB20). (a) Sa- roughness average. (b) Sz - peak-peak height, the difference between the highest and lowest point in surface. (c) Surface skewness, Ssk, or the asymmetry of the height distribution histogram. (d) Surface kurtosis, Sku, or the "peakedness" of the surface topography [15].

Parameter	Information, unit
Load (normally applied), constant or variable	N
Sliding speed	m/s
Pair of materials	Composition, phases, structures
Temperature (environment and in contact)	The second is difficult to measure
Type or relative motion	Sliding, rolling, combined motion, small oscillations, impact
Contact type	Conformal, non-conformal, volumes of the triboelement
Particularities of tribosystem (if the case) abrasive/erosive particles	Material, shape, size and distribution
Contact dynamics	Stiffness, damping, inertial mass

Table 2.
List of important parameters that influence the tribological behavior [42].

indicate extreme holes or peaks on the surface, as for the worn surfaces of composite PBT + 20% glass beads. For v = 0.50 m/s (**Figure 30**) and v = 0.75 m/s, Ssk < 0, reflecting the micro-plowing process. For Sku > 3, all worn surfaces indicated long and narrow valleys, with high peaks, the valley are dominated as result of tearing-off glass beads and maintaining the shape of the extracted beads. Smaller values of Ssk indicate broader height distributions but these polymeric materials have narrow height distribution as all values are above 3 (**Table 2**).

Components with high volume of polymeric material are less heat conductive and prone to have melt/soften contact. The solution given by research and practice: polymeric coatings, thick enough to reduce friction and to bear wear for a specified life and reliability.

During a test, many influencing factors have to be controlled. These can be grouped in

- -mechanical and environmental test conditions (such as contact load or pressure, speed, motion type and environment temperature, composition), and
- -specimen(s) parameters (such as material composition, microstructure, volume, shape and their initial surface finish).

Some of them could be monitored during the test (as friction force), some only at before and after test. For polymers, investigations must be done just after the test as the specimens could age and thus, altering the information.

Researchers have to prioritize what factors are kept constant and what factors will vary on ranges of interest.

A full program of testing under all combinations of these factors would be time-consuming and costly, and may not be required. Often a single factor can be identified as "key" to the material response, and in this case a good approach is to set all the other factors at constant values and vary the chosen factor in a controlled way in a series of tests. Test campaign must promote an objective, to establish variables (materials, working regime parameters, environment) and the most relevant results to be given, non-destructive investigation in order to understand and direction the damage processes during testing.

Tribologists is now using mapping technique when two (or more) factors are changed in a controlled way (normally more coarsely than in parametric studies), the parameter of interest being the friction coefficient, wear or wear rate, temperature or durability till a particular value for wear temperature etc. are reached. The mathematical model for building the map surface is very important. For instance, maps in **Figure 23** are built with double spline curves, enforcing the obtained values from the tests to be on the surface. Sharp peaks or deep zones on the maps could indicate a qualitative change in tribological processes (change in wear process balance, tribochemical reactions induced by temperature threshold etc.)

The mapping technique is efficient for determining the overall behavior of a material or a tribosystem as it provides useful data about the position of transitions in wear behavior for a systematic test campaign. This comes at the expense of a reduction in the detailed knowledge of the variation of friction and wear with any one factor, but once the regime of interest is better defined through the use of maps, then a more detailed parametric study can be conducted.

5. Characteristic mechanisms in the superficial layers of contacts implying polymeric materials

Initially, PTFE was simply used as bushes, seals, but its low mechanical characteristics make the researchers for materials to use it as matrix in composites [9, 39], adding material in other polymers, and even metallic sintered materials, more rigid and less prone to wear.

Burris and Sawyer studied the blend PEEK + PTFE [49]. PEEK has wear resistance, mechanical strength and a higher working temperature as compared to other polymers, but a high friction coefficient in dry regime $\mu = 0{,}4$ and low thermal conductivity. PTFE has a high wear rate, and the fact that has the lowest friction coefficient in similar conditions does not recommend it to be used simple, without blending with another polymer or reinforcements. A qualitative model of a polymeric blend could be modeled as in **Figure 31a**.

Many researchers and producers of polymeric materials recommend only 5–20% PTFE [46, 50, 51], experiments done by Burris and Sawyer [49] obtained an optimum for the wear rate using the blend 30% PEEK +70% PTFE and, thus, underlined the necessity of testing new formulated materials for tribological applications.

Under 20% PEEK, wear has a sharp evolution, explained but not enough PEEK for creating a harder matrix for the soft polymer, thus the last one is easy to be deformed, abraded; the wear is supported by PTFE and not by the harder material (which has a higher wear resistance. The transfer process is more intense, and the wear debris have higher volumes. The authors suggest that preferentially lose of PTFE make the tribolayer grows rich in PEEK and the wear is reduced. At higher

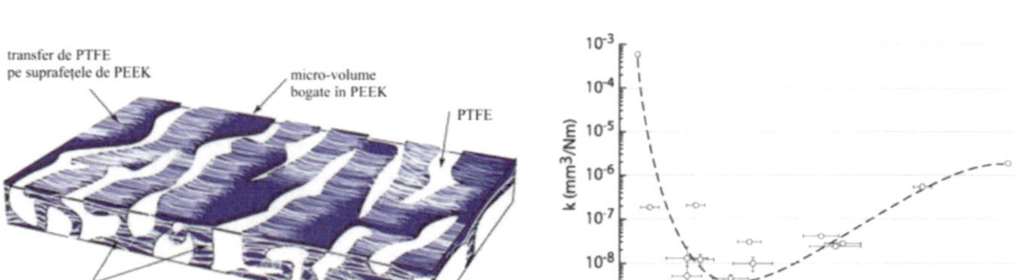

Figure 31.
Contact surface 6,35 mm x 6,35 mm, F = 250 N and alternating sliding on 25.4 mm, v = 0.05 m/s, dry sliding on stainless steel AISI 304. (a) Model proposed by [49]. (b) Wear rate as a function of PEEK concentration.

concentration of PEEK, the wear is dominated by fatigue cracks and the micro-reservoirs of PTFE are in reduced number and the solid lubrication of PTFE is done only on patches. Wear debris made of PEEK generate a more intense abrasive wear, even as third body, care damage the transfer films on both surfaces in contact.

A similar tribological behavior was noticed by Tomescu [9], when a composite copper + PTFE was tested in dry and water lubrication regime.

6. Characteristic mechanisms in the superficial layers of contacts implying polymeric materials

Neale admitted that wear is a complicated process and even if the mechanisms could be described, there are combinations and transitions among them that make them difficult to be understood yet and reduced [52]. Four main wear mechanisms are discussed in literature [23, 46]: abrasion, adhesion, fatigue and tribo-corrosion, with particular, mixt variants (thermal and tribofatigue, fretting etc.).

Aspects of wear mechanisms with different adding materials in polymers are well described and interpreted in [3, 8, 20, 46]. A particular wear process of polymeric materials is the so-called delamination, that is a combined process of sub-layer crack, plastic deformation and material removal (**Figure 32**).

Forms of abrasive wear are micro-cutting, plowing and micro-cracking with material remove are particularized for polymers that are visco-plastic materials.

Adhesion has particular aspects in tribosystems with polymers, including polymeric transfer on the counter surface, especially when this is made of steel.

As Stachowiak and Batchelor [46] mentioned, this transfer has two extreme consequences:

- beneficial, when the transfer film is thin and transform the moving contact in polymer-polymer,
- not beneficial, with lump or insular transfer, that change too much the surface topography.

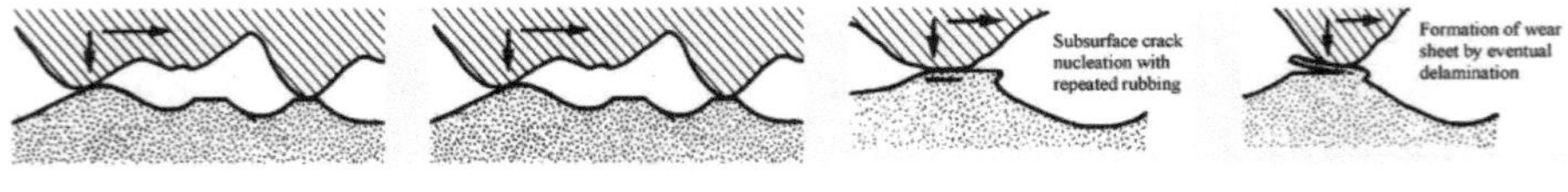

Figure 32.
Wear deterioration of a polymeric body in sliding against a harder material, also known as delamination [35].

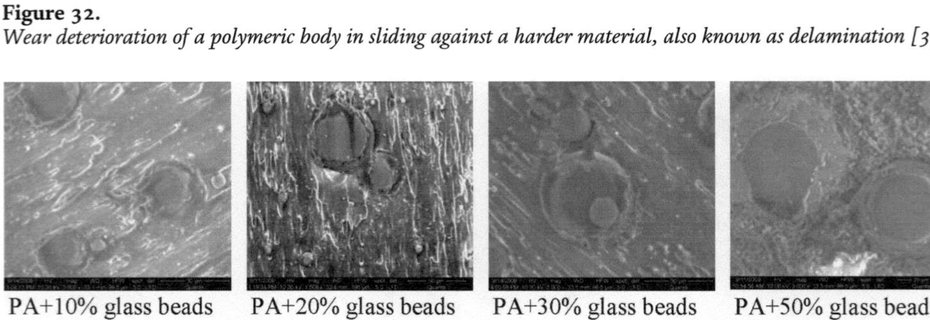

PA+10% glass beads	PA+20% glass beads	PA+30% glass beads	PA+50% glass beads
F = 15 N, v = 1.5 m/s	F = 10 N, v = 1.5 m/s	F = 15 N, v = 1.5 m/s	F = 10 N, v = 1.5 m/s

Figure 33.
SEM images on tribolayer generated from composites with PA6 matrix and different concentrations of glass beads [11].

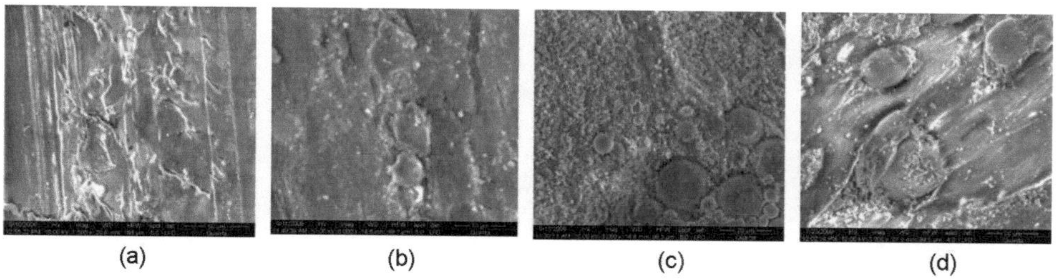

Figure 34.
SEM images for tribolayers: PA disk (a) and for the composite with 50% glass beads (b, c and d), dry sliding on steel (no gold coating of the samples) for SEM investigation. (a) v = 0.5 m/s, p = 1 MPa. (b) v = 0.5 m/s, p = 2 MPa. (c) v = 1 m/s, p = 1 MPa. (d) v = 1,5 m/s, p = 1 MPa [11].

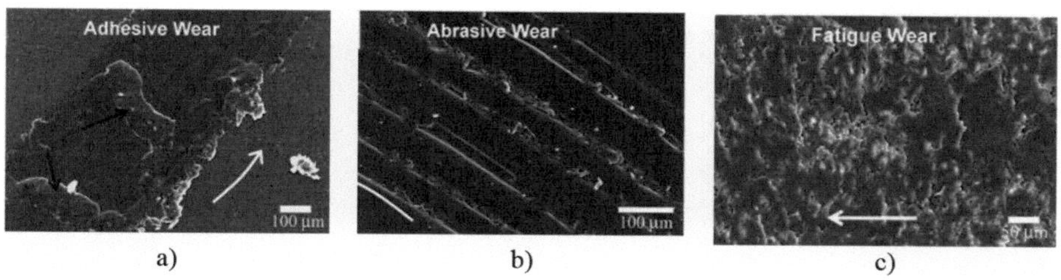

Figure 35.
Typical aspects of the failure mechanisms in sliding on steel in dry regime (a) adhesive wear, (b) abrasive wear, (c) fatigue wear [11].

The solution of reducing wear of polymers is to add materials that keep the polymer into a network (random or organized) to minimize the polymer volume implied in the local deformation and detaching small wear particles instead of big ones.

The research has to establish an optimum concentration of constituents that allow for having a better tribological behavior (reduced wear, permissible working temperature, low power loss due to friction and to keep the functions of the systems in an reliable range).

For instance, Maftei [11] elaborated composites with glass beads in a polyamide matrix with concentration between 5% wt and 50% wt and tested them on pin-on-disk tribotester. SEM investigation revealed agglomerated glass beads, a very thin soften layer of polymer that cover like a blanket the glass beads, justifying the still low friction coefficient. The next figures (**Figure 33** and **Figure 34**) point out differences between wear mechanisms for PA6 (a) (abrasive, fatigue with small cracks) and the composite (detaching smaller polymer debris, al lower sliding velocity the soften layer does not exists and polymer is deformed by the random small movements of the beads in the matrix, at higher velocity (d) several beads roll in the superficial layer as the polymer is less viscos.

Typical aspects of the failure mechanisms in polymer sliding against harder bodies are described in [53–55]: abrasive wear, adhesion wear (with transfer) and fatigue wear (**Figure 35**).

The geometry of the reinforcement makes the wear mechanism to be different for the same fibers, if the matrix is different, as one may see in **Figure 36**. The first line of SEM images is for the matrix of PA6, more ductile than PBT - the matrix of the composite in the second line of SEM images. All tests are done on block-on-ring tester, in dry regime. A more ductile matrix is easier worn and torn-off, the fibers remaining to bear the load and there visible the deformations (flows) induced by a higher load on the fiber ends. In a PBT matrix, more rigid than PA6, the transfer on the steel counterbody is less and the fibers are scratched under higher load.

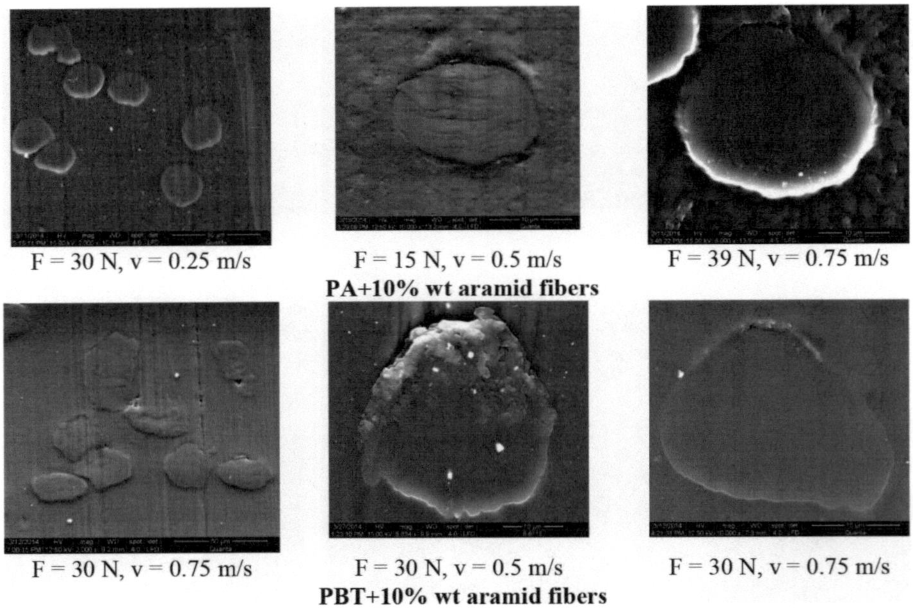

Figure 36.
Block-on-ring, L = 5000 m (thin gold coating of the samples) [16].

Composites with reinforcing particles or fibers: dynamic wear process, in stages: 1 - low wear of polymer and enrichment of the superficial layer in harder materials, 2 - too much hard particles or fibers within tribolayer, the result being big wear particles torn up in bigger conglomerate, 3 - leveling the rough surface after detaching hard particles/fibers by the help of plastic matrix (friction coefficient has high oscillations and the process is repeating.

Friction materials, as for brake pads, need special attention as they have to fulfill requirements as constant friction coefficient and controllable wear (linear would be better). Manoharan et al. [55] presented a study for a composite containing nine major ingredients, including epoxy resin, reinforcement, solid and liquid lubricants etc. (this pointing out the complexity of a composite destinated for brakes). Tests done on disk-on-plate tribotester, in the presence of third body (sand), revealed that wear volume loss of composite brake pad increases with increasing sliding distance and load, but wear rate increases with applied load and decreases with increasing sliding distance. Glass fibers and hard particle fillers were effective in reducing wear rate of the composite. It is reasonable to deduce that binders would increase the adhesion of glass fibers, SiC into the formaldehyde matrix. When the load is increased, microcracks are formed, followed by fragmentation in composite brake pad. Plowing, cracking and accelerated breakage of fibers in composite are evident under higher load. This study is here given in order to underline the necessity of testing new formulated friction materials, no theoretical model being able to reliably predict the tribological behavior in terms of values for wear, friction and durability.

Samyn et al. [56] presented a useful review on tribology of polyimides. Temperature modifies the tribological behavior of this polymer by chemical effects.

The tested sintered polyimides show two sliding regimes: between 100°C and 180°C, friction is high and wear rate increases, with a discontinuous minimum at 140°C. Raman spectroscopy motivated that hydration generates a reversion of polyimide into a precursor. A maximum hydrolysis intensity at 140°C explains the minimum wear rate with acid groups acting as a lubricant. From 180–260°C, friction decreases and wear rate become stable at mild loads, with a maximum value for

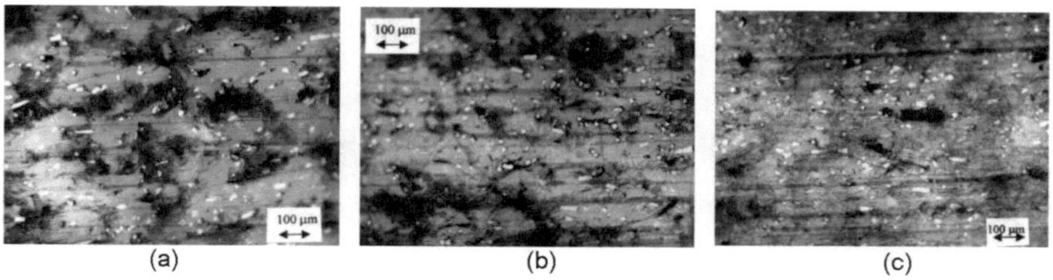

Figure 37.
Images of the partial bearings made of PTFE + short glass fibers with different concentrations, test conditions: v = 2.5 m/s and p = 4.6 MPa, water lubrication, L_x = 10,500 m [8]. (a) 15% glass fibers. (b) 25% glass fibers. (c) 40% glass fibers.

the wear rate at 180°C. Wear rates increase at high loads, but brittleness is not obvious till 150 N, at high temperatures. A discontinuous platelet transfer film develops above 180°C.

Thermoplastic polyimides show three sliding regimes that are related to a combination of chemical and thermal effects.

- at 100 to 120°C, friction increases and is higher and wear rates are lower as compared to sintered polyimides; a thin transfer film develops; dark wear particles were produced by hydrolysis,

- at 120 to 180°C, friction decreases and a transition to high wear rates is initiated; a patchy-like transfer film develops and the polymer surface becomes irregular and opaque due to softening and chemical modification; wear debris become brittle and act as an abrasive,

- at 180–260°C, friction increases and overload wear results from melting; a thick transfer film develops, and the polymer surface smoothens. Roll-like debris are visually observed as an indication for melting. Raman investigation indicates thermal decomposition of aromatic structures into amide monomers on the polyimide surface, weakening strength and producing higher wear.

And study point out the importance of test parameters, here the two polymers, the temperature and the load. Such a study could be done for each polymer of interest, with particular values for the test parameters, as they do not have a pattern due to their diversity in chemical structures and molecular organization.

Agglomeration of reinforcement fibers of particles are observed even in lubricated system with polymer composites sliding against steel. A suggestive model of reinforcements agglomeration in the superficial layer of polymeric composites, due to preferential wear of the polymer matrix has been described by Blanchet and Kennedy [10] from 1992, and then developed by Han and Blanchet in 1997 [57] and experimental results given in **Figure 37** sustained their model. Each worn surface after sliding in water has a similar concentration in short glass fibers, even if initially the concentrations were different.

7. Tendency in using polymeric materials and conclusions

New development in processing polymer-based materials (here including polymers, polymer blends, polymer composites and stratified materials based on polymeric fabrics) make easier to replace metallic parts with ones made of polymer-based materials, at a convenient price.

Test campaign are running faster as the market obliged the designers and producers to give more reliable products and the new achievement in monitoring and investigating the tribological behavior help them to understand and formulate new and adequate materials.

An obvious tendency for these materials is using them as coating, thick enough to fulfill an imposed reliability and durability.

New technologies allow for a better dispersion of the constituents, making the resulting materials more predictable [58–60].

8. Conclusion

Testing is very important when using polymer-based materials. New recipes of polymer-based materials has to follow the logical chain of testing, meaning laboratory specimen - component - partial system - entire system, in order to avoid catastrophic failure of the entire system. Even if it is difficult to imagine now new tribological parameters to be monitored or calculated, variant versions could be adapted for particular applications.

Acknowledgements

This work has been supported by the European Social Fund through the Sectoral Operational Programme Human Capital 2014-2020, through the Financial Agreement with the title „Burse pentru educatia antreprenoriala in randul doctoranzilor si cercetatorilor postdoctorat (Be Antreprenor!)" "Scholarships for entrepreneurial education among doctoral students and postdoctoral researchers (Be Entrepreneur!)", Contract no. 51680/09.07.2019 - SMIS code: 124539.

Author details

Lorena Deleanu[1*], Mihail Botan[2] and Constantin Georgescu[1]

1 "Dunarea de Jos" University, Faculty of Engineering, Department of Mechanical Engineering, Galati, Romania

2 National Institute for Aerospace Research "Elie Carafoli" (INCAS), Bucharest, Romania

*Address all correspondence to: lorena.deleanu@ugal.ro

References

[1] Patnaik A, Satapathy A, Chand N, Barkoulad N M, Biswas S. Solid particle erosion wear characteristics of fiber and particulate filled polymer composites: A review. Wear. 2010; 268:249–263. DOI: 10.1016/j.wear.2009.07.021

[2] Crawford R J. Plastics Engineering. 3rd ed. Oxford: Butterworth-Heinemann; 2002. DOI: 10.1016/B978-0-7506-3764-0.X5000-6

[3] Sinha S K, Briscoe B J. Polymer Tribology. London: Imperial College Press; 2009

[4] Leblanc J L. Filled Polymers. Science and Industrial Applications. Boca Raton: CRC Press; 2010. DOI: 10.1201/9781439800430

[5] Moore D F. Principles and Applications of Tribology, Oxford: Pergamon Press; 1975. DOI: 10.1016/C2013-0-02605-9

[6] Mathew M T, Srinivasa P P, Pourzal R, Fischer A, Wimmer M A. Significance of tribocorrosion in biomedical applications: overview and current status. Advances in Tribology. 2009; 2009:250986, DOI: 10.1155/2009/250986

[7] Myshkin N K, Petrokovets M I, Kovalev A V. Tribology of polymers: adhesion, friction, wear, and mass-transfer. Tribology International. 2005; 38:910–921. DOI: 10.1016/j.triboint.2005.07.016

[8] Briscoe B J, Sinha S K. Tribology of Polymeric Solids and Their Composites. In: Stachowiak G W, editor. Wear – Materials, Mechanisms and Practice. Chichester: Wiley; 2005. p. 223-268. DOI:10.1002/9780470017029

[9] Tomescu (Deleanu) L. Contribution on studying the superficial layers of composites with PTFE matrix, on sliding tribomodels (in Romanian) [thesis]. Galati: ||Dunarea deos|| University of Galati; 1999.

[10] Blanchet T A, Kennedy F E. Sliding wear mechanism of polytetrafluor-ethylene (PTFE) and PTFE composites. Wear. 1992; 153:229–243. DOI: 10.1016/0043-1648(92)90271-9

[11] Maftei L. Contribution on studying the tribological behavior of composites with polyamide and micro glass beads (in Romanian) [thesis]. Galati: ||Dunarea de Jos||University of Galati; 2010.

[12] Pei X, Friedrich K. Sliding wear properties of PEEK, PBI and PPP. Wear. 2012; 274–275:452–455. DOI: 10.1016/j.wear.2011.09.009

[13] Hanchi J, Eiss Jr N S. The Tribological Behavior of Blends of Polyetheretherketone (PEEK) and Polyetherimide (PEI) at Elevated Temperatures. Tribology Transactions. 1994; 37:494–504. DOI: 10.1080/10402009408983322

[14] Padhan M, Marathe U, Bijwe J. Tribology of Poly(etherketone) composites based on nano-particles of solid lubricants. Composites Part B: Engineering. 2020; 201:108323. DOI: 10.1016/j.compositesb.2020.108323

[15] Georgescu C. The evolution of the superficial layers in wear and friction processes involving composite materials with polybutylene terephthalate (in Romanian) [thesis]. Galati: Dunarea de Jos University of Galati; 2012.

[16] Botan M. Caracterizarea mecanică și tribologică a unei clase de compozite polimerice [thesis]. Galati: Dunarea de Jos University of Galati; 2014.

[17] Sharma M, Bijwe J. Influence of fiber–matrix adhesion and operating parameters on sliding wear performance

of carbon fabric polyethersulphone composites. Wear. 2011; 271:2919–2927. DOI: 10.1016/j.wear.2011.06.012

[18] Duan Y, Cong P, Liu X, Li T. Friction and Wear of Polyphenylene Sulfide (PPS), Polyethersulfone (PES) and Polysulfone (PSU) under Different Cooling Conditions. Journal of Macromolecular Science, Part B. 2009; 48:604-616, DOI: 10.1080/00222340902837899

[19] Bahadur S, Sunkara C. Effect of transfer film structure, composition and bonding on the tribological behavior of polyphenylene sulfide filled with nano particles of TiO2, ZnO, CuO and SiC. Wear. 2005; 258:1411–1421. DOI: 10.1016/j.wear.2004.08.009

[20] Friedrich K, Zhang Z, Klein P. Wear of Polymer Composites, p. 269–290, Stachowiak G.W., (editor), Wear – Materials, Mechanisms and Practice, John Wiley & Sons Ltd, England, 2005.

[21] Kurdi A, Kan W H, Chang L. Tribological behaviour of high performance polymers and polymer composites at elevated temperature. Tribology International. 2019; 130:94–105. DOI: 10.1016/j.triboint.2018.09.010

[22] Unal H, Mimaroglu A. Influence of test conditions on the tribological properties of polymers. Industrial Lubrication and Tribology. 2003; 55: 178–183. DOI: 10.1108/00368790310480362

[23] Tudor A. Frecarea şi uzarea materialelor. Bucharest: Bren; 2002

[24] Kiran M D, Govindaraju H K, Jayaraju T, Kumar N. Review-Effect of Fillers on Mechanical Properties of Polymer Matrix Composites. Materials Today: Proceedings. 2018; 5:22421–22424. DOI: 10.1016/j.matpr.2018.06.611

[25] Kmetty A, Bárány T, Karger-Kocsis J. Self-reinforced polymeric materials: A review. Progress in Polymer Science. 2010; 35:1288–1310. DOI: 10.1016/j.progpolymsci.2010.07.002

[26] Mao K, Greenwood D, Ramakrishnan R, Goodship V, Shroutib C, Chetwyn D, Langlois P. The wear resistance improvement of fibre reinforced polymer composite gears. Wear. 2019; 426–427:1033–1039. DOI: 10.1016/j.wear.2018.12.043

[27] Friedrich K. Polymer composites for tribological applications. Advanced Industrial and Engineering Polymer Research. 2018; 1:3-39. DOI: 10.1016/j.aiepr.2018.05.001

[28] Fusaro R L. Self-lubricating polymer composites and polymer transfer film lubrication for space applications. Tribology International. 1990; 23:105-122. DOI: 10.1016/0301-679X(90)90043-O

[29] Godet M. The third-body approach: A mechanical view of wear. Wear. 1984; 100:437–452. DOI: 10.1016/0043-1648(84)90025-5

[30] Liu S, Dong C, Yuan C, Bai X. Study of the synergistic effects of fiber orientation, fiber phase and resin phase in a fiber-reinforced composite material on its tribological properties. Wear. 2019; 426–427:1047–1055. DOI: 10.1016/j.wear.2018.12.090

[31] Increasing Demand for Milled Short Glass Fibers from LANXESS [Internet]. 2013. Available from: http://textilesupdate.com/increasing-demand-for-milled-short-glassfibers-from-lanxess-narrow-fiber-length-distribution-high-purity [Accessed: 2020-08-29]

[32] Sharma S, Bijwe J, Panier S. Exploration of potential of Zylon and Aramid fibers to enhance the abrasive wear performance of polymers. Wear. 2019; 422–423:180–190. DOI: 10.1016/j.wear.2019.01.068

[33] Nanomaterials and related products [Internet]. Available from: http://www.plasmachem.com/download/PlasmaChem-General_Catalogue_Nanomaterials.pdf [Accessed: 2016-02-20]

[34] Zhang X, Wang J, Xu H, Tan H, Ye X. Preparation and Tribological Properties of WS2 Hexagonal Nanoplates and Nanoflowers. Nanomaterials. 2019; 9:840. DOI: 10.3390/nano9060840

[35] Evans D C, Lancaster J K. The Wear of Polymers. In: Scott D, editor. Treatise on Materials Science and Technology. New York: Academic Press; 1979. p. 85-139. DOI: 10.1016/S0161-9160(13)70066-8.

[36] Myshkin N K, Pesetskii S S, Grigoriev A Y. Polymer Tribology: Current State and Applications. Tribology in Industry. 2015; 37:284-290.

[37] Xian G, Zhang Z. Sliding wear of polyetherimide matrix composites I. Influence of short carbon fibre reinforcement. Wear. 2005; 258:776–782. DOI: 10.1016/j.wear.2004.09.054

[38] Khedkar J, Ioan Negulescu I, Meletis I E. Sliding wear behavior of PTFE composites. Wear. 2002; 252:361–369. DOI: 10.1016/S0043-1648(01)00859-6

[39] Şahin Y. Dry wear and metallographic study of PTFE polymer composites. Mechanics of Composite Materials. 2018; 54:403–414. DOI: 10.1007/s11029-018-9751-7

[40] Biswas S K. Friction and wear of PTFE - a review. Wear. 1992; 158:193-211. DOI: 10.1016/0043-1648(92)90039-B

[41] Czikos H. Tribology – A System Approach to the Science and Technology of Friction, Lubrication and Wear. New-York: Elsevier Scientific Publishing Company;1978.

[42] Hutchings I, Gee M, Santner E. Friction and Wear. In: Czichos H, Saito T, Smith L, editors. Springer Handbook of Materials Measurement Methods. Heidelberg: Springer;2006. p. 685-710. DOI: 10.1007/978-3-540-30300-8_13

[43] Ashby M F. Materials Selection in Mechanical Design. 3rd ed. Oxford: Butterworth-Heinemann; 2005. 624 p.

[44] Jones W R Jr., Hady W F, Johnson R L. Friction and wear of poly(amide-imide),polyimide and pyrone polymers at 260°C (500°F) in dry air [Internet]. 1971. Available from: https://core.ac.uk/download/pdf/80652276.pdf [Accessed: 2020-09-02]

[45] Blunt L, Jiang X, Leach R, Harris P, Scott P. The development of user-friendly software measurement standards for surface topography software assessment. Wear. 2008; 264: 389–393. DOI: 10.1016/j.wear.2006.08.044

[46] Stachowiak G W, Batchelor A W. Engineering Tribology. Third ed. Oxford: Butterworth-Heinemann; 2006. 832 p. DOI: 10.1016/B978-0-7506-7836-0.X5000-7

[47] Archard J F. Contact and Rubbing f Flat Surface. Journal of Applied Physics. 1953; 24:981–988. DOI:10.1063/1.1721448

[48] Friction and Wear of Polymers [Internet]. 2005. Available from: http://www.appstate.edu/~clementsjs/polymerproperties/$p$lastics_$f$riction$5f$w$ear.pdf [Accessed: 2020-10-25]

[49] Burris D L, Sawyer W G. A low friction and ultra low wear rate PEEK/PTFE composite. Wear. 2006; 261:410–418. DOI: 10.1016/j.wear.2005.12.016

[50] Wear resistant polymers and composites [Internet]. 2012. Available from: https://www.rtpcompany.com/

products/wear-resistant-2/ [Accessed: 2012-04-08]

[51] Composites [Internet]. 2012. Available from: https://www.solvay.com/en/search?f%5B0% 74 5D=fchemicalcat%3A9696&f%5B1% 755D=fsection%3AProducts [Accessed: 2012-05-21]

[52] Neale M J, Gee M. A Guide to Wear Problems and Testing for Industry, London and St. Edmunds: Professional Engineering Publishing Limited; 2000. 157 p. DOI: 10.1016/B978-0-8155-1471-8.X5002-6

[53] Dasari A, Yu Z-Z, Mai Y-W. Fundamental aspects and recent progress on wear/scratch damage in polymer nanocomposites. Materials Science and Engineering: R: Reports. 2009; 63:31–80. DOI: 10.1016/j.mser.2008.10.001

[54] Abdelbary A. Wear of Polymers and Composites. Cambridge: Woodhead Publishing; 2014. 256 p. DOI: 10.1016/C2014-0-03367-9

[55] Manoharan S, Suresha B, Bharath P B, Ramadoss G. Investigations on Three-Body Abrasive Wear Behaviour of Composite Brake Pad Material. Plastic and Polymer Technology (PAPT). 2014; 3:10-18.

[56] Samyn P, Schoukens G, Verpoort F, Van Craenenbroeck J, De Baets P. Friction and Wear Mechanisms of Sintered and Thermoplastic Polyimides under Adhesive Sliding. Macromolecular Materials and Engineering. 2007; 292:523–556. DOI: 10.1002/mame.200600400

[57] Han S W, Blanchet T A. Experimental Evaluation of a Steady-State Model for the Wear of Particle-Filled Polymer Composite Materials. Journal of Tribology. 1997; 119:694–699. DOI: 10.1115/1.2833871

[58] Gong D, Xue Q, Wang H. Physical models of adhesive wear of polytetrafluoroethylene and its composites. Wear. 1991; 147:9-24. DOI: 10.1016/0043-1648(91)90115-B

[59] Gong D, Xue Q, Wang H. ESCA study on tribochemical characteristics of filled PTFE. Wear. 1991; 148:161-169. DOI: 10.1016/0043-1648(91)90214-F

[60] Nunez E E, Gheisari R, Polycarpou A A. Tribology review of blended bulk polymers and their coatings for high-load bearing applications. Tribology International. 2019; 129:92-111. DOI: 10.1016/j.triboint.2018.08.002

Chapter 5

Tribological Study of the Friction between the Same Two Materials (RAD Steel)

D. Kaid Ameur

Abstract

These demands more and more severe of the organs of friction lead to operating temperatures of more and more high which result in particular a degradation of materials. This is reflected by decreases of performance that could jeopardize the safety (fall of the coefficient of friction) and penalize the economic balance (increase the wear). Our study highlights the interactions between the thermal, tribology, and physicochemistry and has been designed to respond to the following three objectives: (1) characterize at the macrolevel the phenomena of thermal localization and identify their influence on the coefficient of friction, (2) correlate to the local scale these phenomena to the physical mechanisms of friction, and (3) to identify the consequences of the degradation of the material with the temperature, based on the coefficient of friction and the physical mechanisms of friction.

Keywords: tribology, friction, materials, steel, degradation

1. Introduction

This study clearly showed the influence of the degradation of the material on the tribological behavior. However, the diversity of the elements in the presence complicates the interpretation of results.

Finally, the important role of the oxidation has also been highlighted in this study, by its contribution to the degradation of the material on the one hand, to the training of oxides present in the third body, on the other hand. This influence of oxidation could be investigated through the realization of tribological tests under a controlled atmosphere. As an exploratory measure, development on the triometer of an experimental device designed to deprive the contact of the oxygen in the ambient air has been undertaken in this spirit.

2. Proprieties

RAD (Aubert & Duval): A ball bearing steel, and as such previously only used by forgers, it is available in bar stock now (**Figure 1**). It is similar to 5160 (though it has around 1% carbon vs. 5160 ~ 0.60%), but holds an edge better. It is less tough than 5160. It is used often for hunting knives and other knives where the user is

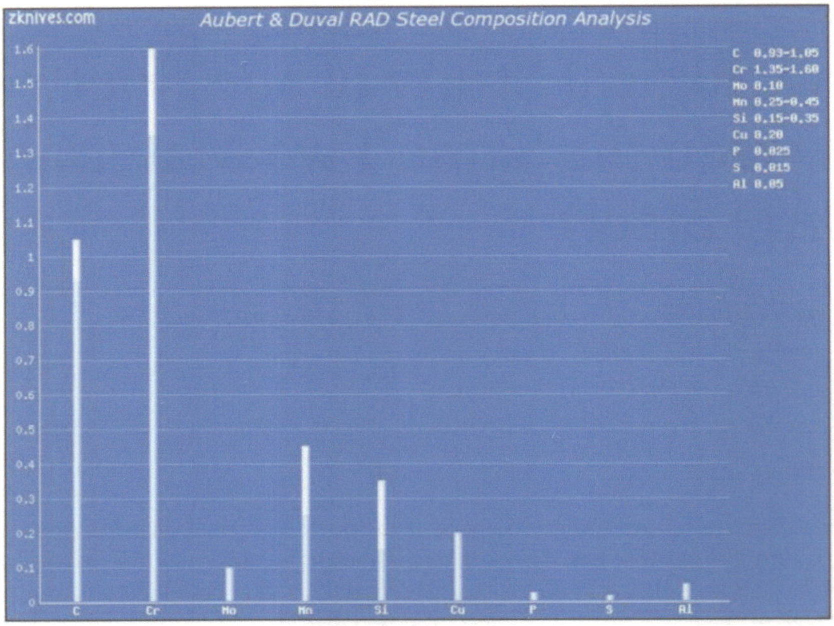

Figure 1.
RAD (Aubert & Duval) steel composition analysis.

willing to trade off a little of 5160's toughness for better wear resistance. However, with the continued improvement of 52,100 heat treat, this steel is starting to show up in larger knives and is showing excellent toughness. A modified 52,100 under the SR-101 name is being used by Jerry Busse in his Swamp Rat knives. The German equivalent 1.3505 has been discontinued [1].

It is used in precision ball bearings and many industrial applications. The balls made of this kind of material feature an excellent surface finish, considerable hardness, and a high load-carrying capacity, as well as excellent wear and deformation resistance. Chrome steel balls (**Figure 2**) are through hardened in order to achieve the maximum mechanical strength [2].

Diameters: 0.025–250 mm.

Precision grades: ISO 3290 G3–5–10–16–20–28–40–100–200–AFBMA G500/G1000.

Equivalent materials to international standards: AFN 100C6–B.S. EN 31–JIS G4805–SUJ2–ASTM 100C6.

Through hardness index:

Up to 12.7 mm HRC 62/66.

From 12.70 to 50.80 mm HRC 60/66.

From 50.8 to 70 mm HRC 59/65.

From 70 to 120 mm HRC 57/63.

Mechanical properties:

Critical tensile strength: 228 kgf/mm^2.

Compression strength: 207 kgf/mm^2.

Modulus or elasticity: 20,748 kgr/mm^2.

Specific weight: 7830 kgf/mm^2.

Chemical compositions, %:

C: 0.90–1.10; Si: 0.15–0.35; Mn: 0.25–0.45; P: 0.025–max; S: 0.025–max; Cr: 1.30–1.60.

Figure 2.
Chrome steel balls.

3. Characterization of the tribological behavior of the dry contact

3.1 Analysis of friction, μ, and μe

The friction coefficient is determined from the tangential force, Q*, measured during the test by a force sensor, and the normal force applied, P, measured by a sensor to gauge the constraints (**Figure 3**) [3].

The value of the coefficient of friction conventional said during each cycle is given by the following expression [4]:

$$\text{ì} = Q * / P \tag{1}$$

While the average value of the coefficient of mechanical friction during the entire test is expressed by the following expression [5]:

$$\bar{\mu} = \frac{1}{N_t} \sum_{i=1}^{N_t} \mu(i) \tag{2}$$

where N is the number of cycles (**Figure 4**).

$$\mu_e = \frac{E_d}{4 * P * \delta_g} \tag{3}$$

where E_d is the dissipated energy in Joules, δ_g is the amplitude of slipping in micrometers, and P is the normal force applied in Newtons.

The average value of the energy coefficient of friction during the entire test is expressed by the following [4]:

$$\bar{\mu}_e = \frac{1}{N_t} \sum_{i=1}^{N_t} \mu_e(i) \tag{4}$$

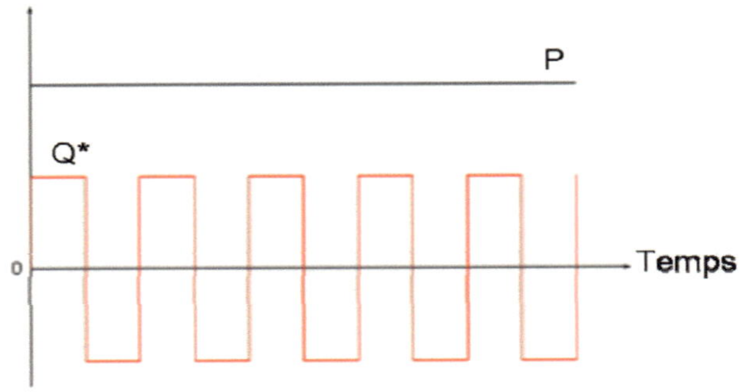

Figure 3.
Illustration of the evolution of the normal force and tangential force during the test (ideal situation of a rolling contact infinitely rigid) tangential as a function of time.

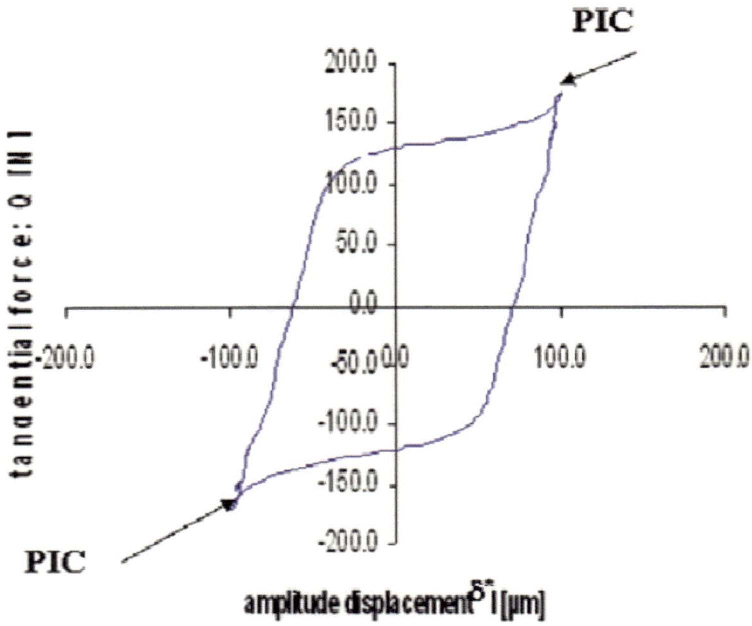

Figure 4.
Cycle of fretting during the test.

To study the evolution of the coefficient of friction during the test, we used a sphere contact/plan of steel 100Cr6 on the machine I, and the diameter of the SPHERE chosen is of 25.4 mm, and the movement is alternative straight with an amplitude of displacement of ±100 mm to a frequency of 10 Hz. A normal load of 86 N is applied during the test friction, which corresponds to a pressure of Hertzian contact maximum in the early test (without wear) of 1.1 GPa. The frequency of the movement and the frequency of the normal load are measured, recorded, and regulated during the test. The tangential force of contact is measured and recorded during the friction (**Figure 5**).

3.2 Evolution of the coefficient of friction

The evolution of the friction coefficient, μ, conventional and the energy coefficient, μe, in function of the cycles of fretting is presented in **Figure 6**. We

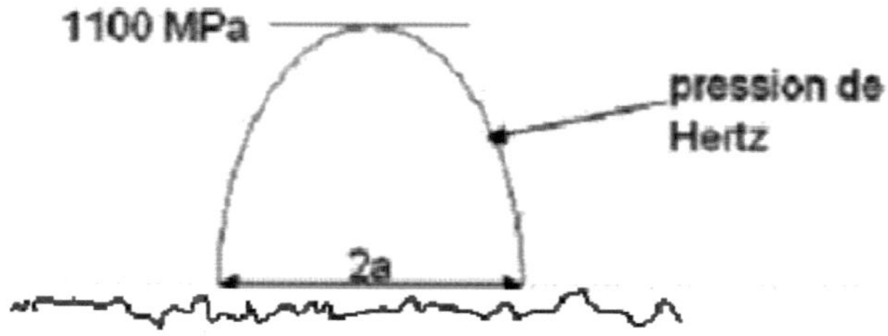

Figure 5.
A mapping of the initial contact sphere/plan for a load applied normal, 86 N (a is the radius of the contact terrestrial, we will appoint subsequently by a_H).

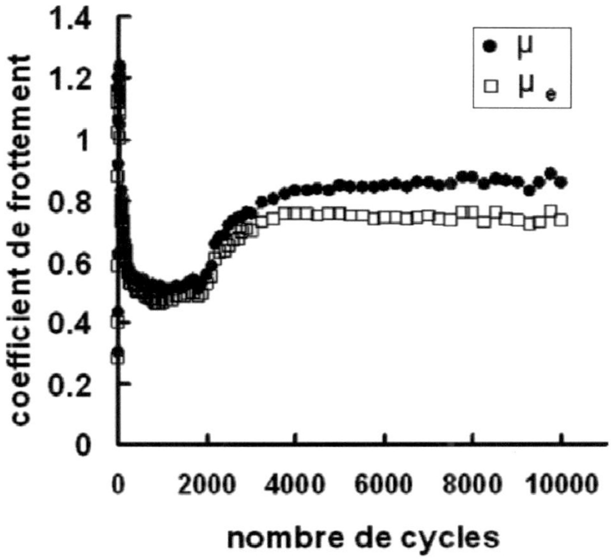

Figure 6.
Evolution of the coefficient of friction is conventional and of the energy coefficient. The friction in function of the number of cycles of fretting of the contact sphere/plan.

note that the evolution of μ and μe presents a similar evolution [6, 7]. The two curves are superimposed up to 2000 cycles, and while the contact stabilizes, we note a significant differentiation of μ and μe. The value of μ stabilizes around 0.85, while μe stabilizes at around 0.75. The values' averages over the whole of the test are, respectively, μ = 0.77, and μe = 0.71. Subsequently, it conducts tests that are interrupted following the evolution of the coefficient of friction (**Figure 7**, six conditions of the number of cycles of four cycles to 10,000 cycles) to appraise the structure of the traces of wear of the plan and of the sphere in function of the establishment of the contact. It still maintains the amplitude of slip, δg, and the amplitude of oscillation, δ*, as constant while all the tests are carried out.

The evolution of the coefficient of friction reveals three phases. During the first phase, we note an increase in the coefficient of friction up to a value of 1.2 (up to 20 cycles); then the second phase corresponds to a fall of up to 0.45 (up to 300 cycles); and then the third phase presents two parts: the first, an increase very progressive of the friction coefficient of 0.45–0.75 (of the 300th–3000th cycle). And finally, the second part of the third phase corresponds to a stabilization of the coefficient of friction around 0.85 (ranging from the 3000th–10,000th cycle).

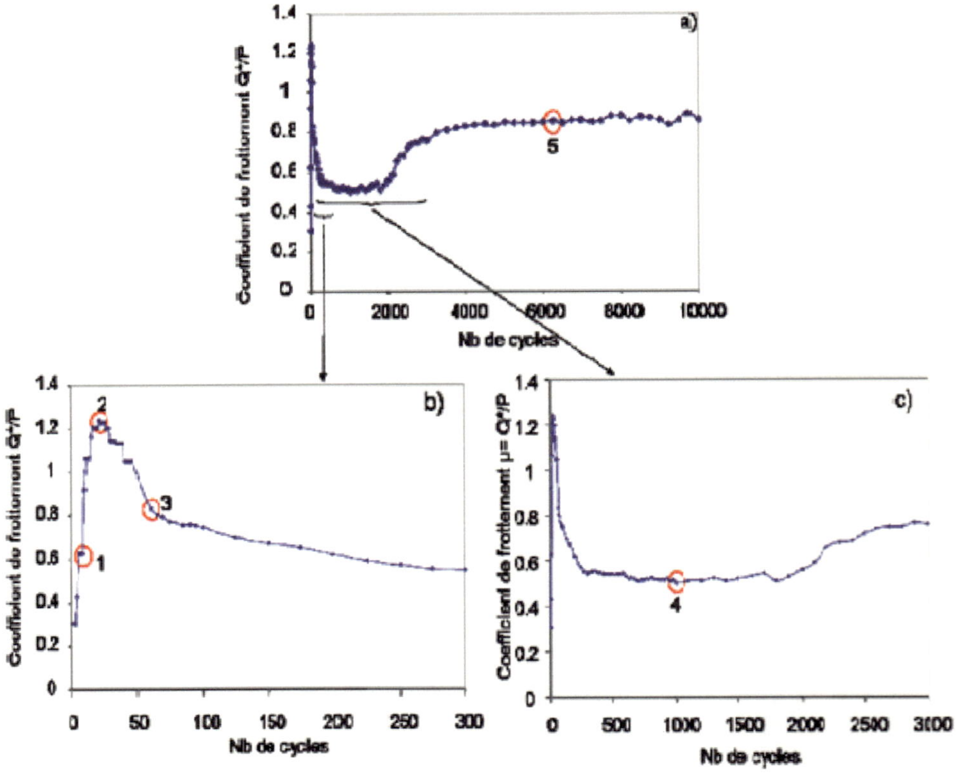

Figure 7.
Evolution of the coefficient of friction in function of the number of cycles for the torque 100C6 sphere/plan (F = 86 N, δ = 100 µm, F = 10 Hz, and R = 12.7 mm). (a) Test INTEGER, (b) zoom up to 300 cycles, and (c) zoom up to 3000 cycles, and the red circles indicate the tests interrupted for the expertise of traces of fretting.*

3.3. Analysis of the trace of wear

Several technics have been used to understand the evolution of the coefficient of friction. In function of the number of cycles: optical microscopy, the electronic microscopy to Sweep (SEM), as well as the EDX analysis [8, 9].

The observation of the traces of wear in the optical microscope after different durations of friction allows us to understand the evolution of the creation of debris and oxides in function of the time. The two surfaces, that of the sphere and that of the plan, are observed. The coloration of the trace of wear gives indications on the presence of oxides.

The SEM is used in order to achieve chemical analyses in the trace of wear and outside of the trace. These analyses confirm about the presence of oxides.

During the first phase, the increase, very brutal, of the coefficient of friction is associated with a contact metal/metal. We interrupt the test in the fourth cycle for observing the surface (**Figure 8**). We note that there is a large surface not worn to the inside of the contact and that we have a wear at the edge of the contact which is more important on the sphere on the plan.

The surface does not oxidize debris. In contrast the surfaces. The darkest shows transfers metal/metal, which explain the very strong increase in the coefficient of friction. Then interrupts our test to 20 cycles, and when the increase of the coefficient of friction reaches its maximum to observe the evolution of the trace of wear (**Figure 9**), we note the existence of a few debris at the edge of the contact, as shown in **Figure 9(c)**, where they are compacted, favoring the phenomena of accession at the edge of the contact, as shown in **Figure 9(d)**.

Analyses of EDX in the trace of wear have intended to assert or non-The presence of oxides in the trace of wear.

Figure 10(a) shows an image SEM of the trace of wear on a sample after 20 cycles of friction. An EDX analysis of the trace of wear (**Figure 10(b)**) shows the detectable elements as in **Table 1**.

Table 1 shows that there has been very little of oxides in the trace of wear to 20 cycles. **Figures 9** and **10** confirm a generalization of the metal-metal interactions after only 20 cycles, which allows us to explain the very high value of the coefficient of friction. In the second phase during the fall of the coefficient of friction, it stops

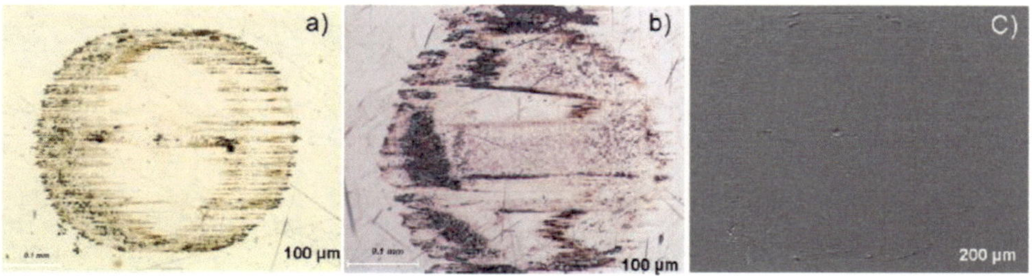

Figure 8.
Traces of wear on the samples to the fourth cycle. (a) Image under the microscope optics of the plan, (b) image in the optical microscope of the SPHERE, and (c) image to the SEM of the plan [10].

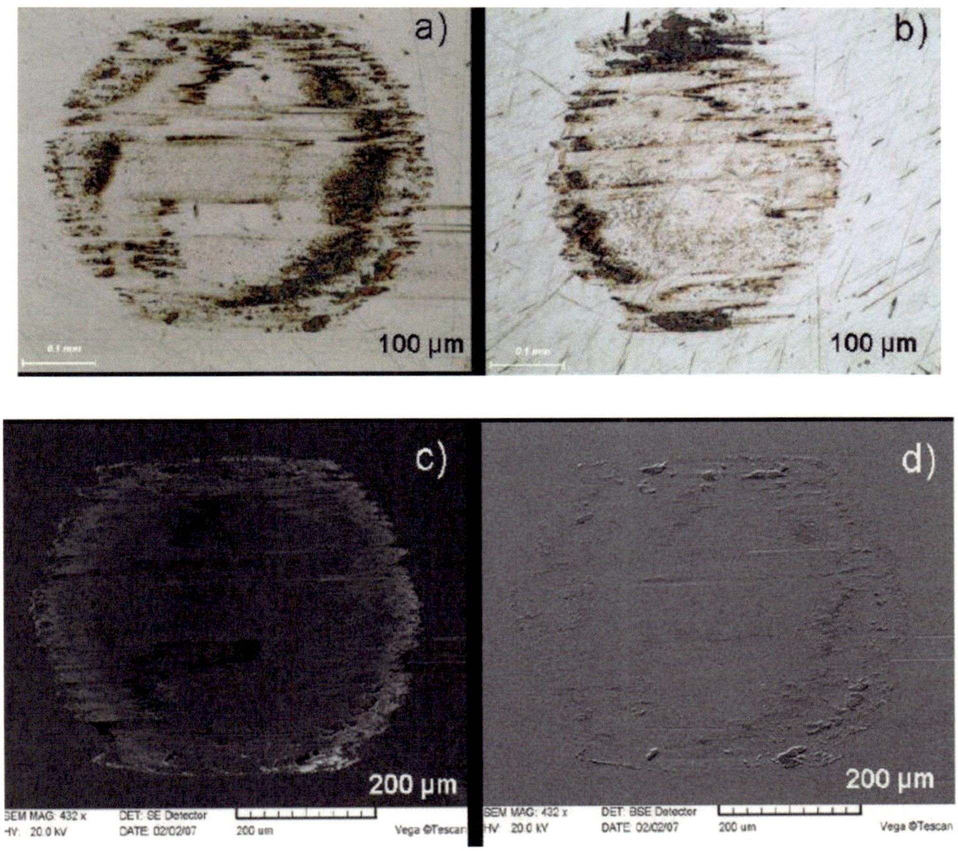

Figure 9.
Traces of wear on the samples in the twentieth cycle. (a) Image under the microscope optics of the plan, (b) image in the optical microscope of the SPHERE, and (c) and (d) image in the WPM in traces of wear [10].

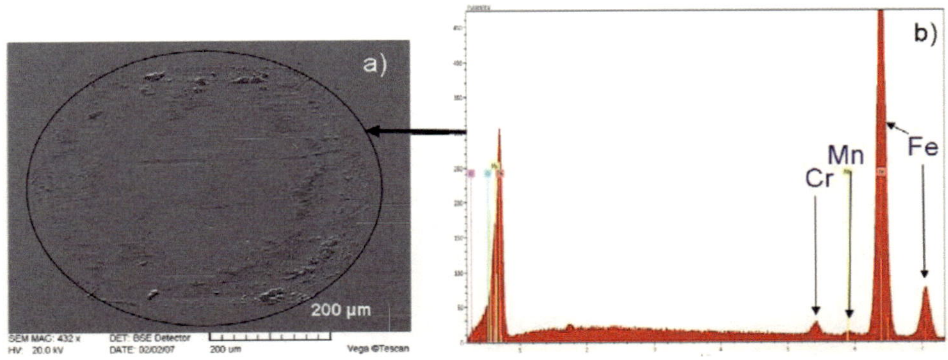

Figure 10.
Chemical analysis of the trace of wear on a sample after 20 cycles of friction. (a) Image SEM of the trace of wear and (b) EDX analysis of the trace.

Item	Atom (%)
Carbon	16.22
Oxygen	0.004
Chrome	1.76
Iron	82

Table 1.
Elements detected in the trace of wear to 20 cycles.

the test at the 60th cycle. The coefficient of friction is of the order of 0.8. **Figure 11** shows the different surfaces observed.

We note that there is a material removal to the inside of the contact (**Figure 11a**, **b**, and **d**). The fall of the coefficient of friction in the second phase is due to the presence of debris that are very oxidized (**Figure 11c**; the debris of white color observed using the WPM, which are ejected at the edge of the contact). After cleaning, it shows the existence of oxides on the worn surface, however, the EDX analysis shows results similar to those observed after 20 cycles with a few traces of oxygen (0.004%) on the surface worn. To locate these traces of oxides, it carries out analyses following a line called "Line Scan." These analyses are carried out line by line on the surface of the traces worn globally. One of these lines shows the presence of a few traces of oxide on the compacted materials at the edge of the trace of wear, as shown in **Figure 12**.

In conclusion, it could be deduced that after 60 cycles, of the first oxidized debris are trained and allow a partial reduction of the coefficient of friction. These debris are however few and are only little members in the interface (because they are easily eliminated). This allows you to explain that the coefficient of friction at this stage remains relatively high. Their accumulation in edge of contact suggests a change of load transfer and flotation a dela pressure on the edges of the contact. A fourth test is conducted and interrupted after 1000 cycles. This condition of solicitation corresponds to the coefficient of friction the lowest ($\mu = 0.5$). To interpret the tribological behavior, it is interesting to compare the optical observations and SEM (**Figure 13**) and the analysis of the surface of the contact (**Figure 14**).

Figure 13 shows that the material removed is distributed over the entire surface of the sphere and the plan [**Figure 13(a)–(c)**]. **Figure 13c** confirms the presence of a large quantity of oxides on the surface, before cleaning, and mainly distributed on the periphery of the contact. By contrast, after cleaning,

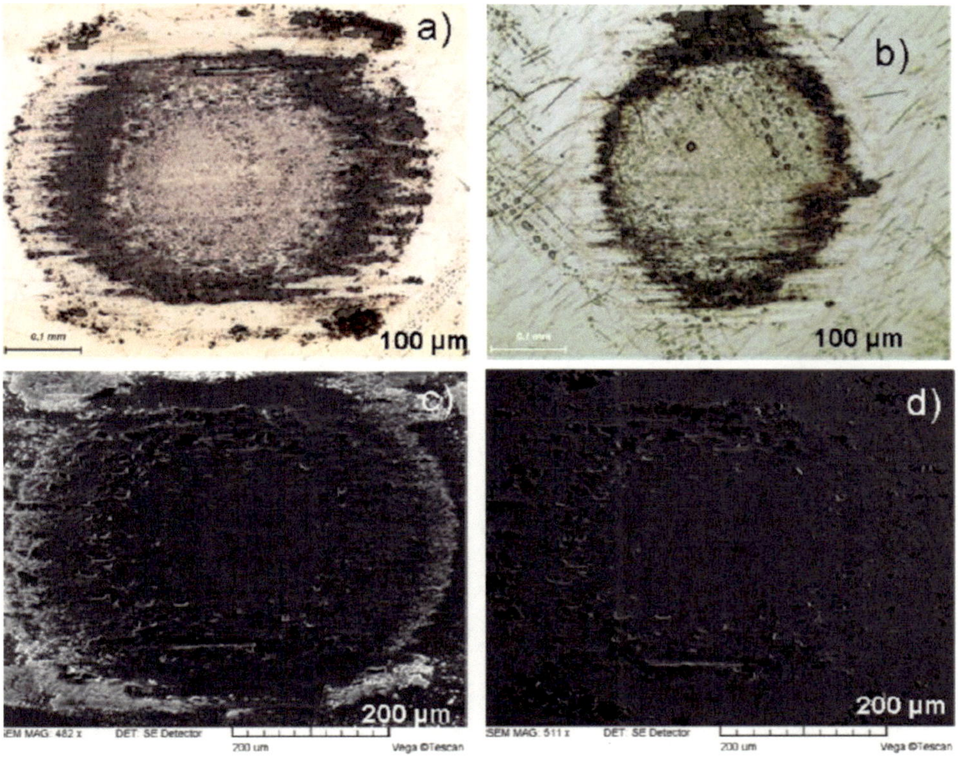

Figure 11.
Trace of wear after 60 cycles of friction. (a) Optical image of the plan, (b) image perspective of the SPHERE, (c) image in the WPM before cleaning, and (d) image to the SEM after cleaning [10].

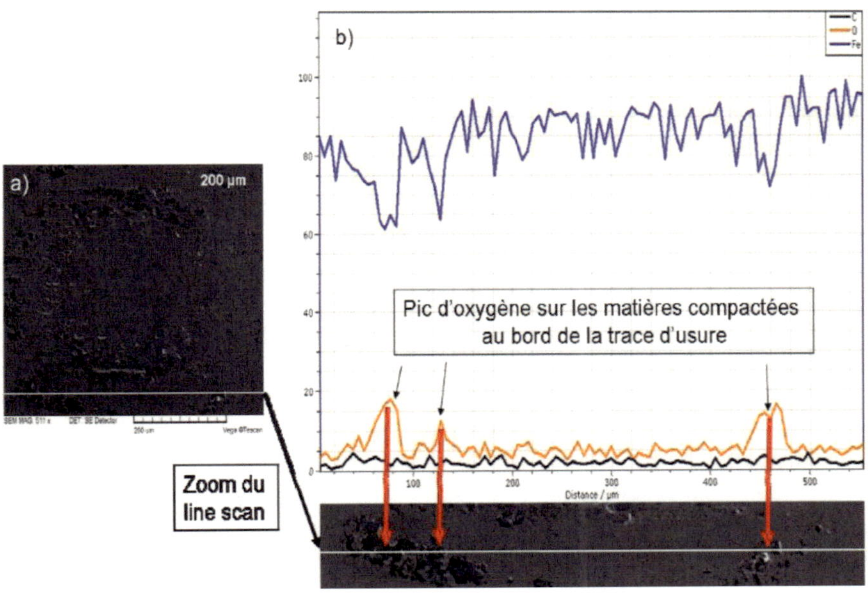

Figure 12.
Analysis of the line scan of the trace of wear on a sample after 60 cycles of friction. (a) Image SEM of a line scan of a plan and (b) zoom of the line scan with the results of EDX of this line scan.

the quantity of debris oxidized is much more low. The EDX analysis (**Figure 14**), however, confirms the presence of a large quantity of oxygen from the acceding debris in the interface.

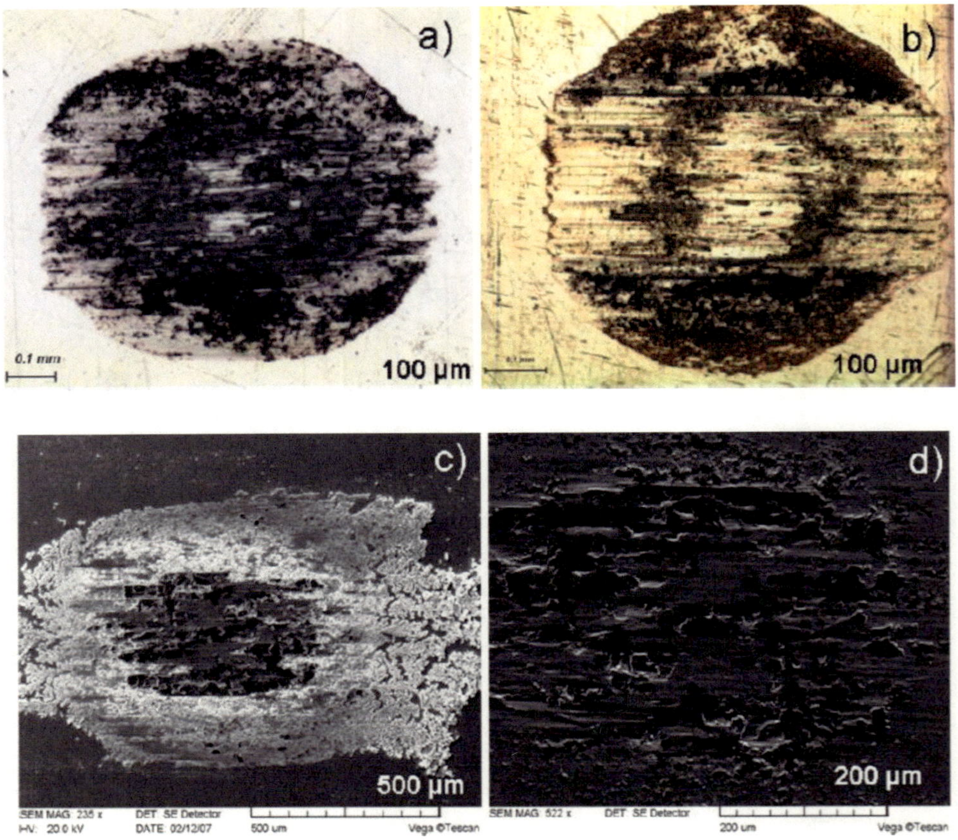

Figure 13.
Trace of wear on a sample after 1000 cycles of friction. (a) Image optics of the plan, (b) optical image of the SPHERE, (c) image to the SEM plan before cleaning, and (d) image to the SEM plan after cleaning [10].

Figure 14 also confirms the existence of oxidized debris on the sphere, especially at the edge of the contact. The concentration in oxides is confirmed by the semi-quantitative analysis (**Table 2**).

What we can remember is that the interface associated with the lowest coefficient of friction corresponds to a structure with a lot of debris oxidized very pulverulently because they are very easily eliminated (comparison of observations of SEM before and after cleaning); mainly located on the outer edges of the contact, this bed of debris very accommodating; and allows to obtain a coefficient of friction that is relatively low. The presence of debris more compacted and members indicates the activation of a process of "Mécano alloying" and the creation at the level of the first body of compacted debris more members. The fifth point analyzed corresponds to a test duration of 6000 cycles. It corresponds to the stabilized state, of contact with a value of the coefficient of friction of the order of 0.8, which is significantly greater than the point of intermediate 5.

Figure 15 illustrates the structure of the interface. It will be noted that these observations have been carried out after cleaning of surfaces.

One can conclude that a layer of compacted oxide is adherent and is distributed on the whole surface of the contact (**Figure 15d** and **e**). It confirms the very large quantity of oxides associated to acceding debris by the semiquantitative analysis of EDX (**Table 3**).

Figure 16 confirms that a similar pattern to that observed on the plan is activated on the sphere. In effect, even after cleaning of surfaces, there is a very large

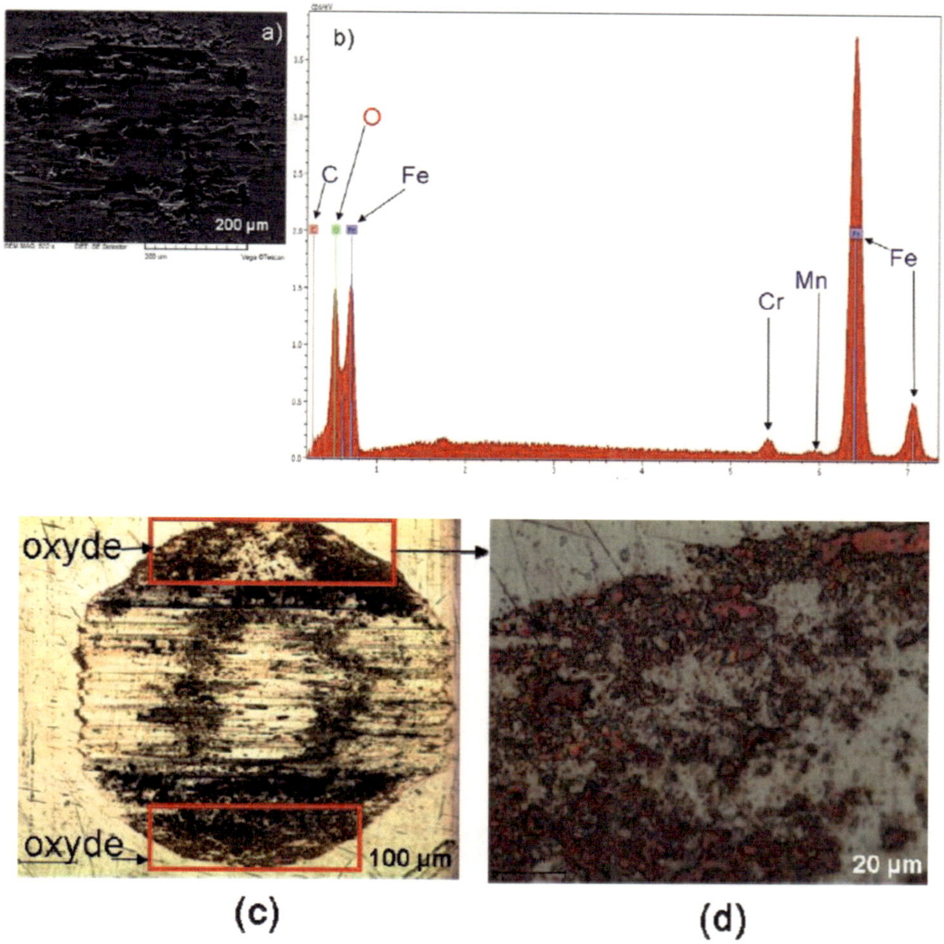

Figure 14.
Chemical analysis of the trace of wear on a sample of the plan and of the SPHERE after 1000 cycles of friction. (a) Image SEM of the trace of wear of the plan, (b) EDX analysis of the trace, (c) optical image of the SPHERE, and (d) zoom of the oxidized portion of the SPHERE [10].

Item	Atom (%)
Carbon	16.9
Oxygen	15.6
Chrome	1.5
Iron	65.8

Table 2.
Elements detected in the trace of wear to 1000 cycles.

quantity of oxides acceding compacted on the surface. In conclusion, it is shown that the process already observed after 1000 cycles of compaction of the third body is becoming widespread. The interface thus evolves to a structure, little member of a third cohesionless body toward a third body compacted, very adhesive.

This evolution of the rheology of the third body, which gives rise to a 3rd Corps member and less complacent than that the bed of cohesionless debris, may explain the increase in the friction coefficient of the second minimum at $\mu = 0.5$, up to the bearing stabilized at $\mu = 0.8$.

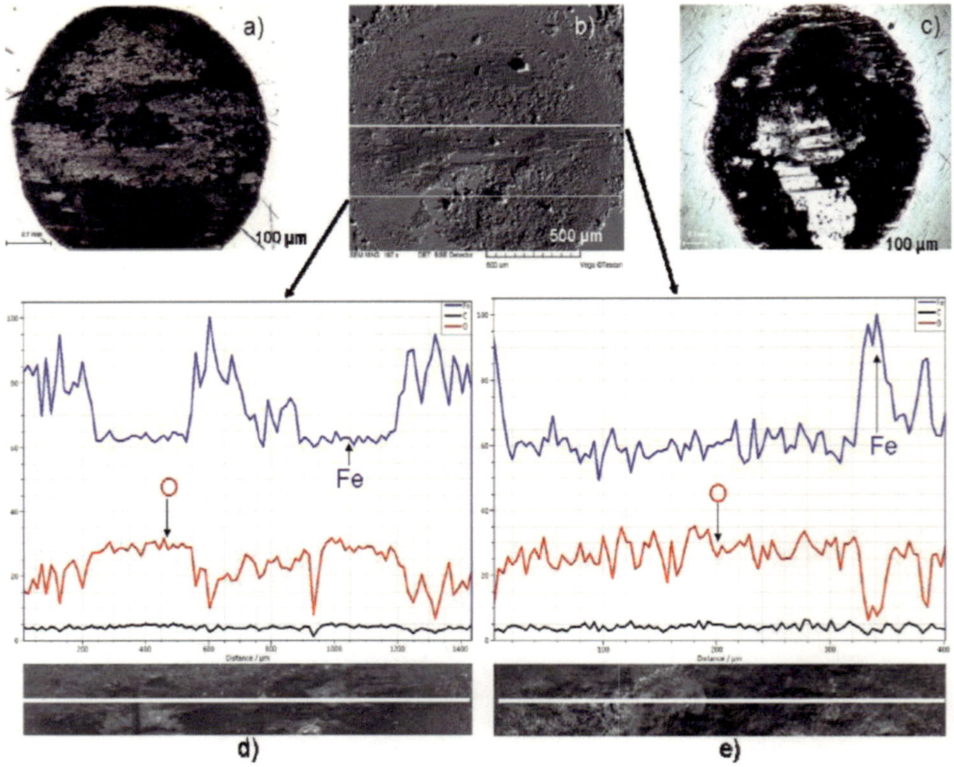

Figure 15.
Traces of wear and chemical analysis of traces of wear on a sample after 6000 cycles of friction. (a) Optical image of the plan, (b) image SEM of the plan, (c) image perspective of the SPHERE, (d) zoom of analysis scan line at the edge of the contact, and (e) zoom of analysis line scan in the middle of the trace of wear.

Item	Atom (%)
Carbon	10.6
Oxugéne	28.3
Chrome	1.4
Iron	59.6

Table 3.
Elements detected in the trace of wear to 6000 cycles.

4. Discussion

The coupled analysis between the evolution of the coefficient of friction and the structure of the interface allows us to establish the tribological scenario next (**Figure 17**).

The very low coefficient of friction observed at the beginning of the test corresponds to an interface involving the native oxides (A). Very quickly, the layer of oxides is eliminated and there is a sharp increase in the coefficient of friction related to activation of interactions metal/metal (B). The interactions metal/metal generalize and generate a maximum shear in the interface (C). The very high stresses generated in the interface lead to formation of debris that oxidize. The interface is more accommodative and the coefficient of friction tends to decrease (D).

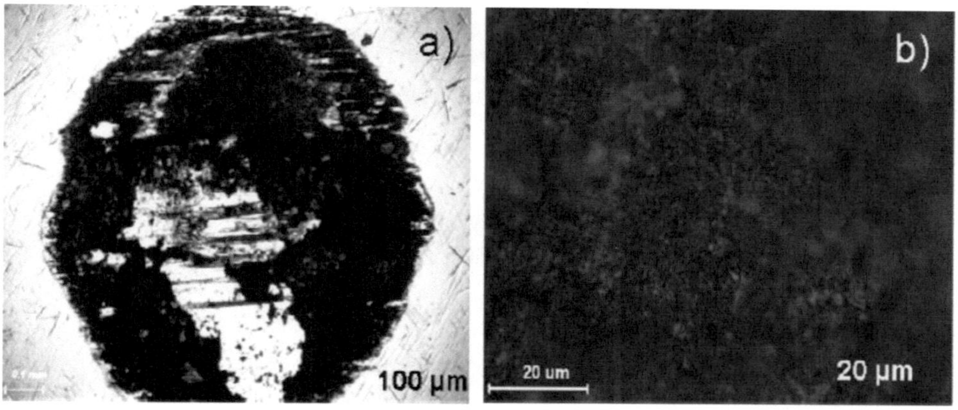

Figure 16.
Traces of wear on the samples after 6000 cycles. (a) Image under the microscope perspective of the SPHERE and (b) zoom of the trace of wear [10].

The training of cohesionless debris generalizes and finally écrante interactions metal/metal. Fully accommodated by a third body, very cohesionless, divided on the periphery of the contact, the interface then presents a second minimum of its coefficient of friction (E).

Under the mechanical action of the loading of fretting, the bed of debris then tends to be compressed. It becomes more adherent and generalizes on the whole of the interface. Less complacent than the bed cohesionless debris (Step e), it presents a coefficient of friction higher (F) of the order of (μ = 0.8). If the evolution between the steps a and e is classically described in the literature, the transition between E and F that we have analyzed clearly shows the influence of rheology in the interface.

Having analyzed the response of the friction coefficient of the dry contact, we will discuss in the next chapter, the response by report to the wear.

5. Conclusion

The objective of this study was to analyze the behavior in fretting of the steel contact. The expertise of the industrial systems in fact appear of slip conditions, total inducing large amplitudes of slip, thus favoring the degradation of the assembly by wear. There is a real need in the industry to have predictive methods about the mechanisms of wear, to limit maintenance inspections while ensuring an optimum level of security.

For the first time, the coupled analysis of the evolution of the coefficient of friction and the structure of the interface allows us to divide the evolution of the coefficient of friction into six phases, appointing A, B, C, D, E, and F and describing the scenario of the evolution of the interface as well as the role of the debris and oxides in the contact.

The elimination of the layer of native oxides (A and B) generates the interactions of metal/metal, which promotes a maximum shear in the interface (C). The training of debris that oxidize tends to decrease the coefficient of friction (D and E). Under the mechanical action of the loading of fretting, the bed of debris then tends to be compact and becomes more adherent, and it generalizes on the whole of the interface. This third compacted body, less complacent than the bed of the cohesionless debris, may cause an elevation of the coefficient of friction (F).

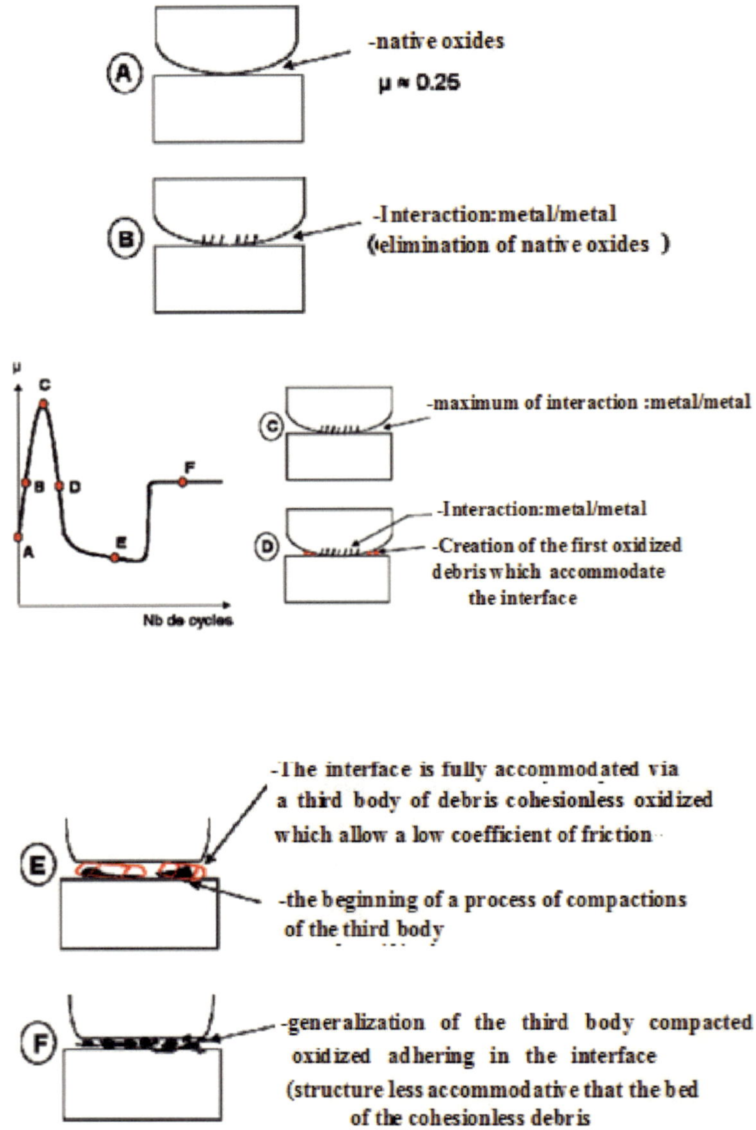

Figure 17.
Illustration of the scenario describing the evolution of the interface and that associated to the coefficient of friction.

By that result, we sought to quantify the kinetics of wear. The successive damage the contact has been formalized through the approaches of Archard and the dissipated energy. The evolution of the coefficient of friction for different sizes of contact shows that the more the contact, the lower the coefficient of friction of the stabilized phase (F). It can assume that larger contact facilitates the trapping of oxidized debris in the contact.

The interface will be more accommodative and will induce a coefficient of friction that is more low. By elsewhere, a greater amount of energy will be dissipated in the third body and not to the level of the first body for the creation of new debris so that the kinetics of wear will also be lower for the great contacts.

Author details

D. Kaid Ameur
Laboratoire de Génie Industriel et du Développement Durable (LGIDD), Centre Universitaire de Relizane, Bormadia, L'Algérie

*Address all correspondence to: djilalikaidameur@gmail.com

References

[1] (Aubert § Duval RAD Steel)

[2] CHROME_STEEL_BALLS_AISI_52100_UNI100C6-htm_s13lccgu-NormeAFNOR Française

[3] Berthier Y. Background on friction and wear. In: Lemaître Handbook of Materials Behavior Models.. Section 8.2. Academic Press; 2001. pp .676-699

[4] Bettge D, Starcevic J. Topographic properties of the contact zones of wear surfaces in disc-brakes. Wear. 2003;**254**:195-202

[5] Bulthé A-L, François M, Desplanques Y, Degallaix G. Comportement d'un couple disque-patin sous sollicitations de freinage et observations in situ du 3ème corps. In: 17ème Congrès Français de Mécanique Troyes. 2005

[6] Bulthé A-L, Desplanques Y, Degallaix G, Berthier Y. Mechanical and chemical investigation of the temperature influence on the tribological mechanisms occurring in OMC/cast iron friction contact. In: 12th Nordic Symposium of Tribology, NordTrib 2006 Helsingor (Danemark). 2006

[7] Cho MH, Kim SJ, Kim D, Jang H. Effect of ingredients on tribological characteristics of a brake lining: An experimental case study. Wear. 2005;**258**:1682-1687

[8] Copin R. Etude du comportement tribologique de couples de matériaux industriels sur tribomètre reproduisant les conditions de freinage ferroviaire [thèse]. Université des Sciences et Technologie de Lille; 2000

[9] Desplanques Y, Roussette O, Degallaix G, Copin R, Berthier Y. Analysis of tribological behaviour of pad-disc contact in railway braking. Part 1: Laboratory test development, compromises between actual and simulated tribological triplets. Wear. 2006. In press

[10] Merhej R. Matériaux [PhD thesis]. France: Ecole Centrale de Lyon; 2008

Chapter 6

Tribology of Ti-6Al-4V Alloy Manufactured by Additive Manufacturing

Auezhan Amanov

Abstract

In this study, the influence of ultrasonic nanocrystal surface modification (UNSM), which was applied as a post-additive manufacturing (AM), in terms of surface, tensile and tribological properties of Ti-6Al-4V alloy by selective laser melting (SLM) was investigated. Ti-6Al-4V alloy was subjected to UNSM at room and high temperatures (RT and HT). It was found that the UNSM enhanced the strength and reduced the roughness of the as-SLM sample, where both increased with increasing UNSM temperature. The UNSM bore influence on tribological properties, where the friction coefficient of the as-SLM sample reduced by about 25.8% and 305% and the wear resistance enhanced by about 41% and 246% at RT and HT, respectively. These are essentially attributed to the enhanced strength, smoothed surface and expelled pores from the surface. Based on SEM images, the damage caused by abrasive wear was the most observed in the wear track of the as-SLM sample than was caused by the highest wear rate. The UNSM temperature-dependent wear mechanisms were comprehensively investigated and elaborated based on the obtained experimental data and observed microstructural images. Indeed, a further investigation is required to improve the characteristics of as-SLM Ti-6Al-4V alloy to the wrought level due to the replacement possibility.

Keywords: roughness, strength, tribology, additive manufacturing, ultrasonic nanocrystal surface modification

1. Introduction

Ti-6Al-4V alloy is a metallic material that attracts much attention from many researchers due to its biocompatibility, good corrosion resistance and high specific strength. Due to these excellent properties, various components are made of Ti-6Al-4V alloy for biomedical and aerospace applications such as medical implants, aerospace crafts, gas turbines etc. [1, 2]. Recently, the advancement of additive manufacturing (AM) has been revelation for manufacturing customized parts with complex geometry in small volume, which is suitable for biomedical and aerospace industries. Selective laser melting (SLM) is one of AM that capable of processing a wide range of metals, alloys and metal matrix composites [3]. Therefore, this method is commonly used for fabrication of Ti-6Al-4V alloy components of those industries.

Attempts for improving their properties have been a critical subject for researchers for overall properties of SLM fabricated Ti-6Al-4V alloy parts [4]. It has been reported that SLM fabricated components of nickel-based superalloys and titanium-based alloys are known to have several issues, such as cracking due to thermal stress by rapid heating and quenching [5, 6], poor surface finish and voids inside the material [6, 7], tensile residual stress, which causes deformation, cracks, and worsening of fatigue strength [4, 8], and columnar grain structures, which cause anisotropy in mechanical properties [9, 10]. To cope with these issues, various post-processing techniques have been applied such as surface milling for removing rough surface layers and possible cracks [8, 11], heat treatment for increasing toughness by phase transformation [12–15], laser polishing for decreasing surface roughness [16, 17], etc. Laser polishing reduced the roughness of laser AM TC4 and TC11 by about 75% and at the same time enhanced the surface micro-hardness of TC4 and TC11 by about 32% and 42%, respectively [18]. The other post-processing method that usually performed is hot isostatic pressing (HIP), which is capable of reducing the porosity and tensile residual stress of SLM fabricated Ti-6Al-4V alloy [19]. Post-heat treatments and HIP can solve some of those issues of SLM fabricated Ti-6Al-4V alloy, but the strength is significantly reduced. While laser polishing can increase the micro-hardness, but cannot solve the porosity-related issues and improve ductility.

For solving the aforementioned issues of metal AM, the application of surface treatment or modification technologies has been proposed, e.g., shot peening (SP), laser shock peening (LSP), and ultrasonic nanocrystal surface modification (UNSM). The application of SP showed a remarkable improvement in fatigue behavior of AlSi10Mg alloys fabricated by AM in the high cycle fatigue region [20]. A previous study reported that LSP can induce a deep level of compressive residual stress, which significantly improves the fatigue life of 316 L stainless steel fabricated by AM [21]. It was also reported that LSP provided more fatigue life improvement than SP, which is largely attributed to the depth of compressive residual stress.

UNSM is one of the mechanical surface modifications that utilizes an ultrasonic vibration energy to improve mechanical properties, tribological behavior, corrosion resistance and fatigue strength of various materials including AM fabricated materials. It was found earlier reported that the fatigue strength of SUS 304 shaft was improved by approximately 80% and the surface hardness is enhanced by both the grain refinement and the martensitic transformation after treatment by UNSM [22]. Furthermore, UNSM induced enhancement of surface hardness, compressive residual stress and grain refinement that resulted in improvement of fretting wear and frictional properties of commercially pure Ti and Ti-6Al-4V alloy [23]. The application of UNSM to AM fabricated materials has also been investigated. For example, Zhang et al. reported that electrically-assisted UNSM reduced the porosity and surface roughness, and enhanced the surface hardness of 3D printed Ti-6Al-4V alloy [24]. In our previous study on SLM fabricated 316 L stainless steel, it was reported that UNSM improved the mechanical properties, tribological behavior and corrosion resistance [25]. In general, UNSM reduces the surface roughness, increases the surface hardness, refines grain size and induces high compressive residual stress with the depth of hardened layer in the range of ~0.1 to ~0.3 mm [23, 25]. UNSM temperature-dependent surface hardness and phase transformation of wrought Ti-6Al-4V alloy were reported earlier [26], but the influence of UNSM temperature-dependent mechanical and tribological properties of AM fabricated Ti-6Al-4V alloy was not investigated yet. Therefore, in this study, the synergy effect of UNSM and local heat treatment (LHT) on the improvement of tensile and tribological properties of SLM fabricated Ti-6Al-4V alloy is investigated. The improvement in tensile and tribological properties after UNSM at different temperatures was compared with the results of the as-printed Ti-6Al-4V alloy.

2. Experimental procedure

2.1 Sample preparation

In this study, the sample made of Ti-6Al-4V alloy were fabricated by SLM (EOS M290, Germany) under the parameters listed in **Table 1**. **Figure 1** shows that the powder was spherical with a diameter of about 30–40 μm. Detailed information of SLM can be found in our previous study [25]. The hardness and yield strength of the as-printed samples were about 380 HV and 820 MPa, respectively. As for chemical composition, the content of Al and V was about 5.8 and 4.2 in wt.%, while the rest was Ti, respectively. The sample fabricated by SLM was used in as-printed state, which is hereinafter referred to as as-SLM, while the UNSM-treated samples at 25°C and 800°C are hereinafter referred to as UNSM-25 Degree-C and UNSM-800 Degree-C, respectively.

2.2 Application of UNSM technology

A UNSM is a cold-forging process that uses a tungsten carbide (WC) tip with a diameter of 2.38 mm to strike the sample surface at 20 kHz, which results in elasto-plastic and surface severe plastic deformation (S^2PD), and heating, whereas forming a nanostructured layer at RT and HT. Due to a small radius of the tip, the contact area with a sample is relatively small causing high contact pressure up to 30 GPa. Advantages of UNSM over other surface peening technologies for particular AM materials are that it smooths out the surface, which is usually rough after AM and also increases the strength simultaneously. Moreover, it somehow shrinkages pores due to the compressive strike. The samples were treated by UNSM using the following variables listed in **Table 2** at 25 and 800°C. The combination of UNSM and LHT was described in our previous study [27]. The main variables are important, while force being the most important because it's magnitude determines the intensity of strain hardening. The force is directly proportional to the surface hardness, the grain size, strain-hardened layer and the compressive residual stress. The roughness is inversely proportional to the force, while it is directly proportional to the feed-rate.

Laser power, W	Scan speed, mm/s	Hatching spacing, mm	Nominal layer thickness, μm
300	900	0.12	50

Table 1.
SLM parameters.

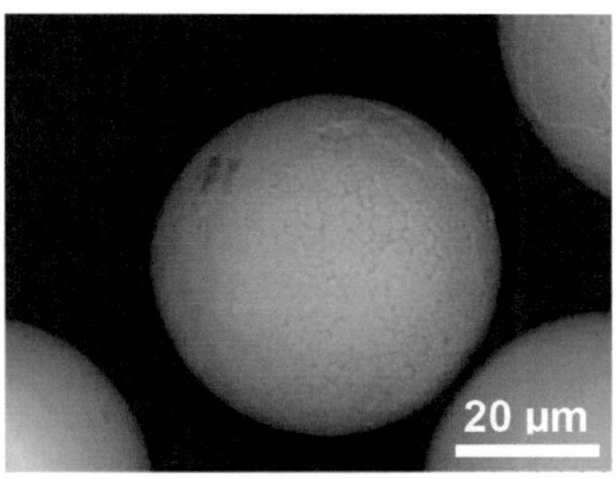

Figure 1.
SEM image of a single powder showing its shape and diameter.

Frequency, kHz	Amplitude, µm	Speed, mm/min	Force, N	Feed-rate, µm	Ball diameter, mm	Ball material	Temperature, °C
20	30	2000	40	10	2.38	WC	25, 800

Table 2.
UNSM treatment parameters.

Roughness data were obtained measured by non-contact laser scanning microscope (LSM: VK-X100 Series, Keyence, Japan). Hardness data were collected by hardness tester (MVK-E3, Mitutoyo, Japan) at a load of 300 gf. X-ray diffraction (XRD) was performed with a Cu Kα radiation (k = 1.54056 Å), a tube current of 40 mA and a voltage of 30 kV over the range of 30–90 with a scanning rate of 100/min by Bruker D8 Advance X-ray diffractometer. Compressive residual stress induced after UNSM was measured by portable device (µ-360 s, Pulstec, Japan), which is a nondestructive method. Tensile-induced fracture and wear mechanisms were investigated by SEM (JEOL, JSM-6010LA, Japan) and chemistry reacted at the contact interface was characterized by energy-dispersive X-ray spectroscopy (EDX: JEOL, JED2300, Japan).

3. Results and discussion

3.1 Surface topography

Figure 2 shows the top surface LSM images of the samples. In general, SLM fabricated samples demonstrated very rough surface due to the partially unmelted powders and also the presence of gas-induced pores on surface as shown in **Figure 2(a)**. The size of unmelted powders was found to be in the range of 30–40 µm, while the size of gas-induced pores was approximately 12 µm. The surface of the UNSM-25C sample is shown in **Figure 2(b)**. Obviously, the UNSM was able to impinge against those unmelted powders and to expel the pores from the surface, which was flattened at the end with no unmelted powders, pores and even cracks. In order to investigate the amalgamation of UNSM with thermal energy, the sample was treated by UNSM at 800°C. **Figure 2(c)** shows the surface of the UNSM-800C sample. The UNSM leaves the trace in the roughness of the surface that distorts by elasto-plastic deformation and S^2PD resulting in refining grains and creating cracks along the pathway of the WC tip. Doubtless, the unmelted powders and pores were not observed following the UNSM at 800°C, which led to the formation of some UNSM-induced isolated cracks noticed in **Figure 2(c)**. The initiated cracks are the indication of over-peening leading to an excessive strain hardening, where the amalgamated impact of UNSM and thermal energy resulted in surface degradation, and consequently imposing practical limitation in terms of surface quality rather than strength. An appearance of UNSM-induced isolated cracks can be explained in three stages: (1) the stage of strain hardening that consists of an intensive increase in surface roughness; (2) the stage of saturation, where a plastic shearing takes place leading to a reduction in dislocation density and cracks initiated; (3) the stage of surface damage, the integrity of surface is destroyed leading to an appearance of cracks, where the surface roughness increases. Actually, over-peening of surface peening technologies may deteriorate the surface integrity leading to a decrease in fatigue cycles or strength [28]. Over-peeing may cause inversion of stress that can reduce the compressive residual stress induced by UNSM. In this regard, it is always required to optimize the impact of surface peening technologies

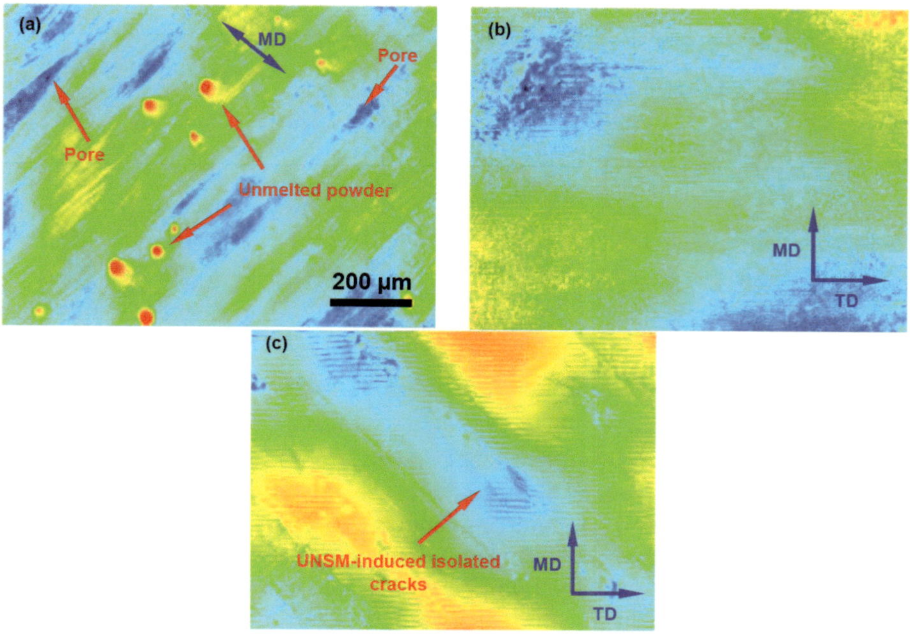

Figure 2.
Top surface LSM images of the as-SLM (a), UNSM-25C (b) and UNSM-800C (c) samples.

in order to avoid such crack initiation with the intention of preventing catastrophic failures of structures. There is still much improvements and optimizations in terms of microstructure- and ductility-based issued of AM materials to be done to be fully replaced with wrought alloys, but a progress in materials science and AM leading to overcome the faced challenge a couple of decades ago. Importantly, there is a way to get rid of from AM-based defects such as pore, unmelted and incompletely melted powders by HIP, which is highly-priced and time consuming that receiving a cautious welcome from various industries.

3.2 Surface integrity

Surface roughness measurement direction (MD) and UNSM treatment direction (TD) for each samples are shown in **Figure 2**. The surface roughness was measured in perpendicular direction to the UNSM TD. As mentioned in the previous sub-section, a rough surface of AM fabricated samples is still considered as one of the main issues. One can be seen that the actual surface contained irregularities in the form of peaks and valleys. **Figure 3(a)** shows the comparison in surface roughness (R_a) profiles of the samples. The as-SLM sample had a roughness of about 9.541 μm, while it was drastically reduced up to 0.892 μm after UNSM at 25°C as shown in **Figure 3(a)**. The roughness of the UNSM-800C sample was 1.237 μm, which is about 7–8 times smoother than that of the as-SLM sample, but it is still rougher than UNSM-25C sample. A bit rougher surface of the UNSM-800C sample than that of the UNSM-25C sample is associated with plastic deformation during the heating that distorted the geometrically pattern resulting in increased the height of irregularities. In general, during the UNSM, a plastic deformation of the top and subsurface layers took place. Expelled pores and disposed peaks and valleys led to the reduction in surface roughness [25]. As those peaks and valleys between irregularities are notches that weaken the surface cause stress concentrations. Feed-rate pathways induced by UNSM and the presence of peaks and valleys are responsible

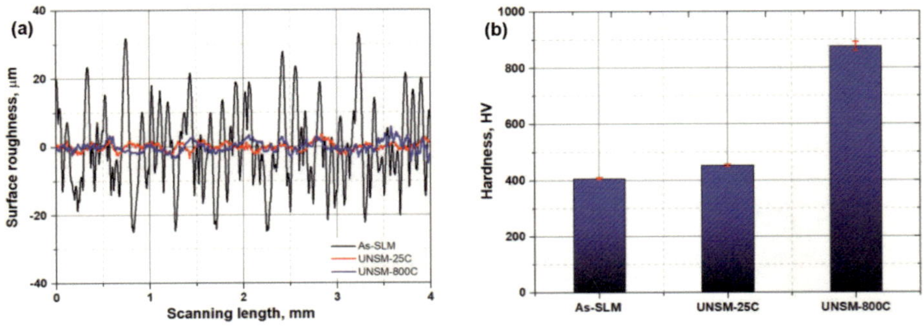

Figure 3.
Surface roughness (a) and hardness (b) of the as-SLM, UNSM-25C and UNSM-800C samples.

for the relatively rough surface of the UNSM-800C sample compared to that of the UNSM-25C sample. The level of plastic deformation increased and lessened flow eliminating with increasing temperature. Smoothed surface by UNSM at 25°C is beneficial to improving main structural properties such as tribology, corrosion and fatigue. Commonly, a surface quality of AM fabricated materials is very rough upon completion of AM [29]. This means that surface is required to be machined or finished. In this study, it is worth mentioning that the surface with no any additional machining or milling was treated by UNSM.

3.3 Surface hardening

Figure 3(b) shows the surface hardness measurement results of the samples. The average surface hardness of the as-SLM sample was approximately 396.4 HV, which increased up to 455.7 and 877.6 HV for the UNSM-25C and UNSM-800C samples, corresponding to a 13.1% and 221.3%, respectively. It is well documented in the literature that the increase in hardness is due to the combination of grain refinement by Hall-Petch expression and increased dislocation density, which are the results of elasto-plastic deformation and S^2PD took place in the top and subsurface layers [30]. The deformation usually refined the grains, which hinder further deformation and gradually diminishes with the depth of the surface layer. Moreover, Zhang et al. reported that the UNSM-induced work-hardening by plastic strain may also play a major role in increasing the hardness [31]. Moreover, in particular for AM materials, the expelled pores after peening technologies may be contributed to the increase in hardness [32].

3.4 Phase transformation

Figure 4 shows XRD patterns of the as-SLM, UNSM-25C and UNSM-800C samples. The change in diffraction peaks and phase transformation were confirmed by the relative intensity and the formation of a new peak after UNSM at 800°C. It was found that the intensity of all alpha phase peaks reduced except for (002) phase after UNSM at both 25 and 800°C. The intensity of alpha (101) phase increased for UNSM-800C sample and reduced for UNSM-25C sample. For as-SLM and UNSM-25C samples, the microstructure exhibited a full α/α`-phase, where α phase resulted from decomposition of α` during the SLM. For UNSM-800C sample, a precipitation of beta (110) phase was detected leading to a microstructure consisting of α and β phases as the temperature was higher than that of Ms. (575°C) [33]. Further, a broadening in full width at half maximum (FWHM) of the α peaks took place after UNSM, which led to the increase in dislocation density [27]. It is also obvious that

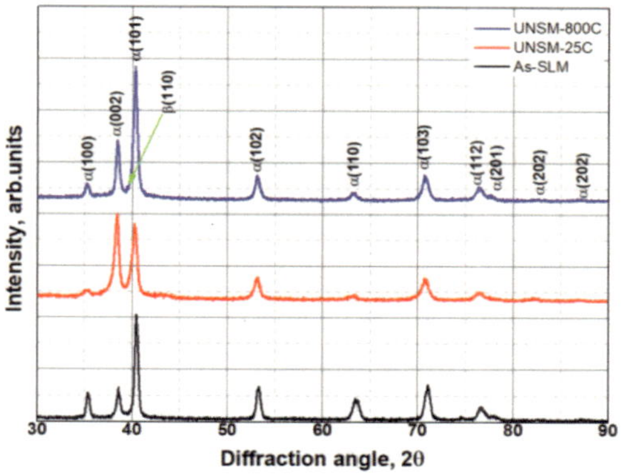

Figure 4.
XRD patterns of the as-SLM, UNSM-25C and UNSM-800C samples.

the UNSM increased the relative intensity of α (101) phase of the UNSM-800C sample as shown in **Figure 4**. Hence, it can be concluded that the UNSM resulted in grain size refinement. As a result, the FWHM values of the α (100) peak for the samples were listed in **Table 3**. It can be seen that FWHM of the as-SLM sample was broadened after UNSM at 25 °C. Furthermore, the FWHM of the UNSM-800C sample was found to be wider than that of the UNSM-25C sample, which signified that the refined grain after UNSM at 25°C was further refined after UNSM at 800°C. In general, it is well documented that the reduction in relative intensity of the peaks is responsible for the grain size refinement, which means that the UNSM refined the coarse grains into nano-grains, and also strain induced lattice distortion [34].

3.5 Tribology

Figure 5 shows the friction coefficient as a function of sliding cycles of the samples. It can be seen that all the samples came into contact with bearing steel (SAE 52100) underwent a running-in and steady-state frictional behavior. As shown in **Figure 5**, the friction coefficient of the as-SLM sample was found to be approximately 0.36 at the beginning of the friction test and increased continuously up to 0.52 for about 2000 cycles, which is considered as a running-in period. Then the friction coefficient continued being stable with a friction coefficient of 0.58 till the end of the test. **Figure 5** also shows the friction coefficient of the UNSM-25C sample. It was found that the friction coefficient demonstrated a similar friction behavior to the as-SLM sample, but the UNSM was able to reduce the friction coefficient in both the running-in and steady-state periods, where the friction coefficient was approximately 0.38 and 0.43, respectively. Overall, the friction behavior of the as-SLM sample was very highly fluctuated, which is associated with the initial rough surface. The frictional behavior of the UNSM-25C sample was relatively lower fluctuated, where the reduced surface roughness after UNSM is responsible for it. In addition,

Samples	FWHM	SD
As-SLM	0.46347	0.00738
UNSM-25C	0.59442	0.00628
UNSM-800C	0.77026	0.01093

Table 3.
Calculated FWHM results based on XRD pattern.

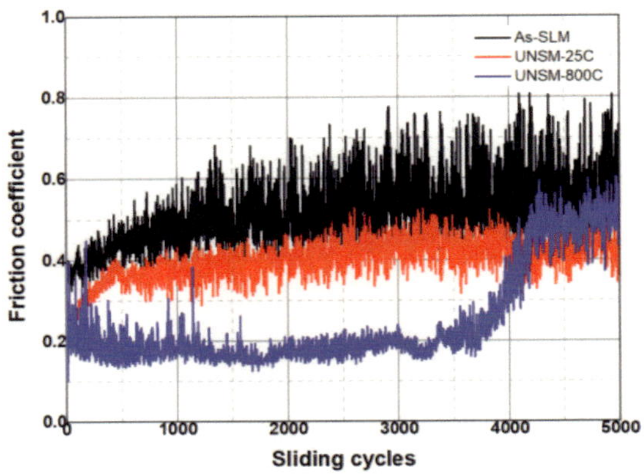

Figure 5.
Friction coefficient of the as-SLM, UNSM-25C and UNSM-800C samples.

as shown in **Figure 5**, the friction coefficient of the UNSM-800C sample was found to be approximately 0.19 at the beginning of the friction test and then continued to be stable for about 3600 cycles, which is considered as a running-in period. Then the friction coefficient gradually increased up to 0.51 and subsequently approached a stable friction coefficient till the end of the test. From the tribological tests, it was obvious that the as-SLM and UNSM-25C samples demonstrated a similar friction behavior, but the UNSM-800C sample extended the running-in period. Essentially, a lower friction coefficient was dominated by initial roughness of the samples, while an increase in hardness of the UNSM-800C sample, which came into first contact with the surface of counterface ball, had harder asperities that could increase the level of plastic deformation. A similar friction behavior was confirmed in the previous study on stainless steel 316 L that the friction coefficient was lower at the beginning of the test due to the initial surface roughness, where the asperities came into contact first and it deformed plastically with continuing reciprocating sliding [25]. Obtained friction coefficient results under dry conditions are in good consistency with the ultra-fined Ti-6Al-4V alloy fabricated by SLM [35].

The wear track dimensions of the samples were measured by 3D LSM as shown in **Figure 6**, which allowed to calculate the wear resistance based on the wear track width and depth dimensions. It can be seen that a significant difference was observed, where the wear track dimensions of the UNSM-800C sample was found to be the shallowest wear track compared to those of the as-SLM and UNSM-25C samples. The maximum peak-to-valley roughness height (R_{max}) of the wear track was about 226.3, 131.6 and 87.8 μm for the as-SLM, UNSM-25C and UNSM-800C samples, while no remarkable distinct in wear track width was observed due to the same contact pressure, respectively. As shown in the inset of **Figure 6**, the wear rate of the as-SLM sample was reduced from 3.57×10^{-8} to 1.48×10^{-8} and 8.70×10^{-9} $mm^3/N \times m$, corresponding to a ~ 41% and ~ 246% enhancement in wear resistance compared to those of the UNSM-25C and UNSM-800C samples, respectively. UNSM eliminated the effect of stress concentration in the inside of wear track, where it's depth did not exceed the thickness of strain-hardened layer containing refined nano-grains and compressive residual stress. Hence, the application of UNSM to the as-SLM sample at 25°C enhanced the wear resistance substantially due to the increase in hardness. Furthermore, a temperature increase of UNSM supplementary enhanced wear resistance. The reduction in surface roughness after UNSM led to the lower friction coefficient, while the increase in surface hardness was responsible for the higher wear resistance compared to that of the as-SLM sample. In addition, an induced compressive residual stress by UNSM

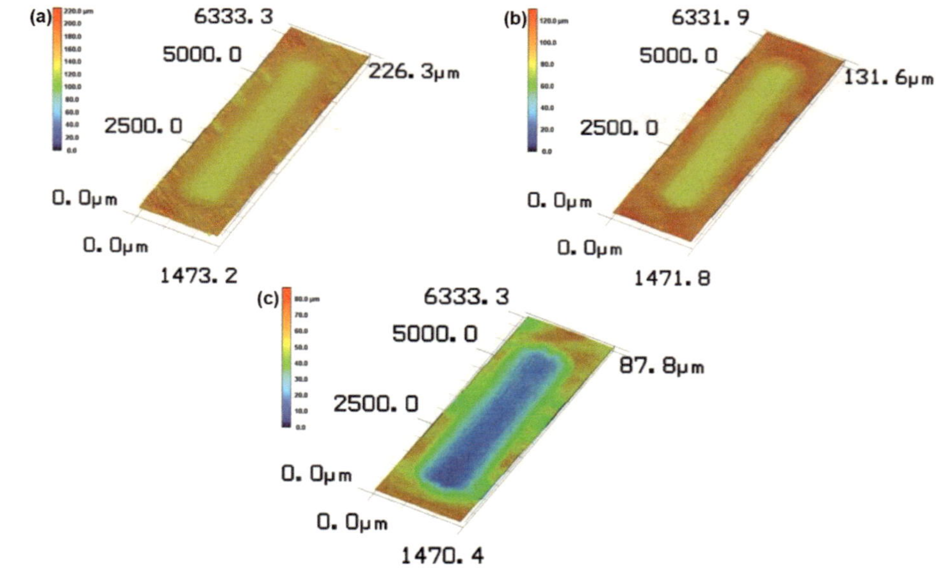

Figure 6.
3D LSM images of the as-SLM (a), UNSM-25C (b) and UNSM-800C (c) samples.

hindered the wear process [36]. As tribology considered as a system of two interacting surfaces, a wear scar of the counterface ball that came into contact with the as-SLM, UNSM-25C and UNSM-800C samples is shown in **Figure 7**. No significant difference in wear scar was observed, but the wear scar of the counterface ball that came into contact with the UNSM-800C sample was relatively smaller than those of the as-SLM and UNSM-25C samples. Beyond the wear scar of the counterface ball that came into contact with the as-SLM and UNSM-25C samples, accumulated debris were attached, while no any debris was found for the UNSM-800C sample. Finally, UNSM ensures improved surface integrity parameters and endurance of

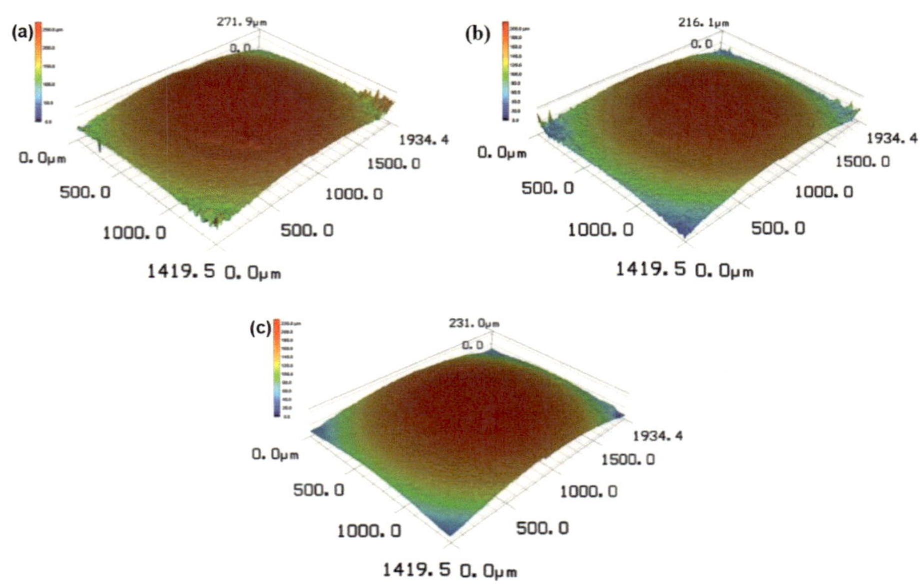

Figure 7.
3D LSM images of the counterface ball that came into contact with the as-SLM (a), UNSM-25C (b) and UNSM-800C (c) samples.

Ti-6Al-4V alloy fabricated by SLM with no subsequent process. Furthermore, it is of interest to note up that the fatigue strength of the UNSM-800C sample may be detrimental due to the presence of cracks on surface (see **Figure 2(c)**) induced by UNSM at HT of 800°C because of the presence of continuous stress leading to a crack propagation [37, 38].

Figure 8 shows the SEM images along with EDX results and oxidation distribution of the samples. It can be realized from SEM image in **Figure 8(a)** that the adhesive wear mechanism was found to be a dominant for the as-SLM sample as it is softer than that of the counterface ball, while a combination of abrasive and adhesive wear mechanisms was dominant for the UNSM-25C sample as shown in **Figure 8(b)**. An increase in temperature of UNSM resulted in changing wear mode as shown in **Figure 8(c)**, where the abrasive wear mechanisms took place for the UNSM-800C sample. Apart from those wear mechanisms, an oxidative wear mechanism came up in all the samples with different oxidation levels as shown in **Figure 8**. For instance, an oxide content over the wear track of the as-SLM sample was about 9.68%, while it was about 9.89% and 11.62% for the UNSM-25C and

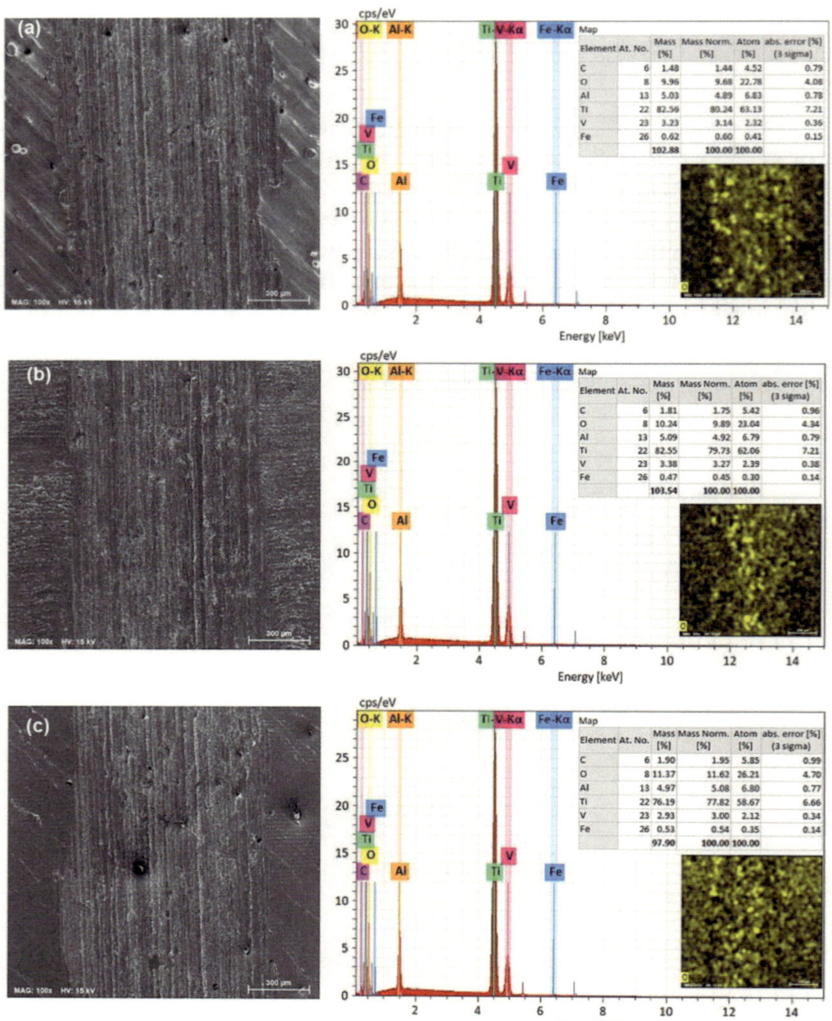

Figure 8.
SEM images along with EDX results of the as-SLM (a), UNSM-25C (b) and UNSM-800C (c) samples.

UNSM-800C samples, respectively. It is clear from oxide distribution mapping (see **Figure 8**) that the level of oxidation of the as-SLM and UNSM-25C samples was nearly consistent, but a relatively high level of oxidation occurred for the UNSM-800C sample. Zhang et al. reported that at a high temperature, a hardness of oxide layer become remarkably higher due to the presence of much oxide [39]. Hence, it is reasonable to hypothesize that it may significantly increase the wear resistance by obtaining low friction coefficient [40]. Typically, there is an advantage to forming an oxide layer between two mating surface as it prevents direct metal-to-metal contact resulting in lower friction coefficient and higher wear resistance. Furthermore, a nearly same amount of Fe, which was transferred from the counterface ball can be seen from the chemical composition table as shown in the inset of **Figure 8**. The presence of Fe along with occurred oxide may react together and form a ferrosoferric (Fe3O4) layer, which provides an advantageous environment for achieving a better tribological behavior and performance [27]. Furthermore, as the surface roughness of the samples deteriorated during the dry tribological tests as shown in **Figure 8**, where the post-test surface roughness of the UNSM-25C and UNSM800C samples was lower than that of the as-SLM sample. Hence, it can be considered that the surface roughness after dry tribological tests is much important than post-surface treatment because the surface during the dry tribological tests comprises a number of grooves of different depth and sharpness – causing local stress concentrations and decreasing the wear resistance.

4. Conclusions

In this study, the influence of UNSM on the surface, tensile and tribological properties of Ti-6Al-4V alloy fabricated by SLM was evaluated. The as-SLM sample had a roughness of about 9.541 μm, which was drastically reduced up to 0.892 and 3.058 μm after UNSM at 25 and 800°C. The average surface hardness of the as-SLM sample was approximately 396.4 HV, which increased up to 455.7 and 877.6 HV for the UNSM-25C and UNSM-800C samples, corresponding to a 13.1% and 221.3% increase, respectively. The surface residual stress of both the UNSM-25C and UNSM-800C samples was transferred into compressive residual stress. The as-SLM sample demonstrated lower YS and UTS than UNSM-25C and UNSM-800C samples, but its elongation was shorter than that of the UNSM-25C sample and longer than that of the UNSM-800C sample. YS and UTS of the UNSM-25C sample was lower and higher than that of the UNSM-800C sample, while the elongation was also longer than that of the UNSM-800C sample. Friction coefficient of the as-SLM sample was reduced by the application of UNSM at 25 °C by about 25.8%, and it further reduced by about 305% increasing the UNSM temperature up to 800 °C. The wear rate of the as-SLM sample was reduced by about 41% and 246% compared to those of the UNSM-25C and UNSM-800C samples, respectively. As a main conclusion, a UNSM at RT and HT may be applied to Ti-6Al-4V alloy fabricated by SLM with the intention of enhancing tensile and tribological properties of various components in aerospace and biomedical applications. Indeed, a further investigation is required to improve the properties and performance of Ti-6Al-4V alloy fabricated by SLM to the wrought level due to the replacement possibility.

Acknowledgements

This study was supported by the Industrial Technology Innovation Development Project of the Ministry of Commerce, Industry and Energy, Rep. Korea (No.

20010482). This research was supported by Basic Science Research Program through the National Research Foundation of Korea (NRF) funded by the Ministry of Education (No. 2020R1I1A3074119).

Conflict of interest

The author has no conflict of interest to declare.

Author details

Auezhan Amanov
Sun Moon University, Asan, South Korea

*Address all correspondence to: avaz2662@sunmoon.ac.kr

References

[1] Bagehorn S, Wehr J, Maier HJ. Application of mechanical surface finishing processes for roughness reduction and fatigue improvement of additively manufactured Ti-6Al-4V parts. Int. J. Fatigue. 2017;102:135-142. DOI: 10.1016/j.ijfatigue.2017.05.008

[2] Gorsse S, Hutchinson C, Gouné M, Banerjee R. Additive manufacturing of metals: A brief review of the characteristic microstructures and properties of steels, Ti-6Al-4V and high-entropy alloys. Sci. Technol. Adv. Mater. 2017;18:584-610. DOI: 10.1080/14686996.2017.1361305

[3] Gu D, Hagedorn YC, Meiners W, Meng G, Batista RJS, Wissenbach K, Poprawe R. Densification behavior, microstructure evolution, and wear performance of selective laser melting processed commercially pure titanium. Acta Materialia. 2012;60:3849-3860. DOI: 10.1016/j.actamat.2012.04.006

[4] Thijs L, Verhaeghe F, Craeghs T, Humbeeck JV, Kruth JP. A study of the microstructural evolution during selective laser melting of Ti–6Al–4V. Acta Materialia. 2010;58:3303-3312. DOI: 10.1016/j.actamat.2010.02.004.

[5] Abe F, Osakada K, Shiomi M, Uematsu K, Matsumoto M. The manufacturing of hard tools from metallic powders by selective laser melting. J. Mater. Process. Technol. 2001;111:210-213. DOI: 10.1016/S0924-0136(01)00522-2.

[6] Mumtaz KA, Erasenthiran P, Hopkinson N. High density selective laser melting of Waspaloy®. J. Mater. Process. Technol. 2008;195:77-87. DOI: 10.1016/j.jmatprotec.2007.04.117.

[7] Ma C, Andani MT, Qin H, Moghaddam NS, Ibrahim H, Jahadakbar A, Amerinatanzi A, Ren Z, Zhang H, Doll GL, Dong Y, Elahinia M, Ye C. Improving surface finish and wear resistance of additive manufactured nickel-titanium by ultrasonic nanocrystal surface modification. J. Mater. Process. Technol. 2017;249:433-440. DOI: 10.1016/j.jmatprotec.2017.06.038.

[8] Shiomi M, Osakada K, Nakamura K, Yamashita T, Abe F. Residual stress within metallic model made by selective laser melting process. CIRP Annals. 2004;53:195-198. DOI: 10.1016/S0007-8506(07)60677-5.

[9] Qiu C, Adkins NJE, Attallah MM. Microstructure and tensile properties of selectively laser-melted and of HIPed laser-melted Ti–6Al–4V. Mater. Sci. Eng. A. 2013;578:230-239. DOI: 10.1016/j.msea.2013.04.099.

[10] Vilaro T, Colin C, Bartout JD, Nazé L, Sennour M. Microstructural and mechanical approaches of the selective laser melting process applied to a nickel-base superalloy. Mater. Sci. Eng. A. 2012;534:446-451. DOI: 10.1016/j.msea.2011.11.092.

[11] Osakada K, Shiomi M. Flexible manufacturing of metallic products by selective laser melting of powder. International J. Mach. Tools Manuf. 2006;46:1188-1193. DOI: 10.1016/j.ijmachtools.2006.01.024.

[12] Lee KA, Kim YK, Yu JH, Park SH, Kim MC, Effect of heat treatment on microstructure and impact toughness of Ti-6Al-4V manufactured by selective laser melting process. Arch. of Metal. Mater. 2017;62:1341-1346. DOI: 10.1515/amm-2017-0205.

[13] Vilaro T, Colin C, Bartout JD. As-fabricated and heat-treated microstructures of the Ti-6Al-4V alloy processed by selective laser melting. Metal. Mater. Trans. A.

2011;42:3190-3199. DOI: 10.1007/s11661-011-0731-y.

[14] Ter Haar GM, Becker T, Blaine DC. Influence of heat treatments on the microstructure and tensile behaviour of selective laser melting-produced Ti-6Al-4V parts. S. Afr. J. Ind. Eng. 2016;27. DOI: 10.7166/27-3-1663.

[15] Vrancken B, Thijs L, Kruth JP, Van Humbeeck J. Heat treatment of Ti6Al4V produced by selective laser melting: Microstructure and mechanical properties. J. Alloys Comp. 2012;541:177-185. DOI: 10.1016/j.jallcom.2012.07.022.

[16] Bhaduri D, Penchev P, Batal A, Dimov S, Soo SL, Sten S, Harrysson U, Zhang Z, Dong H. Laser polishing of 3D printed mesoscale components. Appl. Surf. Sci. 2017;405:29-46. DOI: 10.1016/j.apsusc.2017.01.211.

[17] Rosa B, Mognol P, Hascoët J. Laser polishing of additive laser manufacturing surfaces. J. Laser Appl. 2015;27:S29102. DOI: 10.2351/1.4906385.

[18] Ma CP, Guan YC, Zhou W. Laser polishing of additive manufactured Ti alloys/ Opt. Lasers Eng. 2017;93:171-177. DOI: 10.1016/j.optlaseng.2017.02.005.

[19] Leuders S, Thöne M, Riemer A, Niendorf T, Tröster T, Richard HA, Maier HJ. On the mechanical behaviour of titanium alloy TiAl6V4 manufactured by selective laser melting: Fatigue resistance and crack growth performance. Int. J. Fatigue. 2013;48:300-307. DOI: 10.1016/j.ijfatigue.2012.11.011.

[20] Uzan NE, Ramati S, Shneck R, Frage N, Yeheskel O. On the effect of shot-peening on fatigue resistance of AlSi10Mg specimens fabricated by additive manufacturing using selective laser melting (AM-SLM). Addit. Manuf. 2018;21:458-464. DOI: 10.1016/j.addma.2018.03.030.

[21] Hackel L, Rankin JR, Rubenchik A, King WE, Matthews M. Laser peening: A tool for additive manufacturing post-processing. Addit. Manuf. 2018;24:67-75. DOI: 10.1016/j.addma.2018.09.013.

[22] Yasuoka M, Wang P, Zhang K, Qiu Z, Kusaka K, Pyoun YS, Murakami R. Improvement of the fatigue strength of SUS304 austenite stainless steel using ultrasonic nanocrystal surface modification. Surf. Coat. Technol. 2013;218:93-98. DOI: 10.1016/j.surfcoat.2012.12.033.

[23] Amanov A, Cho IS, Kim DE, Pyun YS. Fretting wear and friction reduction of CP titanium and Ti-6Al-4V alloy by ultrasonic nanocrystalline surface modification. Surf. Coat. Technol. 2012;207:135-142. DOI: 10.1016/j.surfcoat.2012.06.046.

[24] Zhang H, Zhao J, Liu J, Qin H, Ren Z, Doll GL, Dong Y, Ye C. The effects of electrically-assisted ultrasonic nanocrystal surface modification on 3D-printed Ti-6Al-4V alloy. Addit. Manuf. 2018;22:60-68. DOI: 10.1016/j.addma.2018.04.035.

[25] Amanov A. Effect of local treatment temperature of ultrasonic nanocrystalline surface modification on tribological behavior and corrosion resistance of stainless steel 316L produced by selective laser melting. Surf. Coat. Technol. 2020;398:126080. DOI: 10.1016/j.surfcoat.2020.126080.

[26] Amanov A, Urmanov B, Amanov T, Pyun YS. Strengthening of Ti-6Al-4V alloy by high temperature ultrasonic nanocrystal surface modification technique. Mater. Lett. 2017;196:198-201. DOI: 10.1016/j.matlet.2017.03.059.

[27] Amanov A, Pyun YS. Local heat treatment with and without ultrasonic nanocrystal surface modification of Ti-6Al-4V alloy: Mechanical and tribological properties. Surf. Coat.

Technol. 2017;326(A):343-354. DOI: 10.1016/j.surfcoat.2017.07.064.

[28] Bhuvaraghan B, Srinivasan SM, Maffeo B. Optimization of the fatigue strength of materials due to shot peening: A Survey. Int. J. Struct. Chang. Solids. 2010;2(2):33-63.

[29] Iquebal AS, Sagapuram D, Bukkapatnam STS. Surface plastic flow in polishing of rough surfaces. Sci. Rep. 2019;9:10617. DOI: 10.1038/s41598-019-46997-w.

[30] Zhang Q, Xie J, London T, Griffiths D, Bhamji I, Oancea V. Estimates of the mechanical properties of laser powder bed fusion Ti-6Al-4V parts using finite element models. Mater. Design. 2019;169:107678. DOI: 10.1016/j.matdes.2019.107678.

[31] Zhang H, Chiang R, Qin H, Ren Z, Hou X, Lin D, Doll GL, Vasudevan VK, Dong Y, Ye C. The effects of ultrasonic nanocrystal surface modification on the fatigue performance of 3D-printed Ti64. International Journal of Fatigue. 2017;103:136-146. DOI: 10.1016/j.ijfatigue.2017.05.019.

[32] Kahlin M, Ansell H, Basu D, Kerwin A, Newton L, Smith B, Moverare JJ. Improved fatigue strength of additively manufactured Ti6Al4V by surface post processing. Int. J. Fatigue. 2020;134:105497. DOI: 10.1016/j.ijfatigue.2020.105497.

[33] Yang Y, Liu YJ, Chen J, Wang HL, Zhang ZQ, Lu YJ, Wu SQ, Lin JX. Crystallographic features of α variants and β phase for Ti-6Al-4V alloy fabricated by selective laser melting. Mater. Sci. Eng. A. 2017;707:548-558. DOI: 10.1016/j.msea.2017.09.068.

[34] Ye C, Telang A, Gill AS, Suslov S, Idell Y, Zweiacker K, Wiezorek JMK, Zhou Z, Qian D, Mannava SR, Vasudevan VK. Gradient nanostructure and residual stresses induced by ultrasonic nanocrystal surface modification in 304 austenitic stainless steel for high strength and high ductility. Mater. Sci. Eng. A. 2014;613:274-288. DOI: 10.1016/j.msea.2014.06.114.

[35] Li Y, Song L, Xie P, Cheng M, Xiao H. Enhancing hardness and wear performance of laser additive manufactured Ti6Al4V alloy through achieving ultrafine microstructure. Materials. 2020;13:1210. doi:10.3390/ma13051210.

[36] Wang C, Jiang C, Chen M, Wang L, Liu H, Ji V. Residual stress and microstructure evolution of shot peened Ni-Al bronze at elevated temperatures. Mater. Sci. Eng. A. 2017;707:629-635. DOI: 10.1016/j.msea.2017.09.098.

[37] Fotovvati B , Namdari N, Dehghanghadikolaei A. Fatigue performance of selective laser melted Ti6Al4V components: State of the art. Mater. Res. Express. 2018;6(1): 012002. DOI: 10.1088/2053-1591/aae10e

[38] Edwards P, Ramulu M. Fatigue performance evaluation of selective laser melted Ti-6Al-4V. Mater. Sci. Eng. A. 2014;598:327-337. DOI: 10.1016/j.msea.2014.01.041.

[39] Zhang QY, Zhou Y, Wang L, Cui XH, Wang SQ. Investigation on tribo-layers and their function of a titanium alloy during dry sliding. Tribol. Int. 2016;94:541-549. DOI: 10.1016/j.triboint.2015.10.018.

[40] Balla VK, Soderlind J, Bose S, Bandyopadhyay A. Microstructure, mechanical and wear properties of laser surface melted Ti6Al4V alloy. J. Mech. Behav. Biomed. Mater. 2014;32:335-344. DOI: 10.1016/j.jmbbm.2013.12.001.

Chapter 7

A New Concept of the Mechanism of Variation of Tribological Properties of the Machine Elements Interacting Surfaces

George Tumanishvili, Tengiz Nadiradze and Giorgi Tumanishvili

Abstract

The methods of estimation and prediction of tribological properties of the contact zone of interacting elements of machines are characterized by the low informativeness and accuracy that complicates provision of the proper tribological properties and hinders reliable and effective operation of machines. For obtaining more wide information about factors influencing tribological properties of the interacting surfaces, the experimental researches on the high speed (up to 70-m/s) and serial twin-disk machines were carried out. Our researches have shown that with different properties and degrees of destruction of the third body, the coefficient of friction can change up to 10 times or more, the wear rate up to 10^2-10^4 times, etc. This was the basis for a new concept of the mechanism of variation of tribological properties of interacting surfaces. The researches have shown a dependence of tribological properties of the contact zone on the properties and destruction degree of the third body that was assumed as a basis of new concept of the mechanism of variation of tribological properties of these surfaces. The monitoring of the third body destruction onset and development was carried out in the laboratory conditions and a criterion of the third body destruction was developed. The reasons of the negative, neutral and positive friction and mild, severe and catastrophic wear are shown.

Keywords: interacting surfaces, tribological properties, third body, friction coefficient, wear

1. Introduction

Between the interacting surfaces can be continuous or discontinuous third body. Until 70s of the last century the oil layer of hydrodynamic generation existent in the contact zone, was considered as a parameter determining a working capacity of the heavy loaded frictional contact. Many experimental and theoretical works are devoted to study of thickness of this layer [1–5]. An approximate (digital) solution of the elasto-hydrodynamic problem considering thermal processes is given in the work [6], where the temperature, pressure and thickness of the oil layer between the cylinders interacting with the rolling-sliding friction, are determined. However, in spite of many attempts, ascertainment of the reliable relations between the thickness of the oil layer and tribological properties of the contact zone turned out to be problematic [7]. The supplements to the lubricants developed in succeeding years and technical means of study the processes proceeding in the contact zone have radically widened direction of the researches.

The fundamentals of materials science and contact mechanics are developed in works [8–10] and in recent years a new direction of tribology – nano-tribology appeared [11, 12]. New materials were created (graphene etc.) [13]. For tribological modeling are used the methods of mechanics and multiphysics [14–18], methods of finite and boundary elements [19–21], discrete dynamics of dispositions [22], and atomistic methods [23]. However, in spite of this, some engineer aspects of the problems of tribology are not yet properly studied and their solution needs additional researches.

At common operational conditions, various types of boundary films - products of interaction with the environment that prevent the direct contact of rubbing surfaces, cover these surfaces with thin layers. Depending on the friction conditions, properties of the surfaces and environment, these layers may have various tribological properties that will have the great influence on the boundary friction [24–26]. This is confirmed by the results of the experimental researches in the inert gas environment and vacuum, that excludes the possibility of oxidation during friction. Under such conditions, the seizure and intensive wear rate are observed. To prevent these undesirable phenomena, it is necessary to provide the presence of the third body in the contact zone with due properties, control of the friction factor and protection of the third body from destruction.

When the interacting surfaces are separated by the continuous third body, the friction forces mainly depend on the rheological properties of the third body [25, 26] or on the third body viscosity and area of the contact zone: $F = f(\eta, \frac{\Delta v}{\Delta x}, S)$, where S is area of the contact zone; η — viscosity; $\frac{\Delta v}{\Delta x}$- velocity gradient.

Usually, the surfaces are covered with various types of natural and artificial coatings, which represent the components of the third body in the contact zone of the interacting surfaces, are subjected to heavy power and thermal loads. This causes deformations of these coatings, their destruction, activation of the physical and chemical processes proceeding between them and the surfaces and generation of new coatings. Thus, during the interaction of surfaces, the processes of the third body destruction and restoration takes place in the contact zone continuously. When the intensity of destruction of the third body is greater than the intensity of its restoration, the amount of the micro-asperities coming into direct interaction leads to seizure and the wear rate increase because of various kinds of surface damage.

A part of micro-asperities of the heavy loaded interacting surfaces are in direct contact with each other causing their seizure and the remaining part interact with each other through the third body that is schematically shown in **Figure 1**.

For heavy loaded interacting surfaces is typical seizure. This can happen when continuity of the third body is disrupted in individual places; the parts of the direct contact are cleansed from various coatings and boundary layers and are approached to each other at the distance of several atom diameters. As molecular dynamics [27] and atom microscope [28] show, in such conditions they will attract each other generating electron-pair bindings.

Adhesive approach to the friction means invasion of micro-asperities into each other in the contact zone, their close contact without the third body and adhesive scuffing of micro-asperities. The thermal effects accompanying the process have direct influence on the deformation area and value, volume of the deformed

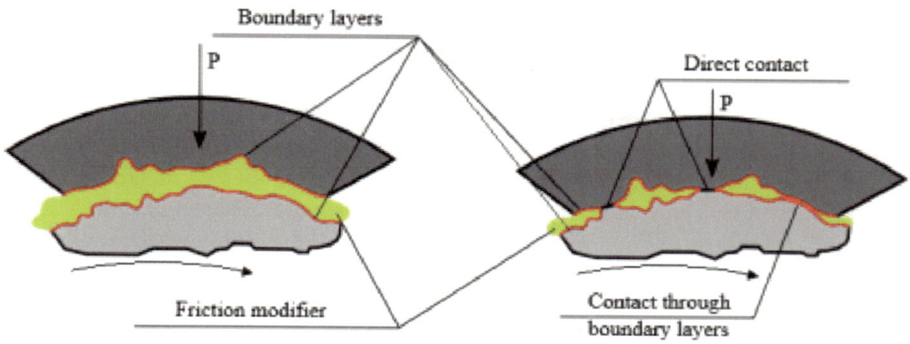

Figure 1.
Types of interaction of the surfaces.

material, variation of the surface structure and physical and mechanical characteristics and damage of various types proceeding simultaneously.

The friction forces between interacting surfaces (at lack of the third body in the places of actual contact) depend on the total area of the actual contacts Ff = ψ (∑τAasp) [24], where τ is effective strength on shear of the actual contact area of interacting surfaces; Aasp – seizure area of the actual contact that depends on the thermal load of the contact zone, thickness of the heated up layer, properties of the surfaces and environment of individual micro-asperities etc.

Hence, the friction forces depend on the contact area in both cases, when the surfaces are separated from each other by the third body fully or partially.

The surfaces are the weakest places of the rigid body from which their destruction begins [29]. Displacement of the coupled places of surfaces relative to each other causes sharp increase of the shear stresses and corresponding deformations, value and instability of the friction forces and rupture of the coupled places. It is possible in this case transfer of the pulled out material from on surface on the other, sharp change of roughness of these surfaces and development of the process of catastrophic wear – scuffing. The shear deformation generated on the surface sharply decreases towards the depth and multiple repetition of such processes results in superficial plastic deformations, lamination and fatigue damage (**Figure 2**) [30, 31].

The damage scales and dominant types in such cases depend on the working conditions. Thus, for providing the interacting surfaces with due tribological properties, their separation from each other by the continuous third body with corresponding properties is necessary.

It should be noted that various types of surface take place simultaneously and proceed with various intensity and a dominant type of damage ascertained visually. The experimental researches have shown that damage intensity and type, of

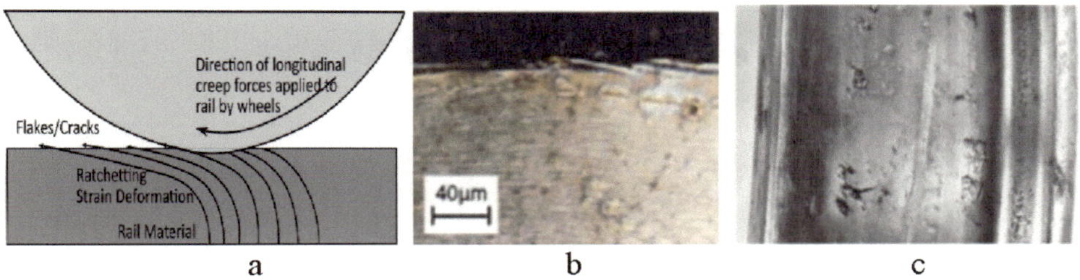

Figure 2.
The scheme of the surface plastic deformation (a); appearance of cracks and lamination (b); appearance of fatigue pits (c).

interacting surfaces are especially sensitive to the relative sliding velocity and shear stresses. Thereat, at low total and relative sliding velocities of the surfaces, when power of the thermal action, velocity and resistance of the shear deformation in the contact zone are comparatively small, stability of the third body and its resistance to scuffing are high and a main type of damage is fatigue wear [4]. With increase of the total and relative sliding velocities of surfaces, thermal load of the actual contact zone and destruction intensity of the third body increases. However, time of action of this load, thickness of the heated up layer and sizes of micro-asperities generated because of the scuffing and subsequent rupture of the seized places, decrease. Such phenomena take place on tread surfaces of the train wheel, near the pitch point of the gear drives, in the rolling bearings etc. (**Figure 3**). At increase of the relative sliding velocity, share of the adhesive wear and scuffing increases and it often becomes a dominant type of damage. For example, a steering surface of the train wheel, places of tooth profile of the gear drive distant from the pitch point, cam mechanisms etc.

For avoiding the above-mentioned non-desirable phenomena, providing the contact zone with the third body having due properties, its protection against destruction and control of the friction coefficient are necessary. However, despite the great number of scientific works this direction could not attract due attention of the scientists until today.

Variation of tribological properties of the surfaces is a result of various mechanical, physical and chemical processes proceeding simultaneously in the contact zone whose essence and mechanism of action are not properly studied [20–22]. This complicates control of the mentioned processes that needs consideration of many factors acting simultaneously. Such factors are:

- Initial tribological properties of the third body and surfaces; influence of interaction of the friction modifier and other materials existent in the contact zone and the surfaces on the properties and stability of the third body and surfaces.

- Structural, physical and mechanical peculiarities and tendency to scuffing of the clean (juvenile) surfaces in the places of the third body destruction; destruction peculiarities of the seized places;

- Influence of the contact zone working conditions on the wear type and rate, variation of the micro- and macro-geometry etc.

a b c

Figure 3.
The damage types: (a) train wheel with fatigue damage of the tread surface and adhesive wear (scuffing) of the flange; (b) gear wheel with the traces of scuffing on the tooth face; (c) inner ring of the rolling bearing with the traces of fatigue damage.

Various interacting surfaces of machines should have different tribological properties: tooth gear drives, cam mechanisms, guides of various types etc., should have stable and as small as possible friction coefficient (≤ 0.1) and friction clutch and brakes – comparatively high and stable friction coefficient (0.25-0.4).

Especially should be noted operational peculiarities of the wheel and rail interacting surfaces. The existent profiles of wheels and rails can be divided into the tread surfaces (which take part in the "free" rolling, traction and braking) and steering surfaces (the wheel flange and rail gauge, which take part in the steering mainly in curves and prevent the wheel-set from derailment). The flange root can roll on the rail corner, and it can take part in traction, braking and steering (**Figure 4**).

But traction (braking) and steering require mutually excluding properties and the "ideal" value of the friction coefficient ($\mu < 0{,}1$) in the contact zone of the flange root and the rail corner is not acceptable for both cases.

As it is seen from **Figure 5**, the power and thermal loads of tread surfaces are relatively low. At working of wheels in the modes of traction and braking, the lateral displacement, rotation about vertical axis and skidding, sliding velocity and distance increase. The flange root and rail corner in the contact zone are characterized by the increased creeping, that at destruction of the third body results in the increased shearing stresses and temperatures.

For interacting surfaces of some mechanisms, such as tooth gear drives, cam mechanisms, wheel and rail etc., the main types of wear are adhesive wear (and its heavy form – scuffing, whose nature is not studied sufficiently and under heavy working conditions it is followed by sharp increase of the friction coefficient instability and wear rate or catastrophic wear) and fatigue wear, that proceed simultaneously and are quite different processes.

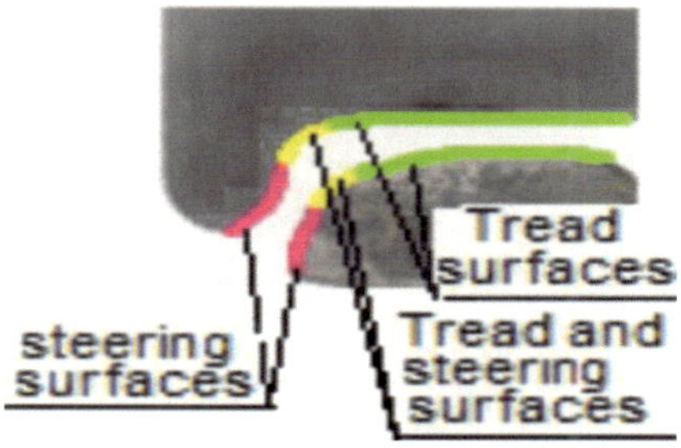

Figure 4.
Components of the wheel and rail interacting surfaces.

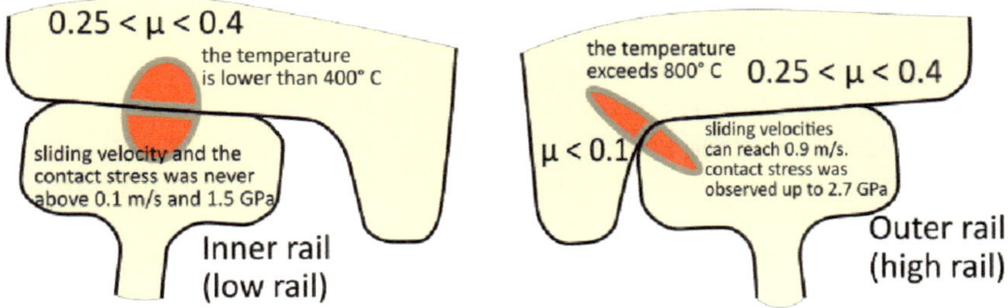

Figure 5.
The ideal values of the friction coefficients and stress distribution in the contact zone of the wheel and rail according to [32] and the thermal loads.

For revealing the factors influencing tribological properties of the interacting surfaces, the experimental researches were carried out on the high-speed and serial twin-disk machines.

2. The experimental researches into variation of tribological properties of the interacting surfaces

2.1 Research into tribological properties on the high-speed twin disk machine

A great number of scientific works are devoted to ascertainment of laws of variation of tribological properties of the interacting surfaces and with perfection of machines, actuality of such works increases. Despite the considerable quantity of works in this direction, the expected results are not obtained yet. The unexpected and catastrophic failure, unlike fatigue, corrosion and other slowly progressing wear types, are subjected some heavy loaded interacting surfaces of gear teeth, cams and followers, sleeve bearings etc. The wheel and rail contact zone is characterized by heavy operational conditions [33] (direct impact of the environmental conditions, high relative sliding and contact stresses) that enhances adhesive and fatigue processes. The wheel and rail contact zone is characterized by the heavy operational conditions (direct impact of the environmental conditions, high relative sliding and contact stresses) that enhances adhesive and fatigue processes raise the problems to be solved for many-sided study of these processes.

For some heavy loaded interacting surfaces of machines are typical unpredictable change of tribological properties and sharp increase of the friction coefficient and wear intensity, so called catastrophic wear. As main cause of the latter is considered the heaviest form of the adhesive wear – scuffing [4] that is not properly studied yet [34] and whose signs are appearance of pits and scratches on the surfaces and transfer of the material from one surface on the other. The various aspects of the complex physical, tribo-chemical and mechanical processes proceeding in the contact zone are not properly studied yet that is accordingly reflected on the operation quality and resource. As an example can be cited interaction of the wheel and rail that occurs on: the tread surfaces during rolling, traction and braking; steering surfaces mainly in curves; flange root and rail corner at rolling, traction, braking and steering. The friction coefficient for wheel-rail interaction can vary in the range 0.05 - 0.8. The values of the friction coefficient for the tread and steering surfaces must be correspondingly in the ranges of 0.25-0.4 and <0.1 [32]. The optimal value of the friction factor for tread surfaces is 0.35 [32] and for steering surfaces - as low as possible. The scuffing on the wheel and rail steering surfaces causes rise of the friction coefficient, energy consumed on rolling, vibrations, nose, wear intensity and probability of derailment.

For more detailed study of the properties and state of the third body in the contact zone we performed the experimental researches on the twin disk machine MT – 1 (**Figure 6**) with the use of existing lubricants and ecologically friendly friction modifiers, developed by us.

The tests were performed at single application of the friction modifier on the rolling surface of the roller. After certain number of revolutions, a thin layer of the friction modifier was destroyed that was revealed by sharp increase of the friction moment and initial signs of scuffing on the surfaces. Without repeated feeding the friction modifier, the damage process was progressed. The rollers with various degree of damage are shown in **Figure 7**: (a) with initial signs of damage; (b) damage in the form of a narrow strip; (c) damage on the whole contacting area.

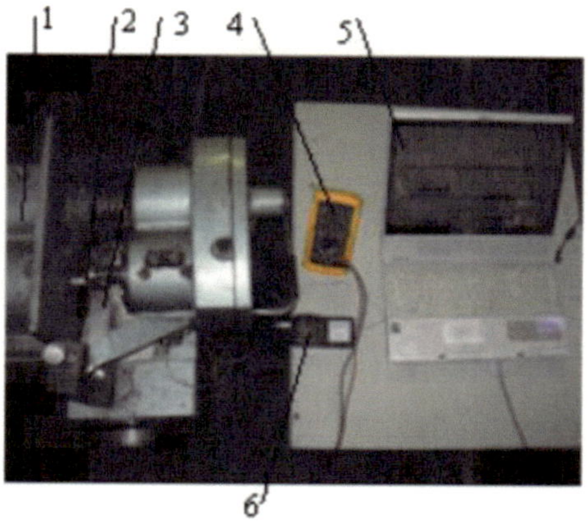

Figure 6.
The twin disk machine model MT1 and measuring means: 1 - twin disk machine, 2 - tribo-elements, 3 - the wear products, 4 - tester, 5 - personal computer, 6 – vibrometer.

Experimental research was performed at rolling of discs with up to 20% of sliding. The rollers had diameters of 40 mm and widths of 10 and 12 mm. The tests were performed at single application of the friction modifier on the interacting surface of the rollers. After certain number of revolutions, a thin layer of the friction modifier (FM) was destroyed that was revealed by sharp increase of the friction moment and initial signs of scuffing on the surfaces. Without repeated feeding of the friction modifier the damage process were progressed. The rollers with various degrees of damage are shown in **Figure 7**: (a) with initial signs of damage; (b) damage in the form of a narrow strip; (c) damage of the whole contacting area.

The graphs of dependences of the friction coefficient and number of revolutions of rollers until appearance of the first signs of scuffing on the contact stress for initial linear contact of disks are shown in **Figure 8**. It is seen from these graphs that for the initial linear contact, when the contact stress is in the range of 0.65-0.77 GPa increase of the contact stress leads to decrease of the friction coefficient. It can also be seen that increase of the contact stress leads to decrease of number of revolutions until the destruction of the third body and onset of scuffing.

The graphs of dependences of the friction coefficient and number of revolutions of rollers until appearance of the first signs of scuffing on the contact stress for initial point contact of disks and anti-frictional friction modifiers are shown in **Figure 9**.

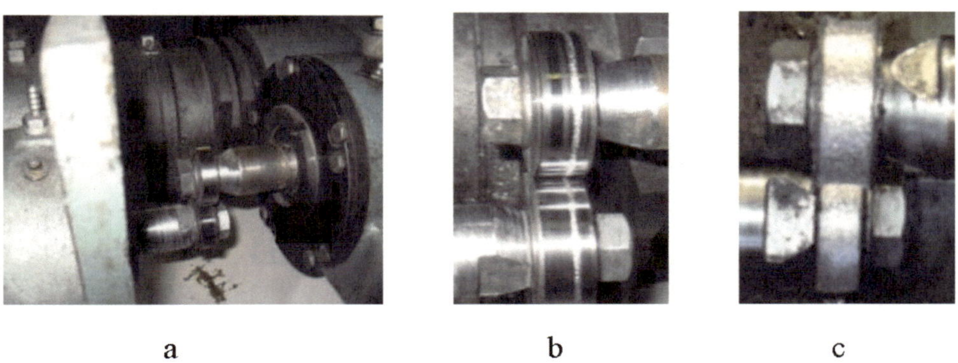

Figure 7.
The stages of damage of the interacting surfaces: (a) damage in the separate points; (b) damage in the form of the narrow strip; (c) damage on the whole area of the contacting surfaces.

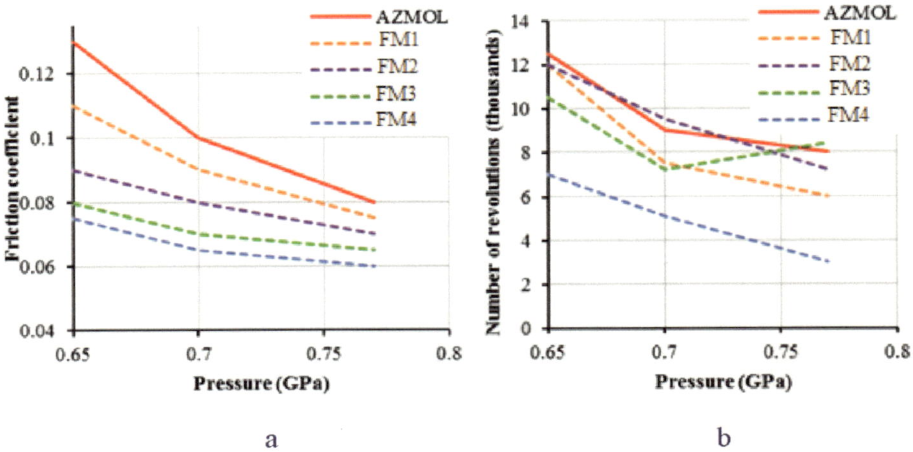

Figure 8.
Dependences of friction coefficients (a) and numbers of revolutions (b) until appearance of the first signs of scuffing on the contact stress for initial linear contact of disks and different anti-frictional friction modifiers.

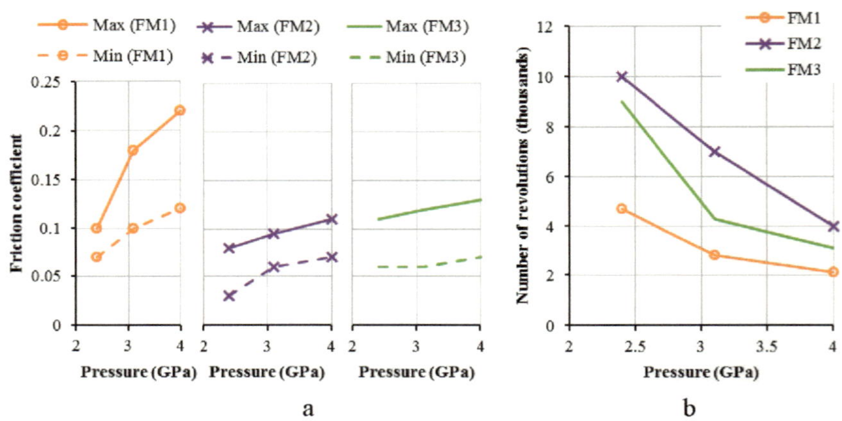

Figure 9.
Dependences of friction coefficients (a) and numbers of revolutions (b) until appearance of the first signs of destruction of the third body (first signs of scuffing) on the contact stress for initial point contact of disks and three different frictional FM-s.

When the contact stress is in the range of 2.42-3.96 GPa the friction coefficient increases with increase of the contact stress. It can also be seen that increase of the contact stress leads to decrease of the number of revolutions until the destruction of the third body and onset of scuffing more intensive than in the previous case.

2.2 Research into tribological properties on the high-speed twin disk machine

At high working velocities, the maximal power and thermal stresses approach to the surfaces and intensity of the third body destruction/restoration and sensitivity of the contact zone tribological properties to working conditions, increase. To promote the mentioned problem, the experimental researches were carried out on the high-speed twin disk machine with independent drive of rollers (**Figure 10**).

During experiments were studied the character of the wear process of working surfaces, influence of various parameters on the lubricant film thickness and friction coefficient at the use of popular mineral lubricants. The researches were

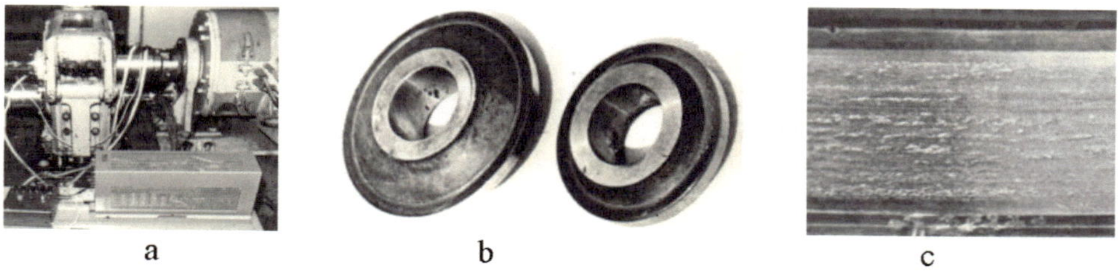

Figure 10.
High speed twin disk machine (a), experimental pieces (b) and a working surface of the roller with traces of scuffing (c) at total speed of rolling 7 m/s, sliding speeds of 3 m/s, linear load 100 N/m.

executed with the use of the high-speed roller machine with independent drive of rollers. Conditions of the experiments and measured sizes were:

- Rolling speed – up to 70 m/s;
- Diameters of rollers 183 mm and 143,3 mm; width of rollers 12 mm and 17 mm;
- Sliding velocity- up to 35 m/s;
- Contact pressure – 5 × 105 – 2 × 106 N/м;
- Dynamic viscosity 49-140 mNs/m^2;

During experiments at the given loading and rolling velocity, the friction torque, sliding velocity and lubricant film thickness were measured. For measurement of speeds of rotation was utilized magnetic pickups, for measurement of the friction torque was utilized the strain gage transducer and contactless skate. The lubricant film thickness was measured by capacitance method [35]. The beginning of scuffing was revealed by the surges of the friction moment. Development of the friction process was accompanied by sharp rise of temperature and characteristic noise. Results of experimental research are shown in **Figure 11**.

The studies have shown that with increase of the rolling speed, the thickness of the lubricating film initially increases (in our case up to 14 m/s) and then decreases slightly.

With increase of the sliding velocity, sharp decrease of the lubricated film thickness is observed. Though measurement of the particularly thin film (boundary

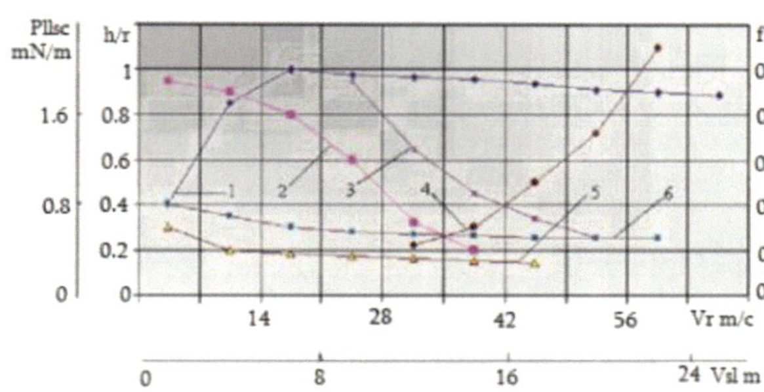

Figure 11.
Dependence of relative lubricant film thickness (h/R), linear scuffing load (P_{llsc}) and coefficient of friction (f) until the appearance of the first signs of scuffing from rolling speed (V_r) and sliding velocity (V_s) at various viscosities (ν) of lubricants: (1) $h/R = \varphi(V_r)$, $P_{ll} = 10^6$ H/m; ν = 157 cSt, (2) $h/R = \varphi(V_{sl})$, $P_{ll} = 2 \times 10^6$ H/m, ν = 157 cSt, V_r = 50 m/s, (3) $P_{llsc} = \varphi(V_{sl})$, ν = 49 cSt, V_r = 50 m/s, (4) $P_{llsc} = \varphi(V_r)$, ν = 157 cSt, V_{sl} = 22 m/s, (5) $f = \varphi(V_{sl})$, $P_{ll} = 1.5 \times 10^6$ H/m, ν = 49 cSt; V_r = 50 m/s, (6) $f = \varphi(V_r)$, $P_{ll} = 10^6$ H/m, ν = 157 cSt.

film) is technically difficult, its presence in the contact zone is indicated by the magnitude and stability of the friction coefficient. Further worsening of the working conditions leads to destruction of the third body in individual places of interacting surfaces.

A particular instability of the friction coefficient was observed at low velocities and high loads: intensive impulses of low frequency were marked and the scuffing marks of significant sizes – scratches and pits were noticed on the rollers surfaces. With increase of the velocity, the time of dwelling of the surfaces in the real contact zone and duration of the thermal impact, values of the amplitude of the friction force variable component decrease; the frequency increases and the individual impulses turn into noise. With further increase of the velocity the friction process is progressed, the temperature on the actual contact area of the interacting surfaces reaches the metal melting point, tonality of the noise rises and turns into whistle and when the frequency exceeds 20 KHz it becomes imperceptible for man.

3. Analysis of results of the experimental researches

The complex physical, mechanical and tribo-chemical processes proceeding in the contact zone of interacting surfaces at direct impact of the environmental conditions raise the problems whose solution demands many-sided approach to these processes. There are many works devoted to these problems [36–38] but they are not solved properly yet. Namely, prediction of the friction coefficient in the contact zone, its control and character of influence of many parameters on its variation are still problematic.

At heavy working conditions, when destruction of the third body is irreversible and scuffing is spread over the factual contact area of the whole surface relative displacement of the surfaces causes sharp increase of the shear stresses, corresponding deformations, values and instability of the friction forces and rupture of the seized places. Strength of the seized places may exceed the strength of the interacting bodies because of which the material pulled out from one surface can form a wear product or can be transferred on the other surface and attached to it that is followed by development of the scuffing process.

Multiple repetition of the shear deformation generated on the surfaces (that sharply decreases towards the depth) causes appearance of cracks on the surfaces, their development and fatigue damage, superficial plastic deformations and lamination. The area of each seized place in the contact zone depends on its power and thermal load; initial micro-geometry of the surfaces; value, velocity and resistance of the deformation etc. Therefore, various working conditions are characterized by corresponding variation of the tribological parameters, namely friction forces, amplitude and frequency of their variable component, wear intensity and roughness of the surfaces. Development of these processes leads to the catastrophic wear due to scuffing.

At low velocities of interacting surfaces, the thermal load of the factual contact zone, velocity of the surface and environment tribo-chemical reaction and resistance of deformation decrease and time of the thermal action and thickness of the superficial heated up layer increase. In such conditions, at destruction of the third body, due to rupture of the seized places, the jerks of low frequency and high amplitudes and sharp instability of the friction coefficient take place and relatively large-size asperities (pits, scratches, asperities, cracks and layers) appear on the surfaces. This is correspondingly reflected on the damage type and roughness of the surfaces.

At high velocities of interacting surfaces, despite several works in this area [39–43], some problems have not yet been resolved. The thermal load of the factual contact zone, velocity of the surface, tribo-chemical reaction of the environment and resistance of deformation increase, whereas time of the thermal action and thickness of the superficial heated up layer decrease. In such conditions, at destruction of the third body, due to rupture of the seized places, the jerks of high frequency and comparatively low amplitudes and instability of the friction coefficient take place and relatively small-size asperities (pits, scratches, asperities, cracks and layers) appear on the surfaces. This is correspondingly reflected on the damage type and roughness of the surfaces. Under the conditions of our experiments at a high rolling speed (more than 40 m/s), traces of fatigue damage and scuffing are not visually observed, however, a high wear rate remains.

Thus, destruction of the third body causes sharp worsening of tribological properties of the interacting surfaces and necessary condition of its avoidance is separation of these surfaces from each other by continuous third body with due properties.

It was ascertained by the experimental researches that destruction of the third body begins in individual points of the factual contact zone that is revealed by appearance of signs of the scuffing in these points. Restoration of the individual damaged points was often observed at unchanged operational conditions but at worsening, the operational conditions the superficial damage quantity increased and multiple damages appeared. At further worsening the operational conditions, a narrow strip of damage is generated spreading afterwards over the whole surface that causes worsening of the tribological parameters and catastrophic wear. The above-mentioned damage stages of the third body are shown in **Figure 12**.

Usually the friction process proceeds at presence of the continuous or discontinuous (restorable or progressively destructible) third body stipulating the character of variation of the friction coefficient. Experimentally it was revealed that to negative friction corresponds the continuous or discontinuous but restorable third body; to neutral friction – multiple seizures of the interacting surfaces and to positive friction – increasing scuffing process that is spread on the whole surface. In **Figure 13** is shown variation of the tractive (friction) force with creep [37].

As it were shown by our experimental researches, at presence of the continuous third body increase of the relative sliding velocity leads to increase of the friction power and contact temperature; decrease of the lubricant viscosity, film thickness and friction force (**Figure 13**, "negative friction"), stable (or smoothly variable) friction torque and low destruction rate of the surfaces. Worsening of the working conditions caused by the partial, non-progressive damage of the third body in the separate unit places corresponds to the separate small impulses of the friction moment. Destruction of the third body in the multiple places leads to the multiple damage of the third body, multiple adhesive junctions of micro-asperities, disruption of these junctions, and a bit little increased impulses of the friction torque and to "neutral friction."

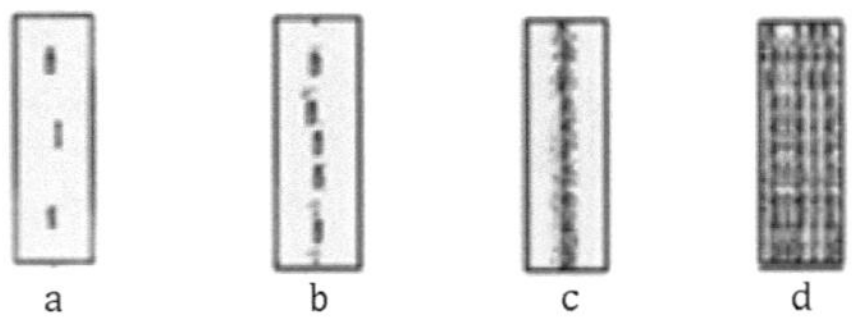

Figure 12.
The damage stages of the interacting surfaces. (a) Unit seizures; (b) multiple seizures; (c) seizures on the narrow strip; (d) seizures on the whole area.

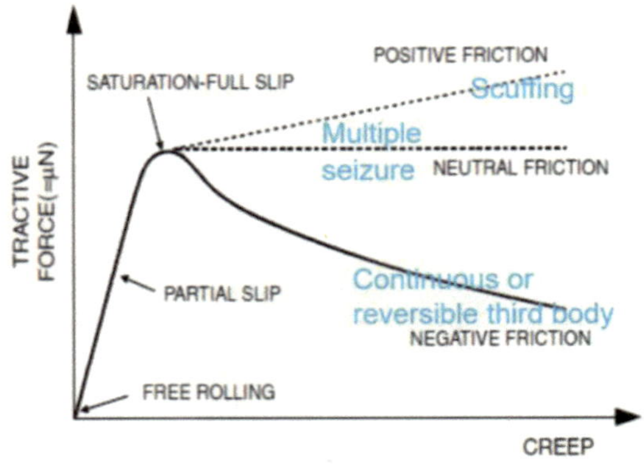

Figure 13.
Friction/creep relationship.

At progressive damage of the third body, the friction torque increases and corresponds to positive friction. Our experimental researches have shown that in other equal conditions the variation of the friction coefficient mainly depends on degree of destruction of the third body. Therefore, preservation of the third body between interacting surfaces and avoidance the scuffing, has a crucial importance for decrease of the friction coefficient, wear rate, etc. This issue became burning especially for wheels and rails in the last 50 years and many works appeared that are devoted to enhancing stability of the wheel flanges against the operational impacts.

In **Figure 14** are shown dependences of the friction factor and various damage types on relative sliding velocity (a) and of the wear rate (types) on slip (b) [32].

Three zones can be distinguished in **Figure 14a**. The low relative sliding velocity, full separation of the interacting surfaces and continuous third body provide high wear resistance of the interacting surfaces and relatively stable friction coefficient (zone 1, **Figure 14a**) that corresponds to "mild" [32] wear rate (**Figure 14b**). In such conditions, the main damage types are the fatigue and plastic deformations.

Small increase of the sliding velocity leads to appearance of small damage sources in multiple places and emergence of small surges of the friction torque (zone 2, **Figure 14a**). The rise of the third body destruction, as well as the magnitude of the friction coefficient and its instability, are clearly reflected in the

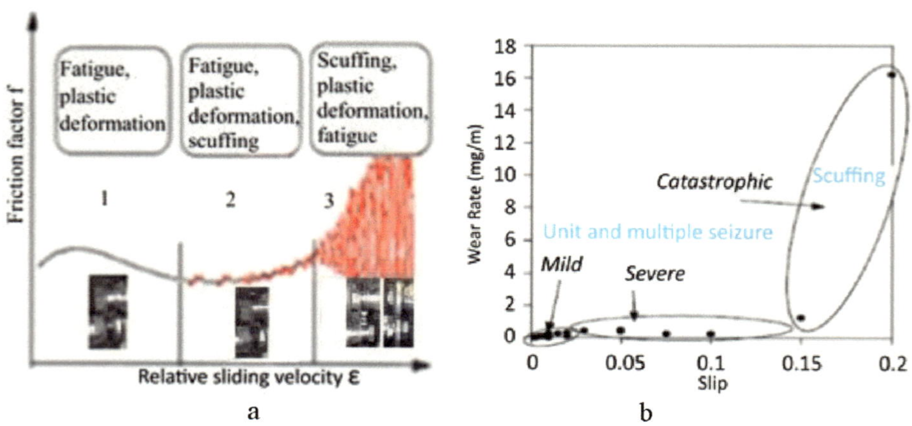

Figure 14.
Dependences of the friction factor and various damage types on relative sliding velocity (a) and of the wear rate (types) on slip (b).

oscillogram of the friction torque and may be predicted on the base of results of the experimental researches. The typical damage types of this zone are fatigue, plastic deformation, adhesive wear and limited rate of scuffing and correspond to "severe" wear rate (**Figure 14b**).

At further increase of the relative sliding velocity, destruction of the third body becomes irreversible and extending and multiple seizures becomes uninterrupted (causing scuffing) and they propagate on the whole width of the interacting surfaces. The typical damage types of this zone are scuffing, plastic deformation and fatigue (zone 3, **Figure 14a**, and "catastrophic" wear rate **Figure 14b**). In this case, the scuffing can be avalanche in nature that quickly disables the machine.

Destruction of the third body makes especially heavy the working conditions of the interacting surfaces and is characterized by increased instability, high wear rate ('catastrophic wear"), vibrations and noise, change of structure and micro-geometry of the surfaces at operation etc.

At low velocities, time of dwelling of individual places of the surfaces in the contact zone and power and thermal actions, variable components of the friction torque and scales of the superficial damage increase and inversely, decrease with increasing speed, although the high wear rate is maintained.

For each operational mode and frictional pairs, this stipulates corresponding micro-geometry and tribological properties.

The methods of calculation of the contact zone power and thermal loads, friction coefficient, wear rate etc., are characterized by low informativeness and precision. This complicates prediction and realization of proper tribological properties of surfaces at various working conditions that prevents machines from reliable and effective operation.

For heavy loaded interacting surfaces are typical destruction of the third body, direct contact of the surfaces and cohesion. The shearing forces, rate of the adhesive and fatigue wear rise sharply in the contact zone at such conditions and friction forces become instable causing the vibrations and noise.

The mentioned types of wear in the contact zone are the results of quite different processes proceeding simultaneously. Besides, identification of the wear type according to the wear signs is often ambiguous that hinders selection of methods for its decrease.

Dependence of tribological properties of the interacting surfaces on the properties of the third body and degree of its destruction were ascertained on the base of results of the experimental researches.

At existence of a continuous third body between interacting surfaces, tribological properties of the contact zone are stipulated by the properties of the third body and at existence of a discontinuous third body, tribological properties of the contact zone are mainly stipulated by the properties of the third body and degree of its destruction.

The signs of onset and development of the third body destruction and a criterion of its destruction are given there. The reasons of the negative, neutral and positive friction, mild, severe and catastrophic wear and types of surface damage at various relative sliding velocities are revealed.

4. Estimation of stability of the third body on the base of EHD theory of lubrication

The most complete mathematical model of lubrication is the elastohydrodynamic (EHD) theory of lubrication [44]. The effectiveness of the EHD theory of lubrication is described by ratio λ or film parameter [45], which is the ratio of film minimum thickness at the Hertzian contact zone to the r.m.s. of the rolling element surface finish:

$$\lambda = \frac{h_{min}}{\sqrt{R_{a_1}^2 + R_{a_2}^2}} \tag{1}$$

where Ra_1 and Ra_2 are the mean roughnesses of the surfaces.

Below are given the integro-differential equations of EHD theory of lubrication with the consideration of the thermal processes that take place in the lubricant film and on the boundaries of surfaces, and the corresponding boundary conditions:

$$\frac{dp}{dx} = 6\mu(V_1 + V_2)\frac{h - h_0}{h^3}, \text{when } x = -\infty, p = 0 \text{ and } x = x_0, p = \frac{dp}{dx} = 0;$$

$$h = h_0 + \frac{x^2 - x_0^2}{2R} + \frac{2}{\pi}\left(\frac{1 - \nu_1^2}{E_1} + \frac{1 - \nu_2^2}{E_2}\right)\int_{-\infty}^{x_0} p(\xi)\ln\left|\frac{\xi - x}{\xi - x_0}\right| d\xi;$$

$$\rho c V\frac{\partial t}{\partial x} = \zeta\frac{\partial^2 t}{\partial y^2} + \mu\left(\frac{\partial V}{\partial y}\right)^2, \text{when } x = -\infty, t = t_0; \tag{2}$$

$$t(x,0) = \left(\frac{1}{\pi\rho_1 c_1 \lambda_1 V_1}\right)^{0,5}\int_{-\infty}^{x} \zeta\frac{\partial t}{\partial y}\bigg|_{y=0}\frac{\partial \varepsilon}{(x - \varepsilon)^{0,5}} + t_0;$$

$$t(x,h) = \left(\frac{1}{\pi\rho_2 c_2 \lambda_2 V_2}\right)^{0,5}\int_{-\infty}^{x} -\zeta\frac{\partial t}{\partial y}\bigg|_{y=h}\frac{\partial \varepsilon}{(x - \varepsilon)^{0,5}} + t_0;$$

$$\mu = \mu_0 \exp\left(\beta p - \alpha\Delta t\right).$$

where p is pressure; V_1 and V_2 – peripheral speeds; μ – dynamic viscosity of lubricant oil in normal conditions; h – clearance; h_0 – minimum clearance; R – radius of curvature; E_1 and E_2 – modulus of elasticity; ν – Poisson's ratio of body materials; t – temperature; ρ, c, ζ, ρ_1, c_1, ζ_1, ρ_2, c_2, ζ_2 – correspondingly density, specific heat capacity and thermal conductivity of lubricant and interacting surfaces; μ_0 – dynamic viscosity of the lubricant; β – piezo coefficient of lubricant viscosity; ζ – lubricant thermal conductivity; α – thermal coefficient of lubricant viscosity; ξ, ε – complementary variables; x_0 – abscissa in the place of lubricant outlet from the gap.

Calculation of the oil film thickness, which separates the bodies, is the main problem of the EHD lubrication theory and there are numerous literature sources about it (Dowson, 1995; Ham rock and Dowson, 1981, etc.). There are various formulas for isothermal and anisothermal solutions for EHD problems describing the behavior of oil film thickness with various accuracies.

The modern friction modifiers contain tribochemically active products that have great influence on their operational properties. The various aspects of properties of these components are not sufficiently studied and they cannot be expressed mathematically. EHD theory of lubrication only considers the mechanical phenomena proceeding in the lubricant film of the contact zone, ignoring other layers.

The thickness of the rough surface boundary layers cannot be measured with the use of the modern methods of measurement of the oil layer thickness. Information about destruction of the boundary layers (and about onset of scuffing as well) can be obtained by sharp increase of the friction torque on the oscillogram. Therefore, onset of the friction torque sharp increase is considered as beginning of the third body destruction.

On the base of system of equations of EHD, theory of lubrication and results of experimental researches considering formula (1), criterion of the third body destruction was developed that has a form:

$$\mathrm{C} = K\left(\frac{R}{\sqrt{{R_{a_1}}^2 + {R_{a_2}}^2}}\right)\cdot\left(\frac{\mu V_{\Xi K}}{P_n}\right)^{0,7}\cdot\left(\frac{P_n\beta}{R}\right)^{0,6}\cdot\left(\frac{\zeta}{\alpha\mu {V_{CK}}^2 {P_{e1,2}}^2}\right)^{e} \leq 1 \qquad (3)$$

As it follows from the formula (2), a criterion of the third body destruction depends on the mechanical and thermo-physical characteristics of interacting surfaces, geometric and kinematic parameters, thermo-physical and tribological parameters of the third body. The properties and stability of the boundary layers are revealed in values of coefficient K and exponent e. The researches have also shown special sensitivity of the third body stability to thermal loads and relative sliding velocities, which must be taken into account to improve working conditions.

The criterion of the third body destruction that is developed on the base of EHD theory of lubrication and results of experimental researches considering stability of the boundary layers has the form:

$$\mathrm{C} = \mathrm{K} \bullet {\mathrm{V}_{\Sigma k}}^{a} \bullet {\mathrm{V}_{sl}}^{b} \bullet {\mathrm{P}_{ll}}^{c} \bullet {\mu_0}^{d} \bullet \mathrm{R}^{l} \bullet \left(\sqrt{R_{a1}^2 + R_{a2}^2}\right)^{f} \bullet \beta^{g} \bullet \zeta^{h} \bullet \alpha^{i} \bullet a^{j} \bullet \mathrm{E}^{n} \leq 1 \qquad (4)$$

where $V_{\Sigma k}$ is a total rolling velocity; V_{sl} – sliding velocity; P_{ll} – linear load; μ - dynamic viscosity of the lubricant; R – reduced radius of curvature of the surfaces; Ra1 and Ra2- average standard deviation of the interacting surfaces; β – piezo coefficient of the lubricant viscosity; ζ– the lubricant thermal conductivity; α – thermal coefficient of the lubricant viscosity; a – thermal diffusivity; The exponents a, b, c, … , n and coefficient K are specified on the base of the experimental data obtained by T.I. Fowle, Y.N. Drozdov, Vellawer, G. Niemann, A.I. Petrusevich, I.I. Sokolov, K. Shawerhammer, G. Tumanishvili and are given in the **Table 1**.

As it is seen from the **Table 1**, destruction of the third body is especially sensitive to the degree b of sliding velocity. It follows from formulae (2) and (3) that with increase of the rolling velocity, radius of curvature, piezo-coefficient of viscosity, heat conductivity factor, thermal diffusivity and coefficient of elasticity, the stability of the third body increases and with increase of the sliding velocity, linear loading, roughness of surfaces and thermal coefficient of viscosity it decreases.

As it was already mentioned, one of the indicators of the third body destruction (scuffing) is appearance of signs of scuffing on the surfaces. According to criteria of destruction of the third body, its destruction is supposed when values of the corresponding criteria are less than 1. K. Schauerhammer experimentally ascertains the conditions of the third body destruction (scuffing) for the gear drive on the gear drive test bench TUME 11 [46]. To predict the destruction of the third body

a	b	c	d	l	f	g	h	i	j	n
0.37 to 0.7	(−0.36) to (−1.32)	(−0.15) to (−0.265)	0.04 to 0.52	0.25 to 0.36	−1	0.6	0.18 to 0.66	(−0.18) to (−0.66)	0.09 to 0.33	0.045 to 0.165

Table 1.
The exponents of formula (3).

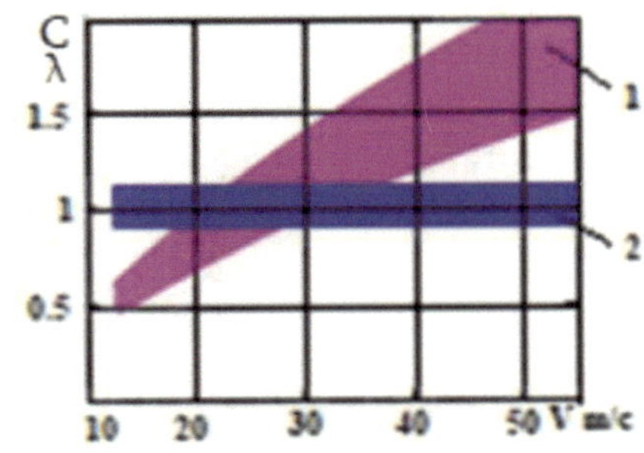

Figure 15.
Dependences of the fields of deviations of the values of λ parameter (1) and criterion C of destruction of the third body (2) developed by us, on the gear wheels circular velocity.

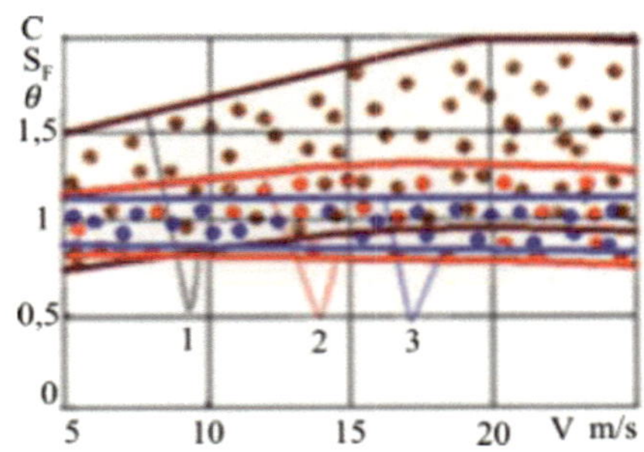

Figure 16.
Dependences of the fields of deviations of the temperature criterion (θ, 1) of H. block, criterion (SF, 2) of G. Niman and Saitzinger and offered criterion (C, 3) of destruction of the third body, on the gear wheels circular velocity.

(scuffing), we used the well-known Dowson and Higginson formulas to determine the lubricating layer parameter (λ) [44, 45] and the criterion C developed by us at the values of the coefficient K = 2.7 and the exponent e = 0.336 in formula (3). Dependences of the fields of deviations of the values of these criteria on the gear wheels circular velocity are shown in **Figure 15**.

As it is seen from the graphs, deviations of the criterion C of destruction of the third body developed by us, are small and constant, while deviations of the parameter λ and its values increase with increase of the gear wheels velocity.

Figure 16 shows the results of similar calculations using the C criterion with the values of the coefficient K = 1.55 and the exponent e = 0.29 in formula (3) and the formulas of H. Block [47] and Niemann G. and Saitzinger K. [48]. The studies were carried out on gear drive test bench FZG for transmissions A, L, N 141, 142, 143, 201, 202, and 203 with lubricant k1.

It is seen from the graphs that deviations of the offered criterion C of destruction of the third body little differ from the unit in the whole range of variation of the circular velocity, while deviations of other criteria significantly differ from the unit and they increase with increase of the circular velocity.

5. Conclusions

- Tribological properties of the interacting surfaces mainly depend on tribological properties of the third body, degree of its destruction, disposition of the surfaces to seizure etc. The researches have shown that the continuous or discontinuous but restorable third body at the initial stage of destruction and progressively destructing third body have quite different properties. In the first case the said properties are stable and depend on the properties of the third body and in the second case, these properties are instable and worsened that are characterized by increasing friction coefficient, catastrophic wear and typical noise.

- Prediction of destruction of the third body is possible in the laboratory conditions by estimation of the friction torque variation and with the use of the criterion of destruction of the third body, with ascertained beforehand values of the experimental coefficients;

- The friction coefficient (negative, neutral and positive), wear rate of the interacting surfaces (mild, severe and catastrophic), damage types (scuffing, fatigue, plastic deformation, adhesive wear) and vibrations and noise generated in the contact zone depend on tribological properties of the third body, its degree of destruction and area of the factual contact zone seized places;

- For the improvement of tribological properties of the interacting surfaces, it is necessary to provide the contact zone with continuous or restorable third body having due tribological properties at the initial stage of destruction.

Acknowledgements

This work was supported by Shota Rustaveli National Science Foundation of Georgia (SRNSFG) under GENIE project CARYS-19-588.

Conflict of interest

The authors declare no conflict of interest.

Author details

George Tumanishvili*, Tengiz Nadiradze and Giorgi Tumanishvili
Institute of Machine Mechanics, Tbilisi, Georgia

*Address all correspondence to: ge.tumanishvili@gmail.com

References

[1] Dowson D. Higginson G. R. Whitaker A. V. Elastohydrodynamic lubrication: a survey of Isothermal solutions. Journal Mech. Eng. Science. Vol. 4, 1962, p. 121-126.

[2] Grubin AN. Fundamentals of hydrodynamic theory of lubrication of heavy loaded cylindrical surfaces. Book: Investigation of the contact of machine components. Central Scientific Research Inst. Tech. & Mech. Eng., (in Russian) 1949, #30, p. 219.

[3] Petrusevich, A. I. Fundamental conclusions from the contact-hydrodynamic theory of lubrication Izv. Akad. Nauk. SSSR (OTN), (in Russian) 1951, 2, 209–223.

[4] Drozdov YN, Pavlov VG, Puchkov VN. Friction and wear in the extreme conditions (in Russian). (in Russian) (1986) Moscow, Mashinostroenie, 224 p.

[5] L. E. Murch, W. R. D. Wilson. A Thermal Elastohydrodynamic Inlet Zone Analysis. J. of Lubrication Tech. Apr 1975, 97(2): 212-216.

[6] Cheng, H. S. A numerical solution of the elastohydrodynamic film thickness in an elliptical contact Trans. ASME F, J. Lubric. Technol., 1970, 92 (1), 155–162.

[7] Finkin E. F., Gu A., Yung L. A critical Examination of the Elastohydrodynamic Criterion for the scoring of gears. Problems of friction and Lubrication, V. 36, #3, 1974.

[8] Greenwood J, Williamson J. Contact of Nominally Flat Surfaces. Proc R Soc. A 1966, 295:300-19.

[9] Persson B, Albohr O, Tartaglino U, Volokitin A, Tosatti E. On the nature of surfaces roughness with application to contact mechanics, sealing, rubber friction and adhesion. Journal of Physics: Condensed Matter 004, 17:R1.

[10] Carpick RW, Salmeron M. Scratching the surface: fundamental investigations of tribology with atomic force microscopy. Chem. Rev. 1997; 97: 1163-94.

[11] Kim SH, Asay DB, Dugger MT. Nano tribology and MEMS. Nano today 2007; 2:22-9.

[12] Bhushan B. Nano tribology and Nano mechanics of MEMS/NEMS and BioMEMS/BioNEMS materials and devices. Microelectronic Engineering 2007; 84:387-412.

[13] Berman D, Deshmukh SA, Sankaranarayanan SK, Erdemir A, Sumant AV. Friction. Macroscale superlubricity enabled by graphene nanoscroll formation. Science 2015; 348: 1118-22.

[14] Popov V. Contact mechanics and friction: physical principles and applications. : Springer Science & Business Media, 2010.

[15] Persson B. Sliding friction: physical principles and applications. : Springer Science & Business Media, 2013.

[16] Franklin FJ, Widiyarta I, Kapoor A. Computer simulation of wear and rolling contact fatigue. Wear 2001; 251: 949-55.

[17] Kalker JJ. Three-dimensional elastic bodies in rolling contact. : Springer Science & Business Media, 2013.

[18] Goryacheva IG. Contact mechanics in tribology. : Springer Science & Business Media, 2013.

[19] Zhu D, Hu Y. A computer program package for the prediction of EHL and mixed lubrication characteristics, friction, subsurface stresses and flash temperatures based on measured 3-D

surface roughness. Tribol Trans 2001; 44:383-90.

[20] Yastrebov VA. Numerical methods in contact mechanics. : John Wiley & Sons, 2013.

[21] Bemporad A, Paggi M. Optimization algorithms for the solution of the frictionless normal contact between rough surfaces. Int. J Solids Structures 2015;69–70:94-105.

[22] Van der Giessen E, Needleman A. Discrete dislocation plasticity: a simple planar model. Modell Simul Mater Sci. Eng. 1995; 3:689.

[23] Yan W, Komvopoulos K. Three-dimensional molecular dynamics analysis of atomic-scale indentation. J Tribol 1998; 120.

[24] Ahmed NS, Nassar AM (2013) Lubrication and Lubricants. In: Tribol. - Fundam. Adv., pp. 55-76.

[25] Yifei M, Turner KT, Szlufarska I (2009) Friction laws at the nanoscale. Nature, Vol 457, 26.

[26] Hou K, Kalousek J, Magel E (1997) Rheological model of solid layer in rolling contact, Wear, 211, 134–140.

[27] Landman U., Luedtke W.D., Burnham N., and Colton R.J. Atomistic Mechanisms and Dynamics of Adhesion, Nano indentation and Fracture, Science, Vol 248, April 1990, p 454-461.

[28] Guo Q., Ross J.D.J., and Pollock H. M., What Part Do Adhesion and Deformation Play in Fine-Scale Static and Sliding Contact?, New Materials Approaches to Tribology: Theory and Application, L.E. Pope, L.L. Fehrenbacher, and Winer W.O., Ed., Materials Research Society, 1989, p 51-66.

[29] V.R. Regel, A. B. Slucker, E. E. Thomashefsky. The kinetic theory of durability of solid bodies (in Russian). Moscow, 1974, p. 560.

[30] F. Braghin a, R. Lewis b, R.S. Dwyer-Joyce b, S. Bruni. A mathematical model to predict railway wheel profile evolution due to wear. Wear 261 (2006) 1253–1264.

[31] Lewis R., Dwyer-Joyce R.S., Bruni S., Ekberg A., Cavalletti M., Bel Knani K.. A New CAE Procedure for Railway Wheel Tribological Design. 14th International Wheelset Congress, 17-21 October, Orlando, USA.

[32] Lewis R, Dwyer-Joyce RS (2004) Wear mechanisms and transitions in railway wheel steels. Proceedings of the Institution of Mechanical Engineers, Part J: Journal of Engineering Tribology, 218(6), 467-478.

[33] ASM Handbook, Friction, Lubrication and Wear Technology,10th ed., Vol. 18, 1992.;

[34] Kenneth C. Ludema. Review of Scuffing and Running in of Lubricated Surfaces, with Asperities and Oxides in Perspective. Wear, 100 (1984) 315 – 331;

[35] Sergovski V. P., Tumanishvili G. I. Capacity method of measurement of the oil film thickness. Collected articles of Institute of Machine Mechanics Book: "Machine Mechanics", Tbilisi, (in Russian) 1979, pp. 119-125.

[36] Magel EE (2011) Rolling Contact Fatigue: A Comprehensive Review, Prescribed by ANSI Std. 239-18 298-102 DOT/FA/ORD-11/24, U.S. Department of Transportation, Office of Railroad Policy and Development Washington, DC 20590. 118 p.

[37] Eadie, D. T., Kalousec, J., and Chiddick, K. C., The Role of High Positive Friction (HPF) Modifier in the Control of Short Pitch Corrugation and Related Phenomena, Proceedings of the

5th International Conference on Contact Mechanics and Wear of Rail/Wheel Systems, Tokyo, pp. 36–41, 2000.

[38] Bolton PJ, Clayton P (1984) Rolling-sliding wear damage in rail and tyre steels. Wear. 93, p. 145 – 165.

[39] Naka N., Eleiche A. M., Suh N. P. Wear of metals at high sliding speeds. Wear 44 1977 104-125.

[40] Donald T. Eadie at all. Effective friction control for optimization of high speed rail operations. Proceedings of the 2010 Joint Rail Conference JRC2010-36010 April 27-29, 2010, Urbana, Illinois, USA;

[41] Magel E. at all. Traction, forces, wheel climb and damage in high-speed railway operations. Wear265 (2008) 1446-1451.

[42] T. Ohyama: Tribological studies on adhesion phenomena between wheel and rail at high speeds. Wear, 144 (1991). 263-27

[43] Gloeckner, P, Sebald, W, Bakolas, V. An approach to understanding micro-spalling in high-speed ball bearings using a thermal elastohydrodynamic model. Tribol Trans 2009; 52: 534–543.

[44] D. Dowson, GR Higginson. The effect of material properties on the lubrication of elastic rollers. Journal of Mechanical Engineering Science. V.2, #3, 1960, p188-194.

[45] Gohar R (2001) Elastohydrodynamics. 2nd ed. World Scientific. London : Imperial College Press, 446 p.

[46] K. Schauerhammer. Untersuchung der Freβtragfahigkeit von Schmierolen im Stirnradgetriebe. – Schmierungs-technik, #10, 1978, s. 297-300.

[47] H. Blok. Les temperatures de surface dans conditions de graissage sous extreme pression – Congr. Mondial du petrole, t. 3, Paris, 1937, s. 471-486.

[48] G. Niemann, K. Seitzinger. Die Erwarmung einsatrgeharteter Zahnrader als Kennwert fur ihre Freβtragfahigkeit. Untersuchungen uber den Einfluβ von Zahnform, Betriebsbedingungen und Schmierstoff und die Erwarmung und die Freβgrenzlast. –VDI-Z, 113, #2, s. 97-105.

Chapter 8

The Effect of Elemental Composition and Nanostructure of Multilayer Composite Coatings on Their Tribological Properties at Elevated Temperatures

Alexey Vereschaka, Sergey Grigoriev, Vladimir Tabakov, Mars Migranov, Nikolay Sitnikov, Filipp Milovich, Nikolay Andreev and Caterine Sotova

Abstract

The chapter discusses the tribological properties of samples with multilayer composite nanostructured Ti-TiN-(Ti,Cr,Al,Si)N, Zr-ZrN-(Nb,Zr,Cr,Al)N, and Zr-ZrN-(Zr,Al,Si)N coatings, as well as Ti-TiN-(Ti,Al,Cr)N, with different values of the nanolayer period λ. The relationship between tribological parameters, a temperature varying within a range of 20–1000°C, and λ was investigated. The studies have found that the adhesion component of the coefficient of friction (COF) varies nonlinearly with a pronounced extremum depending on temperature. The value of λ has a noticeable influence on the tribological properties of the coatings, and the nature of the mentioned influence depends on temperature. The tests found that for the coatings with all studied values of λ, an increase in temperature first caused an increase and then a decrease in COF.

Keywords: physical vapor deposition (PVD) coatings, coefficient of friction, tool life, thermo stability, nanolayers

1. Introduction

The tribological properties are among the most important mechanical characteristics of the coatings, affecting their performance parameters and working efficiency of products. Due to the fact that the coating characteristics vary noticeably with an increase in temperature and often differ radically from the parameters measured at room temperature [1–4], the investigation of the tribological properties at temperatures corresponding to the operating temperature (for example, a temperature in the cutting area during the study of the properties of coatings for cutting tools) is essential. The modern trends in improving the coating properties largely imply more complicated architecture and elemental composition [3–8].

Coatings are being developed with a nanolayer architecture, characterized by a number of significant advantages compared to traditional monolithic coatings [3, 6–8]. In particular, the studies detect improved crack and impact resistance of the coatings with a nanolayer architecture [9–13]. The influence of a nanolayer structure of the coating on its tribological properties is of great importance. The studies have found that the coatings with a nanolayer structure are characterized by a reduced COF, high hardness, and a low level of residual stresses [6], while with a decrease in a thickness of the binary nanolayer λ, the COF decreases and the wear resistance increases [8]. The investigation has also demonstrated that a decrease in the nanolayer period λ leads to an increase in hardness, a decrease in the COF, and an improvement in the resistance to the failure of cohesive bonds between nanolayers [14–17]. In particular, in [15], the studies of the TiAlCN/VCN nanostructured coating with λ = 2.2 nm have found that for this coating, the COF grows with an increase in temperature up to 200°C, but begins to decrease at temperatures above 650°C. The experiments found a decrease in the COF for the CrAlYN/CrN nanolayer coating with λ = 4.2 nm with an increase in temperature up to 650°C [16]. The study of the Ti/TiAlN/TiAlCN nanolayer coating detected its low COF and high wear resistance [17]. The tribological properties of the TiAlN/CNx nanostructured coating with λ = 7 nm were considered in [18]. The tests have found that the TiAlN/CNx nanostructured coating is characterized by small grain sizes and lower surface roughness in comparison with a monolithic coating of a similar composition and also by a lower COF and higher wear resistance.

The influence of the elemental composition of the coatings on their tribological properties was also investigated. The experiments revealed the COF value at room temperature for the coatings of TiN (0.55), TiCN (0.40), TiAlN (0.50), AlTiN (0.7), as well as AlTiN/Si_3N_4 consisting of AlTiN nanoparticles embedded into amorphous Si_3N_4 matrix (0.45) [19]. During the studies focused on the TiAlCrN/TiAlYN and TiAlN/VN coatings, Hovsepian et al. [20] found that the introduction of Y in the coating composition led to a decrease in the COF from 0.9 up to 0.65 during the tests conducted at temperatures ranging from 850 to 950°C. For the TiAlCrN coating, the COF was 1.1–1.6 (at temperatures of 600–900°C), while for the TiAlN/VN coating, the COF was 0.5 (at 700°C). Mo et al. [21] found that the COF of the AlCrN and TiAlN coatings was 0.75 and 0.85, respectively. With an increase in the Cr content in the CrAlN coating, a slight increase in the COF was detected [22]. Nohava et al. [23] studied the properties of the AlCrN, AlCrON, and α-$(Al,Cr)_2O_3$ coatings in comparison with the properties of the TiN-AlTiN reference coating at temperatures of 24, 600, and 800°C. The maximum value of the COF was detected at a temperature of 600°C for all the studied samples, while at a temperature of 800°C, there was a significant decrease in the COF (to 0.6–0.8) to the values lower than those detected at a room temperature (0.3–0.5). Bao et al. [24] found that for the TiCN/TiC/TiN coating, the COF was 0.4–0.5 at a room temperature and 0.6–0.7 – at a temperature of 550°C.

Thus, it can be asserted that:

- for coated products, the COF varies significantly with an increasing temperature;
- in general, the above variation of the COF is initially characterized by its gradual increase with a growth of temperature and then by its more significant decrease with a further increase in temperature;
- the parameters of the nanolayer structure (in particular, the value of the nanolayer period λ) have a noticeable influence on the COF variation.

2. Experimental

2.1 Coating deposition

Coatings were deposited through the filtered cathodic vacuum arc deposition (FCVAD) [5, 7, 25–29] technology. The experiments were carried out using a unit VIT-2, (IDTI RAS – MSTU STANKIN, Russia) with two arc evaporators with a pulsed magnetic field and one arc evaporator with filtering the vapor-ion flow. Furthermore, the complex also included a source of pulsed bias voltage supply to a substrate, a dynamic gas mixing system for reaction gases, a system to control automatically the chamber pressure and a process temperature control system, and a system for stepless adjustment of planetary gear rotation.

Two groups of samples were manufactured:

Group I included samples with three coatings of various compositions, i.e. I-a – Ti-TiN-(Ti,Cr,Al,Si)N coating; I-b – Zr-ZrN-(Nb,Zr,Cr,Al)N coating; I-c – Zr-ZrN-(Zr,Al,Si)N coating.

Group II included samples with the Ti-TiN-(Ti,Al,Cr)N coatings with different values of the nanolayer period λ, formed through varying the turntable rotation frequency n (see **Table 1**).

2.2 Measurement of tribological parameters

During the friction process, a complex system is formed in the actual contact areas. This system possesses a number of specific properties which differ from the properties of the contacting body materials when considered separately, without contact during friction. Apparently, to obtain reliable data correlating with the main factors of friction and wear, it is necessary to assess the properties of the contact area directly. However, there are several reasons which make it difficult to measure the tribological parameters directly during the operation of actual products at elevated temperature of the contact forces. Some specific features of the above-mentioned reasons are as follows:

- forces and temperatures on the contact area are distributed unevenly,
- there is a difficulty in determining the actual contact loads due to the differences in chemical purity and discreteness of contact of the contacting surfaces.

It should be noted that while two surfaces are sliding, the tangential contact force is being affected not only by shear strength of adhesive bonds, but also the deformation component of the friction force [30]. The contacting surfaces (especially those subject to wear) can have significant roughness and be heterogeneous in their physical and mechanical properties due to the polycrystalline structure. Thus, the deformation component of the tangential contact force can have a significant influence on the tribological properties of products, but its direct determination in

Sample	II-a	II-b	II-c	II-d	II-e
Turntable rotation frequency n, min^{-1}	0.25	1	1.5	5	7
Nanolayer period λ, nm	302	70	53	16	10

Table 1.
The samples depending on the turntable rotation frequency.

the friction process is associated with significant difficulties [30]. It is practically impossible to separate the deformation component from the total tangential forces and thus obtain an adhesion component, especially at an elevated contact temperature, and that fact makes it almost impossible to determine the strength of adhesive bonds immediately during the process of metal cutting [31].

Based on the above, it is physical modeling, which makes it possible to determine the tribological characteristics under conditions most closely simulating the conditions of the cutting area, is the most accurate and effective way to find the indicators of the adhesive interaction between the tool and the material being machined.

The proposed method is based on a physical model [25] (**Figure 1**), which reflects with sufficient accuracy the actual conditions of friction and wear at a local contact in the cutting area. In accordance with this model, a spherical indenter 2 made of a coated tool material (imitating an individual asperity of a contact spot of solids subject to friction), compressed by two plane-parallel counterbodies 1 made of the material being machined (with high precision and cleanliness of the contacting surfaces) rotates under load around its own axis. The force F_{exp}, spent on the rotation of the indenter and applied to the cable 3, laid in a groove of the disk 4, is mainly related to the shear strength τnn of adhesive bonds.

The shear strength τ_n of adhesive bonds is found as follows:

$$\tau_n = \frac{3}{4}\frac{F_{exp}}{\pi}\frac{R_{exp}}{r_{ind}^3} \tag{1}$$

where F_{exp} is the circumferential force on the disc, rotating the indenter;
R_{exp} is the radius of the disk in which the indenter is fixed; and.
r_{ind} is the radius of the indent on the samples.

Due to the small dimension of the indenter, it possible to make an assumption about the normal stresses acting on the sphere surface as constant and equal in the entire indentation area (a purely plastic contact).

The above normal stresses are determined as follows:

$$p_n = \frac{N}{\pi r_{ind}^2} \tag{2}$$

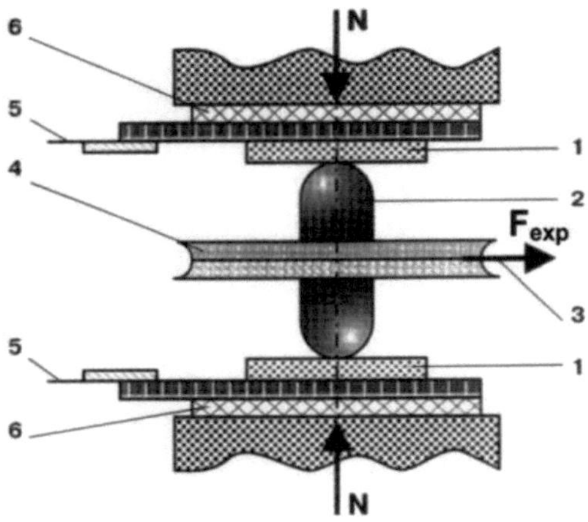

Figure 1.
Friction contact model considered by [3, 4, 25, 31]. 1 – Two plane-parallel counterbodies; 2 – Spherical indenter; 3 – Cable; 4 – Disk; 5 – Copper plates; 6 – Thermal pads; N – Load applied.

The COF consists of two components, i.e. an adhesion component, which results from the solid body molecular interaction in the actual contact area, and a deformation component, which result from the surface layer deformation on solid bodies during the friction process [31]. In [25], the proposed model demonstrated that the forces to rotate the indenter were mainly related to the shear stress of the adhesive (interatomic and intermolecular) bonds, while in this case, no deformation component of the tangential forces was actually detected. Under the conditions of seizure, no deformation or fracture occur on the contact surface during the sliding, while the detected maximum tangential stresses reflect the hardness of near-surface layer of the softest body out of all the contacting bodies. In case when no seizure occurs, the above stresses are related to the dissipation of energy expended to break the bonds formed during the contact of the bodies.

The adhesion (molecular) component of the COF may be defined as follows:

$$f_{adh} = \frac{\tau_n}{p_n} = \frac{3}{4}\frac{F_{exp}}{N}\frac{R_{exp}}{r_{ind}} \tag{3}$$

where N is the load applied (N).

To apply the above method under the conditions of elevated temperatures, a special adhesiometer was developed, which allowed heating the contact area to a temperature of up to 1100°C and providing the typical temperature distribution over the depth of the contacting bodies [25]. Certain shortcomings of the above method include the relatively low rates of deformation and relative sliding. However, in terms of temperature, load, and contact cleanliness, this method is able to simulate well enough the actual conditions of friction and adhesion in the cutting area.

The indenter is a double-sided spherical cylinder with the radius of 2.5 mm and the height of 25 mm made of tool material (carbide).

After the counterbody and the indenter have been installed, the contact area is heated up to the operating temperature, and then the load N is applied, under the influence of which the indenter with the radius sphere r_1 penetrates into the counterbody surface to the depth h. Thus, the plastic contact takes place, and external friction is detected during the punch rotation [31].

The accepted loads in combination with low roughness of the contacting surfaces provide both the required area of actual contact between the indenter and the counterbody and the elimination of the formed oxide and sorbed films and the contacting of metal surfaces close to juvenile. At the same time, the oxygen penetration into the contact area is minimized due to the high density of the contact between the indenter and the counterbody. The experiments were carried out at different values of the contact temperature θ. Thus, the relationships of $\tau_n = f\ (p_r)$ are obtained for different values of θ. The data on the value of τ_n were obtained after the experiment had been repeated three times, with the probable deviation not exceeding 5%.

3. Results and discussion

The microstructures of the coated samples, including I-a – Ti-TiN-(Ti,Cr,Al,Si)N; I-b – Zr-ZrN-(Nb,Zr,Cr,Al)N; and I-c – Zr-ZrN-(Zr,Al,Si)N, are shown in **Figure 2**. The coatings are nanostructured, the thicknesses of the nanolayers are 30–80 nm, and the total coating thickness reaches about 3 μm.

The microstructures of the coated samples under the study are depicted in **Figure 2** [3].

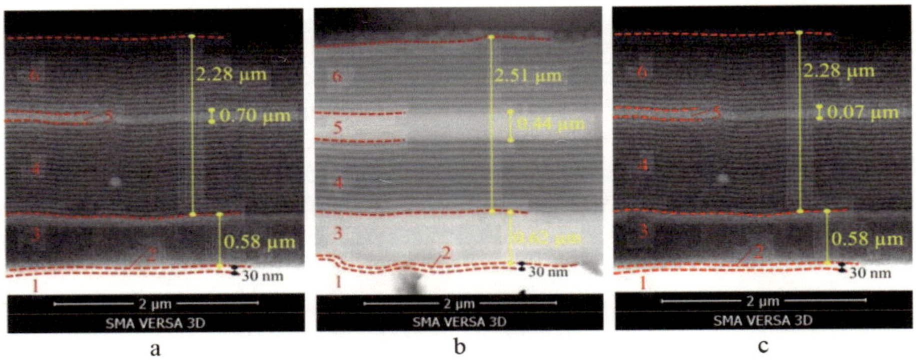

Figure 2.
The structure of the coated samples under the study. a – Ti-TiN-(Ti,Cr,Al,Si)N coating; b – Zr-ZrN-(Nb,Zr, Cr,Al)N coating; c – Zr-ZrN-(Zr,Al,Si)N coating [3]; I-c – Zr-ZrN-(Zr,Al,Si)N coating. 1 –WC-Co substrate; 2 – adhesion sublayer; 3 – transitional layer; 4 – internal zone of wear-resistant layer; 5 – intermediate sublayer; 6 – external zone of wear-resistant layer.

The results of the studies of the nanolayer coating Ti-TiN-(Ti,Al,Cr)N structure (samples of Group II) are presented in **Figure 3** [4].

As seen from **Figure 3**, all the coatings under study have a nanolayer structure. The experiments found the nanolayer thicknesses, which ranged from 10 nm to 302 nm, depending on the coating type. The earlier studies [4,34,35] revealed that each nanolayer of the II-a coating had a complex structure, formed due to the planetary rotation of the toolset during the deposition process [4, 25, 32]. In [4, 25], it is also found that the coatings with the nanolayer thicknesses of 70–10 nm also have a similar complex structure. At the same time, the nanolayers of the II-a and II-c coatings affect the formation of the crystalline structure, and the growth of crystals during the deposition is limited by the boundaries of one nanolayer. The II-d and II-e coatings demonstrate the growth of crystals, which is not limited by the nanolayer boundaries.

The experiments have been conducted at temperatures ranging from 20 to 550°C to investigate the influence of temperature on the tribotechnical properties of tribopairs in the "the material being machined–carbide with wear-resistant complex" interface (shear strength τnn of adhesive bond, normal stress on contact P_{rn}, and relation $\frac{\tau_{nn}}{P_{rn}}$, actually representing the adhesion component of the COF, on which

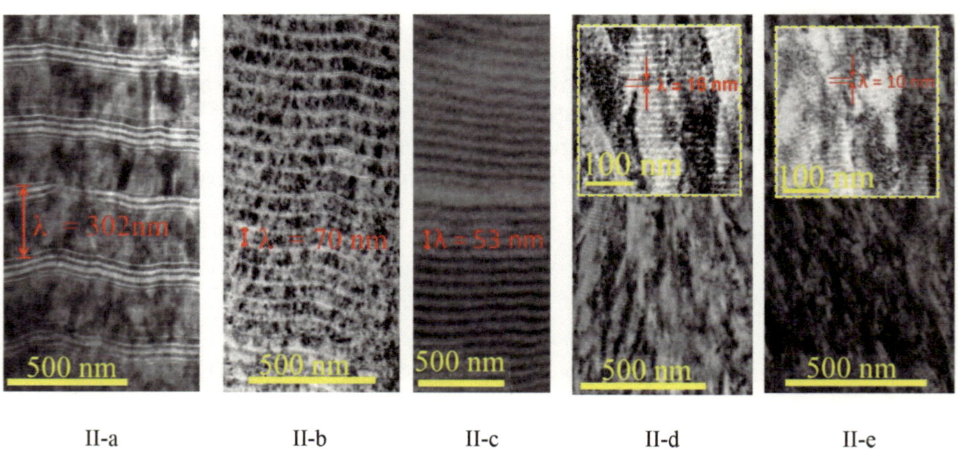

Figure 3.
The nanolayer structure of coatings on samples (TEM) [4]. a – λ = 304 nm, b - λ = 70 nm, c - λ = 53 nm, d - λ = 10 nm.

the deeper deformation of contact layers depends [F1, F2]). **Figures 4** and **5** exhibit curves reflecting the relationship between τ_{nn}, P_{rn}, $\frac{\tau_{nn}}{P_{rn}}$, and temperature. The above relationships demonstrate that the shear stresses τ_{nn} first increase for all the samples under study and then decrease with an increasing temperature. At the same time, for the sample with the Zr-ZrN-(Nb,Zr,Cr,Al)N coating, the shear stresses τ_{nn} initially grow noticeably with an increase in temperature, but when the temperature exceeds 400°C, they begin to decrease, while the process intensity grows with an increase in temperature. In particular, the conducted experiments found that at temperature of 400°C, for samples of carbide WC–Co, the parameter of τ_{nn} was significantly (almost by 2 times) higher for a sample with the Zr-ZrN-(Nb,Zr,Cr, Al)N coating compared to uncoated samples, but at temperature above 400°C, an uncoated sample demonstrated higher τnn. The samples with the Ti-TiN-(Ti,Cr,Al, Si)N coating demonstrated the lowest relationship between τnn and temperature among all the samples under the study. With an increasing temperature, τnn first increases slightly; however, when at temperatures exceeding 300°C, the parameter of τnn starts decreasing with an increase in temperature. It should be noted that at the maximum temperature of 550°C, τnn is approximately equal to τnn at room temperature.

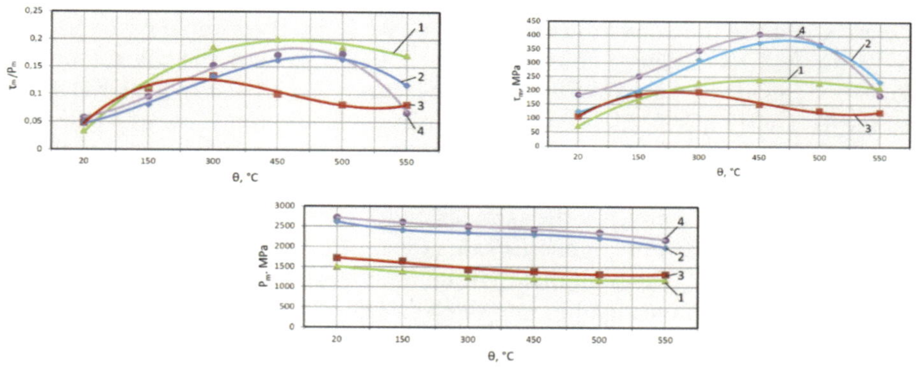

Figure 4.
Influence of temperature on tribotechnical properties of tribopair in the "steel AISI 321–carbide (WC-Co) with coating" interface [3]. 1 – Uncoated sample; 2 – Zr-ZrN-(Zr,Al)N coating; 3 –Ti-TiN-(Ti,Cr,Al,Si)N coating; 4 –Zr-ZrN-(Nb,Zr,Cr,Al)N coating.

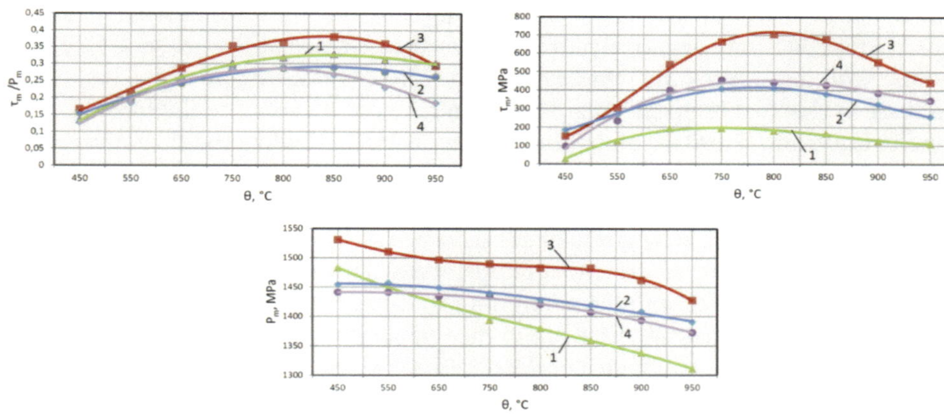

Figure 5.
Influence of temperature on tribotechnical properties of tribopair in the "steel S31600–carbide (WC-Co) with coating" interface [3]. 1 – Uncoated; 2 – Zr-ZrN-(Zr,Al)N coating; 3 – Ti-TiN-(Ti,Cr,Al,Si)N coating; 4 – Zr-ZrN-(Nb,Zr,Cr,Al)N coating.

The specific features of the variation of the adhesion (molecular) component f_{adh} of the COF with an increase in temperature were considered. Coated samples demonstrated noticeably lower values of f_{adh} compared to an uncoated sample, at all temperatures and for all types of carbides. Meanwhile, f_{adh} reaches its maximum at temperature of about 400–450°C, and after that, f_{adh} begins to decrease. The sample with the Ti-TiN-(Ti,Cr,Al,Si)N coating demonstrated the lowest value of the adhesion (molecular) component fadh of the COF. However, when the limit temperature of 550°C is reached, fadh decreases sharply for the sample with the Zr-ZrN-(Nb,Zr,Cr,Al)N coating. Thus, at temperature of 550°C, the samples with the both coatings under consideration demonstrate approximately the same values of f_{adh}, while the substantially lower value of f_{adh} was detected for the uncoated sample. The conducted experiments make it possible to predict that with a further increase in temperature, the sample with the Zr-ZrN-(Nb,Zr,Cr,Al)N coating will demonstrate the minimum value of f_{adh}.

While studying the variation of f_{adh} upon the contact with the counterbody made of S31600 steel (**Figure 5**), it is possible to notice that the dynamics of the change in f_{adh} is similar for all the samples under consideration: the value of f_{adh} first increases, but begins to decrease after a certain temperature is reached. However, for different samples, the value of f_{adh} begins to fall at different temperatures. In particular, for the sample with the ZrN-(Nb,Zr,Cr,Al)N coating, fadh begins to decrease when the temperature reaches 750°C, while for the sample with the Ti-TiN-(Ti,Cr,Al,Si)N coating – at temperature of 850°C. If upon the contact with the counterbody made of American Iron and Steel Institute (AISI) 321 steel, the sample with the Ti-TiN-(Ti,Cr,Al,Si)N coating exhibits the minimum value of f_{adh} in a range from 300 to 550°C, then upon the contact with the counterbody made of S31600 steel, the sample with the Ti-TiN-(Ti,Cr,Al,Si)N coating demonstrates the highest value of fadh among all the samples under consideration. Meanwhile, the remaining samples showed largely the same dynamics of the variation in f_{adh} for counterbodies made of the both materials under consideration.

The results of the investigation into the tribological parameters of the samples II-a – II-e are presented in **Figure 6**.

The investigation of the relationship between the tribological properties and temperature for samples with various coatings demonstrated that with an increase in temperature, the value of f_{adh} varies nonmonotonically and is of extreme nature. Within a temperature range from 500 to 800°C, an increase in the parameters of the fictional contact is related to an increase in the adhesive interaction on the contact surface. At 800°C, the adhesion on the contacting surface of friction is maximum, which can negatively affect the wear resistance of the product. The sample with the II-e coating demonstrates the higher value of f_{adh}, which begins to decrease at elevated temperatures. This phenomenon can relate to the formation of tribological oxide films (titanium and aluminum oxides), while the thicknesses of the coating layers are of key importance.

The sample with the II-d coating demonstrated the most favorable value of f_{adh}, as well as the lowest shear strength τ_n of adhesive bonds. The advantages of this coating are especially clearly demonstrated at temperatures above 600°C. It is important to note that for the samples with the minimum nanolayer thicknesses (II-d and II-e), the value of f_{adh} continuously increases up to the temperature of 800°C and then begins to noticeably decrease, and such a decline is especially clear for the II-e coating). For the coatings with large values of the nanolayer period λ (II-a, II-b, and II-c), a decrease of f_{adh} is observed at temperatures above 700°C, and such a decline intensifies with a decrease in λ. Such a decrease is least pronounced in the II-a coatings with the maximum nanolayer period λ. Thus, it can be assumed that for the coatings with the minimum value of λ (II-d and II-e), the active oxidation begins at

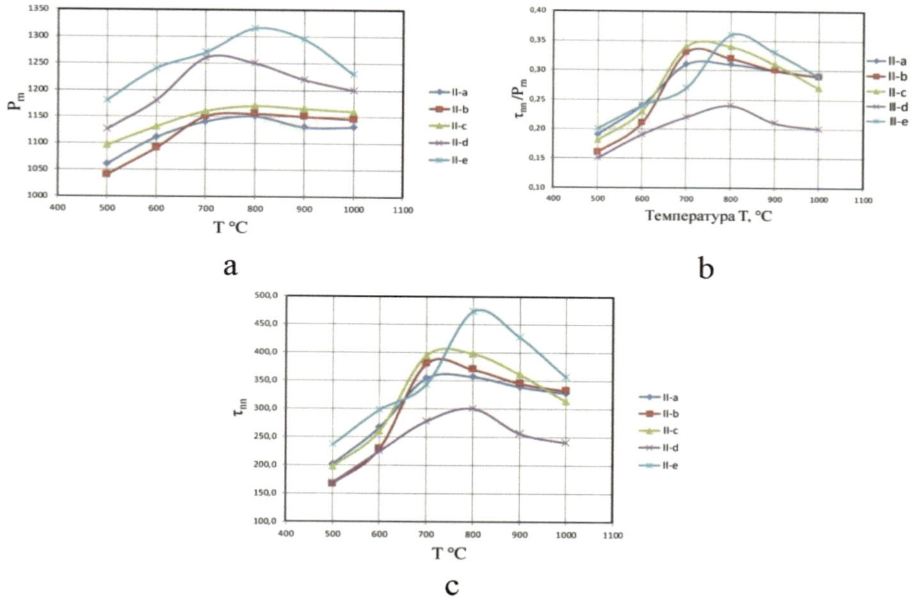

Figure 6.
The results of the investigation into the tribological properties of the samples II in the "steel AISI 1045–carbide (WC-Co) with coating" interface [4]: A – Normal stresses $\mathbf{P_{rn}}$*; b – Strengths on cut* τ_n*; c – Coefficient of friction* $\mathbf{f_{adh} = \tau_n/P_{rn}}$.

temperatures above 800°C, while for the coatings with larger values of λ (II-a, II-b, and II-c), the oxidation processes start already at a temperature of 700°C. Thus, it can be concluded that a smaller value of λ provides better resistance to thermal oxidation, due to an increased number of interlayer interfaces inhibiting the processes of thermal destruction of the surface layer of the coating [4, 11].

There is a significant difference between the tribological characteristics of the II-d and II-e samples. In particular, when the value of λ decreases from 16 nm (II-d) to 10 nm (II-e), the value of f_{adh} increases significantly, and the most significant difference in f_{adh} is demonstrated by the II-d and II-e samples in the temperature range from 800 to 900°C. When the above temperatures are reached, the spinodal decomposition of the (Ti,Al,Cr)N phase begins, accompanied by the formation of a soft hexagonal AlN phase and also a release of pure Ti,Al and Cr [4, 11]. The interlayer interfaces can slow down the formation of the decomposition zone, although, if the value of λ is too small, this effect becomes significantly weaker, and as a result, more intense spinodal decomposition is detected for the II-e sample and less intense – for the II-d sample.

Figure 7 depicts the influence of λ on the adhesion component f_{adh} of the COF. Based on the obtained approximating curves, it is possible to distinguish two distinct extrema of f_{adh}, including the maximum in a range of λ = 70–53 nm and the minimum in a range of λ = 16–10 nm. A decrease in the value of λ from 16 to 10 nm leads to a noticeable increase in f_{adh} at all the temperatures under consideration.

Temperature has a significant influence on the value of f_{adh} and the nature of its variation depending on the value of λ. Based on the data presented in **Figure 7**, three temperature ranges can be distinguished, characterized by the different influence of the nanolayer period λ on the adhesion component f_{adh} of the COF.

When temperatures are within a range from 500 to 600°C, the influence of λ on f_{adh} is not significant, and the maximum values of f_{adh} are typical for coatings with the maximum (302 nm) and minimum (10 nm) values of λ.

At temperatures of 700–900°C, the influence of λ on fadh enhances noticeably. There is a significant decline in the value of f_{adh} with a decrease in λ from 53 to

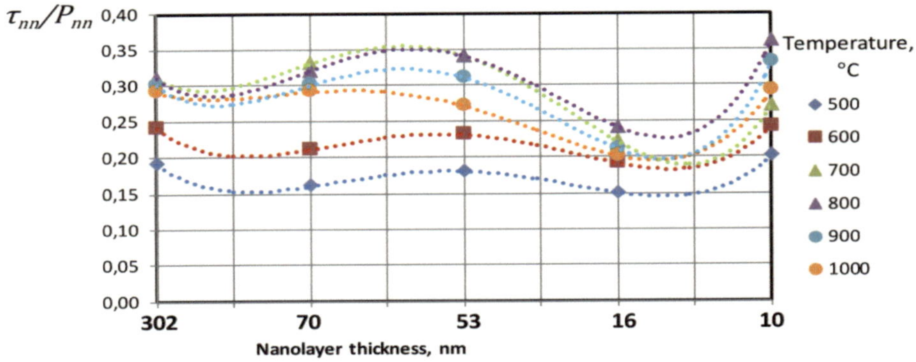

Figure 7.
Relationship between the adhesion component f_{adh} *of the COF and the nanolayer period* λ *at different temperatures [4].*

16 nm, and equally significant growth is detected with a further decrease in λ from 16 to 10 nm. At a temperature of 1000°C, a decrease in λ from 302 to 16 nm leads to a continuous decrease in f_{adh} and then to a noticeable increase at λ = 10 nm.

The specified temperature ranges can be related to the temperature in the cutting area under various conditions of machining. In particular, temperatures within a range from 500 to 800°C are usually detected at relatively low cutting speeds v_c, when the mechanisms of adhesive and abrasive wear play a key role, while oxidation and diffusion processes are relatively weakly expressed [33–35]. A range of temperatures within 800–900°C is typical for significantly high cutting speeds, when oxidation and diffusion processes begin to play an important role, and the destruction of the external layers of the coating begins due to spinodal decomposition [11, 36–38]. At the same time, such elevated temperatures trigger the formation of protective oxide films, which have a positive influence on the tribological parameters of the cutting process. A range of temperatures within 900–1000°C is typical during dry cutting at the highest possible cutting speed. At the temperatures within the above range, the oxidation and diffusion processes prevail, and a coating can fail as a result of active oxidation and spinodal decomposition [11, 33–35].

Thus, at relatively low cutting speeds, the best cutting properties can be demonstrated by a tool with the II-d coating, characterized by the minimal tendency to adhesion to the material being machined in combination with the significantly high hardness, which provides good resistance to both adhesive and abrasive wear. During the cutting at high cutting speeds (and, accordingly, elevated temperatures in the cutting area), the best cutting properties can be expected from a cutting tool with the II-e coating, characterized by better resistance to oxidation and propagation of spinodal decomposition due to the maximum number of interlayer interfaces that restrain the indicated phenomena. The II-e coating is also characterized by the highest hardness. Meanwhile, a significantly high value of the adhesion component of the COF for the II-e coating can even play a positive role in the turning of difficult-to-cut materials. This effect can be associated with a decrease in the contact stresses due to an increase in the contact area of the chips with the rake face of the tool at a significantly lower intensity of the increase in the normal forces, acting on the contact area of the rake face, which in turn reduces cracking and brittle fracture in the coated tool [39]. The above assumptions were confirmed by the conducted cutting tests [3, 4]. The processes occurring in the surface layers of the coatings under the simultaneous action of elevated temperature, oxidation, and diffusion were considered in detail using the example of coating I-a – Ti-TiN-(Ti, Cr,Al,Si)N (**Figure 8**) [39]. The formation of a layer with signs of active oxidation

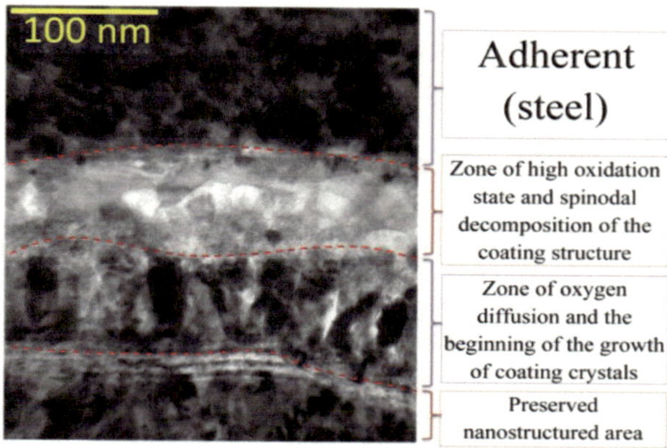

Figure 8.
Structure of the surface layers in I-a – Ti-TiN-(Ti,Cr,Al,Si)N coating at the boundary with the steel adherent (TEM) [39].

can be noticed in the interface area between the steel adherent and the coating. This layer is also characterized by the phenomenon of spinodal decomposition and active diffusion of Fe. The nanolayer structure of the coating was completely destroyed in this area. Below the mentioned layer there is a layer characterized by active growth of coating grains under the influence of the elevated temperature. This layer also contains signs of diffusion of O and Fe, but in much smaller volumes. Finally, below there is an area of the coating with the preserved nanostructure. The area demonstrates no presence of diffused Fe and O.

The study focused on the distribution of chemical elements in the area of the "coating – steel adherent" finds (see **Figure 9**) the presence of diffusing Fe in the surface layers of the coating at a depth not exceeding 200 nm and diffusing O at a depth not exceeding 300 nm from the coating surface. There is also a diffusion of Ti from the coating into the steel adherent to a depth not exceeding 100 nm. Thus, under the influence of such factors as temperature, oxidation, and diffusion of Fe, a

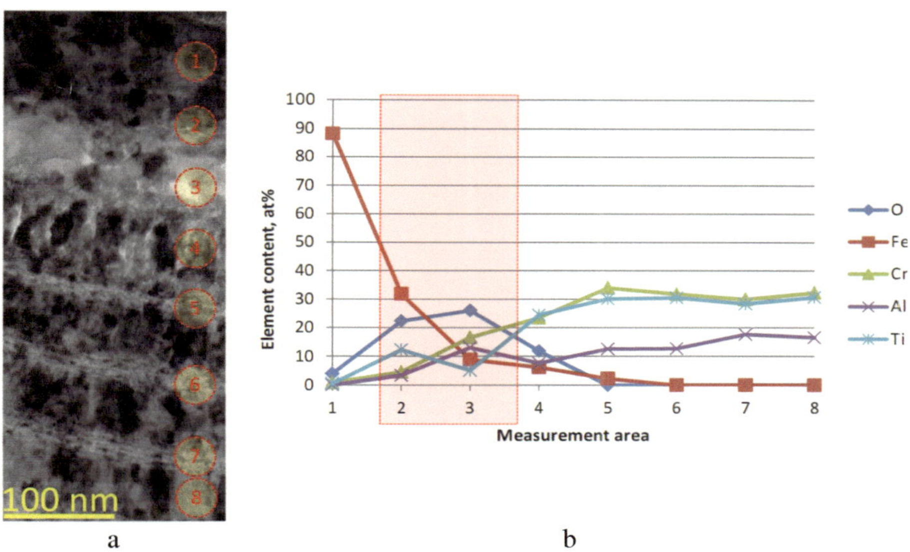

Figure 9.
Study of the oxidation layer. (a) Diagram of research areas (TEM), and (b) distribution of chemical elements by areas.

surface layer with new special properties, influencing, among other phenomena, a decrease in the value of fadh is being formed in the coating.

4. Conclusions

The processes that take place in the cutting area are very complex and difficult for modeling. Till present, there is no any ideal model (both mathematical and mechanical), which would take into account all the factors involved in the cutting process (stresses, temperature, diffusion and chemical processes, etc.), and such a model is unlikely to be created in the near future. Accordingly, an entirely adequate assessment of the working efficiency of a coating can be made only in the course of cutting tests. At the same time, there are several techniques, including those considered in this chapter, which make it possible to predict the performance properties of the coatings with a fairly high probability.

The conducted experiments found the following:

1. For the coatings under study, the adhesion (molecular) component f_{adh} of the coefficient of friction (COF) first grows with an increase in temperature and then begins to decrease noticeably with a further increase in temperature.

2. The temperature at which fadh begins to decrease depends not only on the coating material, but also on the counterbody material. In particular, for AISI 321 steel, the above temperature is 400–450°C, while for AISI 1045 and S31600 steels, it stays within a range of 800–850°C.

3. The samples with the Ti-TiN-(Ti,Cr,Al,Si)N coating showed the smallest value of f_{adh} in the temperature range under study (despite the fact that at room temperature, this value differed little from the data of other samples). The smallest change in f_{adh} at varying temperature was also detected for the samples with the Ti-TiN-(Ti,Cr,Al,Si)N coating. Meanwhile, the presence of Si in the coating composition does not have a noticeable influence on f_{adh}, because the Ti-TiN-(Ti,Al,Cr)N coating demonstrated a result similar to the results obtained during the study focused on the Ti-TiN-(Ti,Cr,Al,Si)N coating.

4. The experiments detected the influence of the nanolayer period λ of the Ti-TiN-(Ti,Al,Cr)N coating on its tribological properties. The tests found that for the coatings with all studied values of λ, an increase in temperature first caused an increase and then a decrease in f_{adh}.

5. Three temperature ranges, characterized by different influence of λ on f_{adh}, are detected. In the temperature range of 500–600°C, the influence of λ on f_{adh} is insignificant. In the temperature range of 700–800°C, the influence of λ on f_{adh} grows noticeably, and there is a gradual increase of f_{adh} with a decrease in λ from 302 to 53 nm, then noticeable decrease in f_{adh} follows a decrease in λ from 53 to 16 nm, and, again, there is a noticeable increase in fadh with a further decrease in λ from 16 to 10 nm. At temperatures of 900–1000°C, there is an almost continuous decrease in f_{adh} with a decrease in λ from 302 to 16 nm, and then a noticeable increase in f_{adh} with a further decrease in λ from 16 to 10 nm.

6. Under the simultaneous action of elevated temperature, oxidation processes, and diffusion of Fe from the steel counterbody, a surface layer with new properties is being formed in the coating, with a positive effect on the decrease in f_{adh}.

Acknowledgements

This study was supported by a grant of the Russian Science Foundation [Agreement No. 18–19–00312 dated 20 April 2018].

Conflict of interest

The authors declare no conflict of interest.

Author details

Alexey Vereschaka[1*], Sergey Grigoriev[2], Vladimir Tabakov[3], Mars Migranov[4], Nikolay Sitnikov[5], Filipp Milovich[6], Nikolay Andreev[6] and Caterine Sotova[2]

1 IDTI RAS, Moscow, Russia

2 Moscow State Technological University STANKIN, Moscow, Russia

3 Ulyanovsk State Technical University, Ulyanovsk, Russia

4 Ufa State Aviation Technical University, Ufa, Russia

5 National Research Nuclear University MEPhI, Moscow, Russia

6 National University of Science and Technology MISiS, Moscow, Russia

*Address all correspondence to: dr.a.veres@yandex.ru

References

[1] Navinsek B, Panjan P. Oxidation resistance of PVD Cr, Cr–N and Cr–N–O hard coatings. Surf Coat Technol 1993;59:244–8

[2] Milosev I, Strehblow H-H, Navinsek B. XPS in the study of high-temperature oxidation of CrN and TiN hard coatings. Surf Coat Technol 1995; 74–5:897–902.

[3] Vereschaka, A., Aksenenko, A., Sitnikov, N., Migranov, M., Shevchenko, S., Sotova, C., Batako, A., Andreev, N. Effect of adhesion and tribological properties of modified composite nano-structured multi-layer nitride coatings on WC-Co tools life. Tribology International 2018;128:313–327

[4] Vereschaka, A., Grigoriev, S., Tabakov, V., Migranov, M., Sitnikov, N., Milovich, F., Andreev, N. Influence of the nanostructure of Ti-TiN-(Ti,Al, Cr)N multilayer composite coating on tribological properties and cutting tool life. Tribology International 2020;150: 106388

[5] Vereshchaka, A.A., Vereshchaka, A. S., Mgaloblishvili, O., Morgan, M.N., Batako, A.D. Nano-scale multilayered-composite coatings for the cutting tools. International Journal of Advanced Manufacturing Technology 2014;72(1–4):303–317

[6] Hovsepian P.E., Ehiasarian A.P., Deeming A., Schimpf C. Novel TiAlCN/VCN nanoscale multilayer PVD coatings deposited by the combined high-power impulse magnetron sputtering/ unbalanced magnetron sputtering (HIPIMS/UBM) technology. Vacuum 2008;82:1312–1317

[7] Vereschaka A.A., Tabakov V., Grigoriev S., Aksenenko A., Sitnikov N., Oganyan G., Seleznev A., Shevchenko S. Effect of adhesion and the wear-resistant layer thickness ratio on mechanical and performance properties of ZrN – (Zr,Al,Si)N coatings. Surf. Coatings Technol. 2019;357:218–234

[8] Zhang Z.G., Rapaud O., Allain N., Mercs D., Baraket M., Dong C., Coddet C. Microstructures and tribological properties of CrN/ZrN nanoscale multilayer coatings. Applied Surf. Sci. 2009;255:4020–4026

[9] Bouzakis K.D., Michailidis N., Skordaris G., Bouzakis E., Biermann D., M'Saoubi R. Cutting with coated tools: Coating technologies, characterization methods and performance optimization, CIRP Ann. Manuf. Technol. 2012;61: 703–723

[10] Zhang S., Sun D., Yongqing F., Hejun D. Recent advances of superhard nanocomposite coatings: A review, Surf. Coatings Technol. 2003;167:113–119.

[11] Vereschaka, A., Tabakov, V., Grigoriev, S., Sitnikov, N., Oganyan, G., Andreev, N., Milovich, F. Investigation of wear dynamics for cutting tools with multilayer composite nanostructured coatings in turning constructional steel. Wear 2019;420–421:17–37

[12] Vereschaka, A.A., Grigoriev, S.N., Sitnikov, N.N., Batako, A.D. Delamination and longitudinal cracking in multi-layered composite nano-structured coatings and their influence on cutting tool life. Wear 2017;390–391: 209–219

[13] Skordaris G., Bouzakis K.-D., Charalampous P., Bouzakis E., Paraskevopoulou R., Lemmer O., Bolz S. Brittleness and fatigue effect of mono- and multi-layer PVD films on the cutting performance of coated cemented carbide inserts, CIRP Ann. Manuf. Technol. 2014;63:93–96.

[14] Araujo J.A., Araujo G.M., Souza R. M., Tschiptschin A.P. Effect of

periodicity on hardness and scratch resistance of CrN/NbN nanoscale multilayer coating deposited by cathodic arc technique. Wear 2015; 330–331:469–477.

[15] Kamath G., Ehiasarian A.P., Purandare Y., Hovsepian P.E. Tribological and oxidation behaviour of TiAlCN/VCN nanoscale multilayer coating deposited by the combined HIPIMS/(HIPIMS-UBM) technique. Surf. Coatings Technol. 2011;205: 2823–2829.

[16] Hovsepian P.E., Ehiasarian A.P., Braun R., Walker J., Du H. Novel CrAlYN/CrN nanoscale multilayer PVD coatings produced by the combined high power impulse magnetron sputtering/unbalanced magnetron sputtering technique for environmental protection of γ-TiAl alloys. Surf. Coatings Technol. 2010;204:2702–2708.

[17] Al-Bukhaiti M.A., Al-Hatab K.A., Tillmann W., Hoffmann F., Sprute T. Tribological and mechanical properties of Ti/TiAlN/TiAlCN nanoscale multilayer PVD coatings deposited on AISI H11 hot work tool steel. Applied Surf. Sci. 2014;318:180–190.

[18] Wang M., Toihara T., Sakurai M., Kurosaka W., Miyake S. Surface morphology and tribological properties of DC sputtered nanoscale multilayered TiAlN/CNx coatings. Tribology Int. 2014;73:36–46.

[19] Coatings Specifications – Coating Guide [Internet], http://www.platit.com/coatings/coating-specifications

[20] Hovsepian P.E., Lewis D.B., Luo Q., Munz W.-D., Mayrhofer P.H., Mitterer C., Zhou Z., Rainforth W.M., TiAlN based nanoscale multilayer coatings designed to adapt their tribological properties at elevated temperatures. Thin Solid Films 2005; 485:160–168

[21] Mo J.L., Zhu M.H., Lei B., Leng Y. X., Huang N., Comparison of tribological behaviours of AlCrN and TiAlN coatings — Deposited by physical vapor deposition, Wear 2007;263:1423–1429

[22] Wang L., Zhang G., Wood R.J.K., Wang S.C., Xue Q., Fabrication of CrAlN nanocomposite films with high hardness and excellent anti-wear performance for gear application, Surface & Coatings Technology 2010; 204: 3517–3524.

[23] Nohava J., Dessarzin P., Karvankova P., Morstein M., Characterization of tribological behavior and wear mechanisms of novel oxynitride PVD coatings designed for applications at high temperatures, Tribology International 2015;81: 231–239.

[24] Bao M., Xu X., Zhang H., Liu X., Tian L., Zeng Z., Song Y., Tribological behavior at elevated temperature of multilayer TiCN/TiC/TiN hard coatings produced by chemical vapor deposition, Thin Solid Films 2011;520:833–836.

[25] Vereschaka A, Tabakov V, Grigoriev S, Sitnikov N, Milovich F, Andreev N, Sotova C, Kutina N. Investigation of the influence of the thickness of nanolayers in wear-resistant layers of Ti-TiN-(Ti,Cr,Al)N coating on destruction in the cutting and wear of carbide cutting tools. Surface & Coatings Technology 2020;385:125402; doi:10.1016/j.surfcoat.2020.125402

[26] Sobol'O.V., AndreevA.A., GrigorievS.N., VolosovaM.A., Gorban'V.F. Vacuum-arc multilayer nanostructured TiN/Ti coatings: structure, stress state, properties. Metal Science and Heat Treatment 2012;54(1–2):28–33

[27] Metel A., Grigoriev S., Melnik Y., Panin V., PrudnikovV. Cutting Tools Nitriding in Plasma Produced by a Fast

Neutral Molecule Beam Japanese Journal of Applied Physics. 2011;50(8):08JG04,

[28] Fominski V.Yu., Grigoriev S.N., CelisJ.P., Romanov R.I., OshurkoV.B. Structure and mechanical properties of W–Se–C/diamond-like carbon and W–Se/diamond-like carbon bi-layer coatings prepared by pulsed laser deposition. 2012;520(21):6476–6483

[29] Fominski V.Yu., Grigoriev S.N., Gnedovets A.G., Romanov R.I. Pulsed laser deposition of composite Mo-Se-Ni-C coatings using standard and shadow mask configuration, Surface and Coatings Technology, 2012;206 (24): 5046–5054

[30] Shuster L.S. Device for investigating adhesion interaction. Patent of Russia 34249 26/03/2003.

[31] Shuster L.S. Adhesive interaction of the cutting tool with the material being processed. Mashinostroenije, Moscow, 1988.

[32] Vereschaka, A.A., Bublikov, J.I., Sitnikov, N.N., Oganyan, G.V., Sotova, C.S. Influence of nanolayer thickness on the performance properties of multilayer composite nano-structured modified coatings for metal-cutting tools. International Journal of Advanced Manufacturing Technology 2018;95(5–8): 2625–2640

[33] Loladze T.N., Nature of brittle failure of cutting tool. Ann CIRP, 1975; 24(1):13–16.

[34] Boothroyd G., Knight W.A. Fundamentals of machining and machine tools, CRC Press, Boca Raton, 2006.

[35] Shaw M.C. Metal Cutting Principles, Clarendon Press, Oxford, 1989.

[36] Choi P.-P., Povstugar I., Ahn J.-P., Kostka A., Raabe D., Thermal stability of TiAIN/CrN multilayer coatings studied by atom probe tomography. Ultramicroscopy. 2011;111(6):518–523.

[37] Povstugar I., Choi P.-P., Tytko D., Ahn J.-P., Raabe D., Interface-directed spinodal decomposition in TiAlN/CrN multilayer hard coatings studied by atom probe tomography. ActaMaterialia 2013;61:7534–7542.

[38] Barshilia H.C., Prakash M.S., Jain A., Rajam K.S., Structure, hardness and thermal stability of TiAlN and nanolayered TiAlN/CrN multilayer films, Vacuum. 2005;77(2):169–179 DOI: 10.1016/j.vacuum.2004.08.020

[39] Vereschaka A, Tabakov V, Grigoriev S, Sitnikov N, Milovich F, Andreev N, Bublikov J. Investigation of wear mechanisms for the rake face of a cutting tool with a multilayer composite nanostructured Cr-CrN-(Ti,Cr,Al,Si)N coating in high-speed steel turning. Wear 2019;438–439:203069 DOI 10.1016/j.wear.2019.203069

Section 2

Friction and Lubrication

Chapter 9

Lubricant and Lubricant Additives

Debashis Puhan

Abstract

Lubricants have been used by humans for thousands of years in their simple machines such as wheel-axle bearings and sledges. Modern machines are much more complicated and are composed of many different machine elements which are in relative motion under varying loads, speeds and temperatures. Industrial lubricants are significant for all kinds of industries whether machine building, chemical, textile, wood, food-processing, automotive, or wind power. Today's lubricants have evolved to a complex mixture of chemical structures that ensure not only lower friction but also provide various other functionality such as lower wear, improved heat transfer, sealing, as well as control of soot, impurities, sludge and deposit formation in the mechanical equipment. Lubricant research and development has become indispensable in automotive engines and drive trains as these have been rapidly advancing towards smaller sizes, increased power, better fuel economy and lesser emissions. Development of lubricant additives and lubricant formulation has led to extended service intervals, enhanced fuel efficiency and improved machine durability. Future trends of lubricant development and use in the Industry 4.0 era and rise of electric vehicles look promising where several stakeholders already have taken their first steps.

Keywords: lubricant, lubrication, lubricant additives, base oil, greases, solid lubricant

1. Introduction

Lubricants have been in use for hundreds of centuries and are essential to our survival. Natural lubricants such as saliva and synovial fluid lubricate the food for easy mastication and reduce wear and tear of our joints respectively. Cooking oils prevent sticking of food onto frying pans and baking trays at the same time as conducting heat. Ancient Egyptians used lubricants to slide large stone blocks for building the great pyramids while the Romans used lubricant on the axles of their chariots [1]. Ancient lubricants were plant and animal based natural oils. With the onset of industrial revolution and our reliance on metal-based machinery and engines, petroleum-based lubricants witnessed a growth.

Modern lubricants are far more complex and perform various other functions in addition to lubricating such as cleaning, cooling, and sealing. The primary function of most lubricants is to reduce friction and this property is known as lubricity. A lubricant can be used in solid form, semi-solid, liquid form or gaseous form. Examples of solid lubricants are graphite and Molybdenum disulphide (MoS_2), semi-solid lubricants are greases, and liquid are automobile engine oil. Depending on the requirements of a said application, the physical state of lubricant is chosen. For example, in space environments where liquid lubrication is not feasible due to vacuum, solid lubricants are chosen. Air bearing are preferred in applications in machine tool applications where precision is of primary importance such as cutting and finishing of optical lenses. Greases are used where a liquid oil would not remain in position due to its tendency to flow or when a sealing action is needed to prevent water-ingress in addition to lubrication. Today's lubricants are designed and packaged to meet specific requirements for specific applications by lubricant formulators. The lubricant for automobile transmission and drive train has different requirements to satisfy compared to lubricant for an internal combustion engine or turbines. Further depending upon the type of turbines viz. gas, steam or hydraulic, the lubricant needs to be designed.

2. Lubricant composition

Typically, industrial lubricants contain 70-90% base oils and the rest is additives [2]. Base oils impart primary vital properties of the lubricant such as viscosity, viscosity stability, thermal stability, solvency, low temperature flow and volatility, oxidation stability. Additives have been used in lubricating oil since the 1920s and the demand for lubrication has resulted in continuous growth in the size of the market (USD 14.35 billion in 2015) with huge investments in research and development to design and formulate superior lubricants that meets present and future environmental regulations and consumer expectations. Despite this, lubricant formulation has mostly remained an art. This is because blending a new formulation for optimizing viscosity and obtaining optimum performance through performance tests for friction and emissions is much easier than testing for parameters, such as impact on engine wear, sludge build-up and piston cleanliness which require long duration engine tests. For example, during the development of Castrol's engine oil, 'Edge with Titanium Fluid Strength Technology', over 2400 unique formulations were engine-tested for an equivalent of 1.9 million miles.

To understand the need for additives, one must understand the implication of the Stribeck curve shown in **Figure 1**. Machine elements such as engine bearings work in hydrodynamic lubrication regime where the major function of the lubricant is to maintain its viscosity at all temperatures while ensuring a thick fluid film to keep the two contacting surfaces in relative motion separated at all loads and speeds. Rolling element bearings work on elasto-hydrodynamic lubrication (EHL) where the contacting surfaces deform elastically and there is a very thin film separation. Cams and tappets work in the boundary lubrication regime where there is substantial metal to metal contact. And in the reciprocating motion of engine piston rings, all four kinds of lubrication regime occur.

Classical lubrication theory assumes that a lubricating oil is a Newtonian fluid with a fixed viscosity and the contacting surfaces to be rigid. George Osborne Reynolds approached the fluid film hydrodynamic lubrication using mathematical and physical approach (1886) to predict friction, film thickness and load carrying capacity [3]. In the real world, oils undergo shear thinning and behave as a non-Newtonian fluid due to heat and pressure developed at the contact. Furthermore, real surfaces are rough and can undergo elastic as well as local plastic deformation under fluid pressure. The EHL theory was developed by H.M. Martin (in 1916) and later by Ertel (in 1939) and Grubin (in 1949), Petrusevich (in 1951), Dowson and Higginson (in 1959), Dowson and Hamrock (in 1977) for predicting traction, load carrying capacity and film thickness in heavily loaded contacts [4]. The non-Newtonian behavior of thin films under high pressure and lower rolling speeds has guided lubricant formulators to consider the shear-stress/shear-strain behavior, pressure-viscosity dependence of the lubricant as these are closely linked to the molecular properties of the oil composition. On the

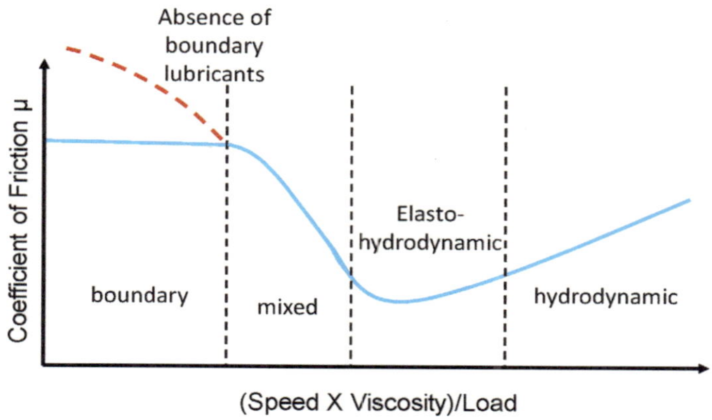

Figure 1.
Stribeck curve.

Stribeck curve, when the speed is very low, the friction is governed by the chemistry of the lubricant molecules i.e. the molecular structure and orientation. It was W. B. Hardy (in 1920), who coined the term boundary lubrication and later with Ida Doubleday (in 1922) established the basic concepts of boundary lubrication theory [5]. The boundary lubrication properties such as friction and film thickness at the interface of two surfaces are affected by the force fields of molecules in relation to their structure and polarity. Hence, formulators add boundary film additives to reduce friction in this regime [6]. Additives must also reduce wear, as in this regime, the two surfaces in relative motion are in contact hence prone to significant wear. Furthermore, additives need to counter the side effects of continuous as well as intermittent use of the lubricant such as they need to control heat, deposit formation, prevent foam formation, prevent fouling and corrosion due to water ingress, prevent wear, increase film strength under concentrated contacts, control greenhouse gas emissions. Therefore, as high as 30% content of modern automotive lubricants are chemical additives while some industrial oils may only contain 1% or less.

2.1 Base oil

A typical petroleum-based oil with no additive is called a base oil or base stock [7]. The generation of base stock starts with the identification and selection of a good petroleum crude followed by atmospheric distillation, vacuum distillation and solvent processing. Solvent processing has two processes viz. solvent extraction, and solvent dewaxing where undesirable molecules are separated where as in hydroprocessing undesirable molecules are converted to desirable ones. Hydroprocessing a general term for conversion of less desirable crude fractions into good quality feedstock using catalyst and hydrogen at high temperature and pressure. This can be categorized as hydrofinishing, hydrotreating and hydrocracking according to increasing order of severity. Another process called hydrodewaxing is required to remove long chain linear paraffins, isomerize straight chains to branched chains which improves the pour point. Only 10 % of crudes are converted to base stocks for lubricants, hence the refinery owners call the shots regarding the choice of crudes to be converted to base oils while balancing the cost, yield and demand relative to all the other refinery products. However, to satisfy lubricant performance demands under severe operating conditions, high quality base stocks are needed. The lubricant manufactures (refinery owner can also be lube manufacturer) buy these base stocks and other chemical compounds and formulate their lubricant for example SAE15W40 or ILSAC GF-5, meeting standards set by Original Equipment Manufacturers (OEMs), professional bodies, and international institutions like American Petroleum Institute (API), International Lubricant Standardization and Approval Committee (ILSAC), Society of Automotive Engineers (SAE) to name a few.

Owing their origin to petroleum crudes, lubricant base stocks are also mixtures of long chain hydrocarbons containing three types of chemical groups. i.e. paraffins, naphthenes and aromatics. The paraffins can be further classified into branched or straight chains. The chain length and branching affects the melting point and crystallization temperature of the paraffins. During the production of lube base stock, most of the unsaturated bonds, paraffin wax and sulfur content are removed, however depending on the severity of hydroprocessing, some wax, unsaturates and sulfur may remain. Base Stocks have been classified into 5 categories by the API according to the presence of saturates, sulfur content and viscosity as shown in **Table 1**. Group I, II and III are derived from petroleum crude while Group IV is reserved for Polyalphaolefins (PAO) which are synthesized from gaseous hydrocarbons. Group V is for all other base stocks that are not included in other four groups such as mineral based napthenics, synthetic esters, polyglycols, silicones, polybutenes, phosphate esters etc. These oils are designed for severe performance requirements.

2.1.1 Application based on properties of base oil

Group I base oils are processed by solvent processing in which wax and multiring aromatics are removed while some sulfur and aromatics remain. These base oils are typically used in marine and diesel engine oils, heat transfer oils, hydraulic oils, conventional greases, industrial gear oils, and machine tool oils. These oils have high solvency, very high viscosity but poor viscosity index. The maximum operating temperature is 93°C.

Group II base stocks have undergone catalytic conversion to remove wax, aromatics and sulfur compounds. These base oils are used as automotive engine oil, automatic transmission fluid, gear oils and turbine oils. The maximum operating temperature of these base stocks is 121°C.

Group III base oils are similar to Group II oils but have a higher viscosity index. The severity of hydroprocessing removes any ring structures, as a result, its solvency is poor. These base oils are used in premium passenger vehicles, automatic transmission fluids, and food grade lubricants. The maximum operating temperature is 121°C.

Group IV oils are synthetically produced Polyalphaolefins from low molecular weight organic material. They have uniform molecular structures. They provide excellent high temperature performance and oxidation stability and therefore used in high performance engine and gear applications, heavy duty industrial

	Type	Saturates (%)	Sulfur (%)	Viscosity index
Group I	Solvent refined	<90	>0.03	80 to <120
Group II	Hydrofinished	≥90	≤0.03	80 to <120
Group III	Hydrocracked	≥90	≤0.03	≥120
Group IV	Polyalphaolefins	—	—	—
Group V	Synthetics and naphthenics and other stocks not included in Group I, II, III or IV	—	—	—

Table 1.
API base stock classification.

compressor, power transmission fluid, hydraulic fluid, heat transfer fluid and in bearings as greases or liquid or lubricants. A wide range of viscosity grades of PAOs can be produced by varying the number of olefin molecules linked together. PAOs have a maximum operating temperature of 132°C.

Group V base stocks can be either naphthenic or synthetic in origin. The naphthenics provide very good low temperature performance and hence are used in applications which operate in a narrow temperature range such as transformer oil, process oils, grease. These oils have high solubility and available in a wide range of viscosities. The synthetics on the other side vary widely in their types and properties. These include polyglycols, silicones, polybutenes, organic esters, phosphate esters and they are used as compressor oils, brake fluids, heat transfer oils, aviation engine oil. Their maximum operating temperature is dependent on the nature of their molecular structure. Silicon oils are known to have a maximum operating temperature of 232°C.

Finally, Group II+ and Group III+ are two types of base stocks that are not included in the original API classification. The plus refers to increase in viscosity index (VI) in the higher limit of the API specification. Group II+ has VI minimum between 110 and 115 and Group III+ has minimum VI somewhere between 130 and 140.

As mentioned previously, mineral oils are mixtures of hydrocarbons containing paraffins, naphthenics and aromatics of various structures and carbon numbers ranging between 20 to 40+, at varying amounts depending on their degree of refining and processing. Typical composition of various API groups have been identified, but their exact structures are still not known. There is a considerable variability in the performance among the mineral origin base oils owing to the source of the crude. Therefore, lubricant manufactures conduct performance tests on their formulations to ascertain satisfactory performance with any new base stock in the same API category.

2.2 Lubricant additives

The primary function of Lubricant additives is to improve the properties of the base stock under different operating conditions and the high performance requirements of any machinery. Lubricant additives are chemical components that need to blend well with the base oil to function as a single fluid. Additive manufactures often sell several additives combined into an additive package and diluted with a base oil at a higher concentration. The additive package is then dosed into the lubricant blend by the lubricant manufacturer at an appropriate treat rate to give the desired performance. The concentration of various additives is constrained by various factors such as their primary function (for example dispersancy, wear protection and so on), their synergetic or antagonistic behaviour with other additives, and regulations set by industry bodies.

In a nutshell, lubricant additives can be categorized into various kinds based on their general roles of performance improvement and service life extension. First category are additives that impart new properties to the lubricant also known as surface protective additives. Examples include antiwear additives, extreme pressure additives, corrosion inhibitors, detergents and dispersants. The second kind of additives enhance the existing properties already present in the lubricant hence known as performance additives. Viscosity index improvers, viscosity modifiers, friction modifiers, pour point depressants belong to this type. The third type of additives known as lubricant protective additives, are the ones that counteract the negative effects or changes that take place during the service life of the lubricant. These include antifoamants and, antioxidants. In this chapter only the major types of lubricant additives are discussed. Other additives such as demulsifiers, emulsifier, biocides may be added depending on the intended applications.

2.2.1 Types of lubricant additives

2.2.1.1 Pour point depressants

As the name indicates, these are additives that reduce the pour point of the lubricant, i.e. the lubricant remains in liquid state and maintains its fluidity (pourability) at lower temperatures than without these additives. Usually as temperature decreases, paraffin molecules in the oil start to crystallize as wax (below 50°C) and the oil loses its ability to flow by gravity or to be pumped under pressure. This also affects the viscosity of the oil. Additives such as alkylaromatic polymers and polymethacrylates prevent wax crystal growth by modifying the interface between the wax and the oil molecules, to a certain extent thus lowering the pour point by about 20–30°F (11–17°C). These are present up to a fraction of a percent in all paraffin-based lubricants that lubricate machine elements such as bearings, gears exposed to cold start and cold (winter) operating temperatures. Modern multi-grade engine oils/motor oils composed of partly synthetic oil and partly mineral oil along with these additives, have pourpoints as low as -32°C.

2.2.1.2 Viscosity index improvers

Viscosity index improvers (VII) also known as viscosity modifiers are additives that prevent the oil from losing its viscosity at high temperatures which is a natural tendency of any liquid. These additives are available in all shapes and sizes and quality [8]. Polymethylmethacrylates, olefin copolymers, hydrogenated poly(styrene-co-butadiene or isoprene), esterified polystyrene-co-maleic anhydride are commonly used VIIs. The large oil soluble flexible polymer molecules uncoil and spread out as temperature increases thereby increasing the viscosity as shown in **Figure 2**. Furthermore, their numerous branches entangle with those of other neighboring molecules. By doing this, these macromolecular structures can trap and control smaller oil molecules, thus increasing the viscosity of the lubricant.

Permanent and temporary shear thinning of VII-thickened formulations can also occur depending upon the quality of the VII. In heavy duty application, due to the large compressive pressure between the two mating surfaces, VII polymer molecules, tend to align with each other and get "squashed" or even get chopped to small pieces under high shear conditions. When the polymer coils elongate and become aligned in the direction of the flow the viscosity temporarily drops resulting in reduced oil film thickness. After the lubricant leaves the contact between

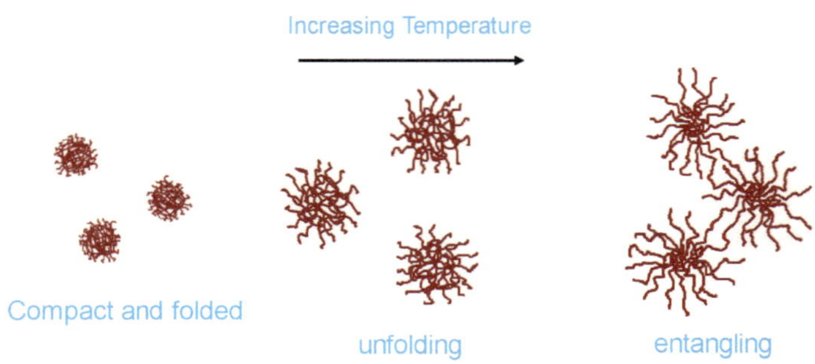

Figure 2.
Mechanism of VII.

the mating parts, the polymer coils return to their original shape and the viscosity of the lubricant returns to normal. This phenomenon is referred to as temporary shear-thinning. However, under further high shear rates, the long and flexible polymer chains can be cut or ruptured or pulled and ripped apart into smaller chains by molecular scission. Unfortunately, once this has occurred, the broken polymer chains cannot re-form into the single large chain and this causes the oil to permanently lose viscosity leading to a reduction in oil film thickness, oil film failure and an increase in wear. This phenomenon is referred to as permanent shear-thinning.

2.2.1.3 Anti-wear agent

Anti-wear additives are additives that prevent two-body wear of the metallic countersurfaces in the boundary lubrication regime where the film thickness is small and there is asperity - asperity contact. These additives are polar in nature which enables them to attach to the metallic surfaces followed by tribochemical or mechanochemical reactions to form an anti-wear film. This newly formed film undergoes wear and formation at the top layers thus protecting the underlying metallic surface. As these additives form films by chemical reactions, they get used up and the amount of antiwear additives present in the lubricant reduces with time. These are typically phosphorous compounds. Zinc dialkyldithiophosphate (ZDDP) is the most common, the most researched and has been used since the 1940s [9]. Its use has been reduced in passenger vehicles in the last decade due to zinc metal causing poisoning of the catalyst in the exhaust gas catalytic convertor. ZDDP also provide antioxidant and corrosion-inhibition properties to the lubricant. Owing to the multi functionality of ZDDP, finding its replacement has been challenging because molybdenum-based additive such MoDTC (molybdenum dithiocarbamate) or MoDDP (molybdenum dithiophosphate) molecules cannot work as antioxidant. On the other hand, ash-less antiwear additives such as hindered phenols and amines are very expensive and are required in larger quantities. Till date, ZDDP is considered as the most cost-effective antioxidant and antiwear additive available, and the alternatives are currently very expensive.

2.2.1.4 Antioxidants

Antioxidants or oxidation inhibitors prevent the oxidation of the components of the base oil there by increasing the life of the lubricant. Oxidation of the lubricant molecules occur at all temperatures but at higher temperatures, it is accelerated. The presence of wear particles, water, and other contaminants also promote Oxidation of the lubricant molecules which then leads to formation of acids and sludge. The acids may further cause corrosion in the metallic parts while the sludge formation increases the viscosity of the lubricant. Almost every lubricating oil and grease contains antioxidants and examples include Zincdialkyldithiophosphates, hindered phenols, sulphurized phenols, and aromatic amines. These compounds decompose peroxides and terminate free-radical reactions that occur in the lubricant. These are sacrificial in nature hence their quantity gets reduced with time.

2.2.1.5 Defoamants

Defoamants or antifoaming agents are additives that prevent the lubricant from forming a foam and speed up the collapse of the foam if it does form. Foaming occurs because of constant mixing of the oil with air or other gases leading to air entrapment. Foam disrupts cooling of parts as it is not a good conductor of heat. It reduces the load carrying capacity and the lubricant flow leading to excessive engine wear. Silicone polymers such as polymethylsiloxane at a few parts per million and organic copolymers such as alkoxy aliphatic acids, polyalkoxyamines, polyethylene glycols, and branched polyvinyl ethers, at higher concentration are widely used in mineral oils. The antifoaming agents are essentially insoluble in the lubricant hence they need to be finely dispersed in the lubricant. These droplets attach themselves to the entrapped air bubbles and aid in forming bigger bubbles (via coalesce). The larger bubbles rise readily to the surface followed by bursting to release the trapped air. Bursting occurs by thinning of the air bubble film as the additive spreads due to its low surface tension.

2.2.1.6 Friction modifiers

Friction modifiers are used in engine oil and transmission oil to alter the coefficient of friction that would be experienced between the sliding parts when only the base oil is present. Friction reduction results in improved fuel economy. Organic and sulfurised fatty acids, amines, amides, imides, high molecular weight organic phosphorus and phosphoric acid esters are added to the range between 0.1 and 1.5% in finished lubricants as friction modifiers. Glyceryl monooleates and Molybdenum compounds such as MoDTC and MoDTP also function as friction modifiers. They preferentially adsorb very strongly on to the metallic surface. The head of the friction modifier is attracted to the metal surface and the long tail with at least 10 carbon atoms remains solubilized in the oil as shown in **Figure 3** [10]. The chemical structure and the polarity of the molecules play a major role in the friction reduction. Ionic lubricants [11], a class of ionic liquids that are room-temperature molten salts consisting of cations and anions are also very good surface additives. The polarity of head group provides for strong surface adsorption. The physical, chemical and tribological properties of ionic liquids can be tailored to suit a wide variety of applications ranging from its use as polymer brushes in biological application, or as water soluble or oil soluble lubricant additive.

2.2.1.7 Detergents

Detergents keep surfaces free of deposits and neutralize corrosive acids formed due to oxidation. These molecules are chemical bases consisting of a polar substrate and a metal oxide or hydroxide [12]. Metallo-organic compounds of calcium and magnesium phenolates, phosphates, salicylate and sulfonates are recommended. Overbased detergents are used in marine engine lubricants to neutralize large amounts of acidic components produced by fuel combustion or oil oxidation. Ash (burning of organometallic species) and soot particles (largely carbon with sulfur adsorbed) is formed by burning of the oil in internal combustion engines. Ash can then form unwanted residues at high temperatures or simply deposit on surfaces. The deposit precursors particles are insoluble in the oil and have greater affinity for detergent molecules. The additive molecules cling to the surface of the particle and envelop it thereby also acting as dispersants and prevent those particles to agglomerate and to later settle as deposits. A detergent additive is normally used in conjunction with a dispersant additive.

2.2.1.8 Dispersants

Dispersants are used mainly in engine oil along with detergents to keep engines surfaces clean and free of deposits [12]. Dispersants keep the insoluble soot particles and the precursors of deposits in the internal combustion engine finely dispersed or suspended in the lubricant even at high temperatures. These suspended particles are subsequently removed by oil filtration or oil change. Thus, dispersants minimize damage to engine surfaces and formation of high temperature deposits. Generally,

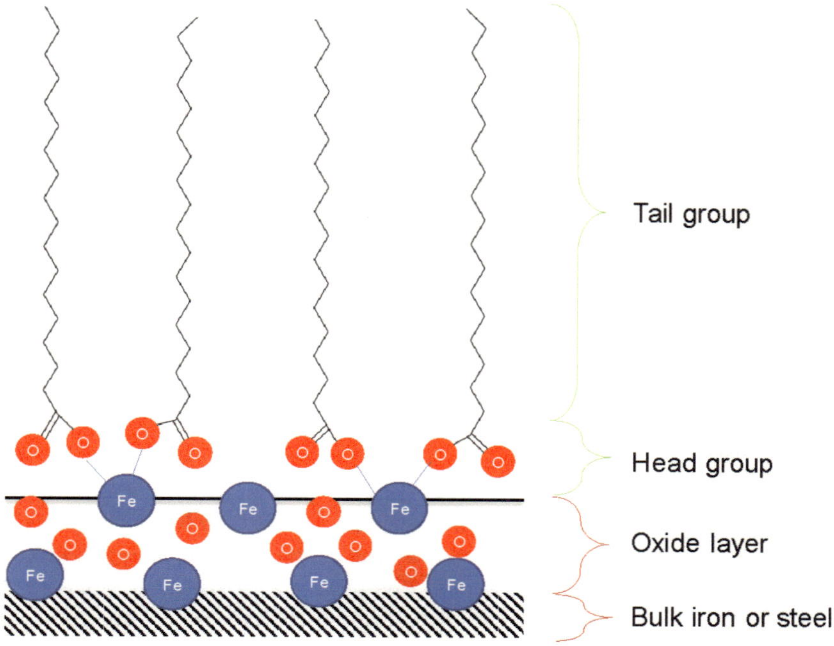

Figure 3.
Adsorption of polar headgroups onto metallic surface.

polymeric and ashless dispersants are used today such as polymeric alkylthiophosphonates, alkylsuccinimides, succinic acid esters/amides, and their borated derivatives as well as organic complexes containing nitrogen compounds.

2.2.1.9 Corrosion and rust inhibitor

Corrosion and rust inhibitors are additives that reduce or eliminate rust (corrosion of iron and steel) and corrosion by neutralizing acids and forming a protective film, either adsorbed or chemically bonded on the metal surfaces. Preferential adsorption of polar constituent on metal surface forms the protective film that prevents corrosive materials such as organic acids from reaching and attacking the metal. These are usually compounds having a high polar attraction towards metal surfaces such as succinates, alkyl earth sulfonates, metal phenolates, fatty acids, amines as well as zincdithiophosphates. Some of these inhibitors are specific to protecting certain metals. Hence, an oil may contain several types of corrosion inhibitors.

2.2.1.10 Extreme pressure (EP) additives

Extreme Pressure additives are required to reduce friction, control wear and prevent severe surface damage in heavy duty application of gears and bearings at high temperatures and pressures. They are also known as antiscuffing additives. They react chemically with metallic surfaces to form a sacrificial surface film that prevents the welding and subsequent seizure of asperities at the metal-to-metal contact. Additionally, they contribute to smoothing of the surfaces as these are formed at contact asperities and the load is then distributed uniformly over a greater contact area, thus reducing the severity of wear and ensuring effective lubrication. Effectiveness of EP additives relies on their reactivity and their ability to readily form thick surface films at high loads and high contact temperatures that are created at the mechanical contacts. These additives usually contain sulfur and phosphorus compounds and chlorine or boron compounds. Ashless EP additives such as dithiocarbamates, dithiophospates, thiolesters, phosphorothioates, thiadiazoles, aminephosphates, phosphites may be preferred in some applications where chlorine may cause corrosion.

Other additives such as demulsifiers, emulsifier, biocides are added to meet specific requirements. Emulsifiers are used as a binder between oil and water molecules in oil-water-based metal-working fluids to help create a stable oil-water emulsion. Without the emulsifier, oil and water will separate out from each other due to differences in specific gravity and interfacial tension. On the other hand, demulsifiers are used to separate oil-water emulsions. Demulsification and removal of the aqueous phase from the oil-based lubricant minimizes harmful effects such as corrosion, foaming and cavitation from occurring. Biocides may be added to water-based lubricants to control the bacterial growth.

2.3 Greases

Lubricating grease are a class of lubricant that do not flow like a fluid but bleed (release oil) when squeezed between contacting surfaces. They have a gel like consistency and can be described as a solid or semifluid like material and in some cases can be used in vertical or overhead applications because they can have good drip resistance due to their Non-Newtonian rheological properties. They are particularly useful in applications that are sealed for life; for example bearings and remote gearboxes. Functionality of greases include sealing out contamination and water ingress, prevent corrosion, compatible with polymers and elastomers, provide antiwear and extreme pressure load protection while reducing friction. Greases have three main components: fluid, thickener and additives [13].

A typical grease consists of 75-95% fluid base stock, 2-25% thickener and 0-25% additives. The base stock is chosen based on the required applications. Hence, the fluid can be petroleum based for most automotive and industrial application, synthetic based for low and high temperature application or may be wax based (no flow) for high load caring capacity. The thickeners are metal soaps which are created using the fundamental reaction of an acid and a base. Calcium based soaps have been the simplest and earliest used thickeners with a maximum operating temperature between 60 and 70°C. In the last decade, calcium sulfonate greases, polyurea greases, aluminum complex greases, lithium complex greases, sodium complex greases, and clay-based greases have received general acceptance due to their higher service temperatures of more than 150°C. However, each having their own set of pros and cons. For example, calcium-based greases do not perform well over a wide range of temperature, sodium based have deteriorated performance in the presence of water. Clay and polyurea based greases are used for high temperature (service temperatures of 190-220°C) and application that have limited relubrication access. The third component of greases are additives. Commonly used additives are listed below.

- Antiwear additives
- Antioxidants
- Extreme Pressure additives
- Friction Modifier
- Rust and corrosion inhibitor

- Tackifiers (adhesive agents)
- Odorants (perfumes)
- Dyes

Tackifiers are additives that increase the adhesive property of the grease or the lubricant. They prevent the lubricant from flinging off the metal surface during rotational movement. To be acceptable to the manufacturers and the end users the greases must be free of offensive odor and have a desirable color. Therefore, odorants and dyes are added to the grease. Although these have little effect on the grease performance, their appeal to senses has an impact on the product selection.

Greases have the unique ability to incorporate liquid as well solid additives. Solid additives such as molybdenum disulphide (MoS_2), graphite, hexaboron nitride and polytetrafluoroethylene (PTFE) in the form of fine dispersed powder (nano and micro particles) have been used in lubricants and greases to provide ultralow friction and wear protection. Solid additives provide a physical separation between two contacting surfaces when fluid is unable to provide load support. The lattice structure of these solid lubricants plays an important role in transferring a thin low shear layer on the metal surfaces especially where load is high, and speed is low.

2.4 Solid lubricants

Solid lubricants on their own are vital to niche applications such as space missions, satellites release and deployment mechanisms. Vacuum and microgravity of space eliminates the use of liquid lubricants. The importance of a thin film of solid lubricant can be emphasized on the fact that the success of an entire mission can be compromised if the, receiver and transmitter antennas or solar arrays packed securely during launch fail to release smoothly with precision due to extreme friction of a deployment mechanism. Solid lubricants can be used at low temperatures as well as at high temperatures where the liquid lubricant may solidify or vaporize respectively. Even at extreme pressures where liquid lubricant will be squeezed out, solid lubricants are used. However, solid lubricants are not limited to extreme conditions [14]. A wide variety of low friction coatings are used in various engineering applications that require high electrical and thermal conductivities, low wear rates and high lubricity at all operating temperatures. Newer engineered coatings have increased complexity and have transitioned from single or multi component structures to nanostructured and functional gradient structures.

2.4.1 PTFE

The most recognized polytetrafluoroethylene (PTFE) coating is Teflon® discovered in 1938 at DuPont. These are highly linear fluorocarbon molecules. They offer a low friction surface with moderate wear. They have low chemical reactivity and low surface energy. PTFE on its own performs best at low loads and wears rapidly at higher loads; hence they need reinforcements to increase strength and load bearing capacity. Such materials are called composite materials.

2.4.2 Carbon based materials

Carbon based materials such as graphite in both micro and nano forms is a popular solid lubricant additive whereas diamondlike carbon (DLC) makes excellent low friction coatings. However, they have different mechanisms of friction reduction. Graphite has hexagonal crystal structure which has the intrinsic property of easy shear. DLC exhibits high hardness and low friction due to an amorphous structure that combines graphitic and diamond phases. They can be doped with hydrogen or nitrogen for achieving desirable properties. Recently, a series of patents on superlubricity of nano-diamonds and graphene films have been filled [15].

2.4.3 MoS_2

Molybdenum disulphide are transition-metal dichalcogenides like Tungsten disulphide (WS_2) work on the mechanism of interlamellar shear between covalently bonded hexagonal basal planes like that of graphite. Their performance is affected by moisture content [16].

Apart from the advantages mentioned earlier, solid lubricants also have a few disadvantages. They have less ability to carry away heat and contaminants away from the contact. They have poor self-healing properties, and they are not easily entrained into tribological contacts.

3. Application methods

After the selection of a suitable lubricant, its application method i.e. its delivery to the mechanical components such as gears, bearings, cams, tappets, chains, guideways and couplings of a machine or engine in correct quantity is considered. Liquid lubricants are applied to the machine elements by two methods. First type is called 'all loss method' while second type is called 'reuse method' [2]. In all loss method a small quantity of lubricant is applied periodically and the lubricant after use, gradually leaks away to waste. In reuse method, elaborate lubricating systems are designed to feed the required quantity of lubricant to various machine elements of the system. The lubricant after leaving the machine components is collected, cooled then filtered and recirculated to lubricate the machine components again. Most open gears, ropes, guideways, chains and rolling element bearings (except sealed for life rolling element bearings) are lubricated by all loss method. Nearly all grease lubricated elements also are lubricated by this method. Some devices used for all loss lubrication method include hand-held oiling device, grease gun, drop feed cups, wick feed cups, wick oilers, pad oilers, mechanical force feed lubricators, airline oilers and automatic or semiautomatic spray units. Reuse method of lubricant application has the oil circulating through a network of pipes that is pressurized by a pump or aided by gravity. However, closed oil sump systems employ splash oiling, bath oiling or ring oiling methods. Centralized lubricant application systems can reduce the quantity of lubricant usage as well as labour costs. Automobiles use an oil mist lubrication system to lubricate various machine elements of the engine and drive train. Considerations with regards to the lubricant characteristics and composition are required while designing a compatible system. Oil condition monitoring and routine checks are vital for the longevity of the machine elements lubricated by any method.

Solid lubricants on the other hand are applied as powder dispersed in a liquid lubricant or in a solid phase matrix. They can be also applied as a thin film coating. The coatings are made by various methods such as dip coating, thermal spraying, and cold. More robust ways of coating includes chemical vapor deposition or physical vapor deposition methods. Electrochemical processes are used for producing coatings of polymer-based solid lubricants and their composites.

Physical and chemical characteristics	Performance test evaluation	Engine test evaluation
Color	Oxidation Tests	Oxidation stability and bearing corrosion protection
Density and API gravity	Thermal Stability	Single cylinder high Temperature tests
Carbon Residue	Foaming Tests	Multi cylinder high temperature tests
Flash point	Corrosion and Rust Protection Test	Multi cylinder low temperature tests
Neutralization Number	EP and Antiwear Test	Rust and corrosion protection tests
Total Acid Number	Emulsion and Demulsibility Test	Oil Consumption rates and volatility
Total Base Number		Emission and protection of emission control systems
Pour Point		Fuel Economy
Sulphated Ash		
Viscosity		
Volatility		

Table 2.
Evaluation techniques and testing.

4. Lubricant analysis

Lubricant analysis primarily refers to the characterization and evaluation of the lubricant for various physical, chemical and performance properties in all stages of its life cycle. For solid lubricants and coatings, the analysis includes determining the composition and structure by using a range of spectroscopic and microscopic observation methods as well as measuring the coating consistency and thickness applying non-destructive evaluation techniques. Standardized tribological lab tests are used to evaluate their performance. Once a candidate material is identified, larger-scale bench testing, such as engine tests, are conducted.

For liquid lubricants, characterization is done to a much larger extent due to the large variety of applications and the implication of chemistry of the formulated oils on the machine elements performance and the overall performance of the entire equipment/machine or engine. The lubricant analysis can be classified as physical and chemical characteristics evaluation, Performance test evaluation and Engine Test evaluation as summarized in **Table 2** [2]. Additionally, end users also develop their in-house in-service lubricant analysis to monitor and maintain the condition of the lubricant. The objective of such analysis is to ensure optimum performance, achieve expected life of the equipment as well as the lubricant flowing through it.

5. Future trends

Current automotive lubricants are optimized for internal combustion engines and drive trains. Electric vehicles (EVs) which use electric motors possess new challenges of lubrication such as high-power density of the small gear box which require efficient cooling. Hydro lubricants and synthetic gear oils are excellent candidates for such application but may pose sealing issues which requires innovative solutions. Lubricants are also required in the rolling element bearings of EVs that must stop electro-erosion caused by high frequency high energy discharges by being conductive. Ionic liquids are having shown good performance in preventing built up of potential [17].

The future for lubricants formulation, manufacturing and end-use is oriented towards efficiency in terms of cost and time, customized and optimized for each individual tribo-system and run reliably for even longer drain interval time. New industrial lubricants must meet stringent regulations and guarantee ecological sustainability, and climate change actions. Hence, new lubricants will contribute to 'Green Tribology'.

Tribology of lubricants plays a significant role in technology and economics of industrial development. As the fourth industrial revolution 'Industry 4.0' (combines automation with the internet of things) is in progress, several concepts can be extended to lubricant and lubricant additive development and evaluation. For example new electronic smart sensors can be used for lubricant analysis and condition monitoring. The information of various performance parameters can be stored as data and transferred to the stakeholders and decision-making points using the information and communication technologies involved in Industry 4.0. such as internet and wireless connections to and via several electronic devices. Continuous monitoring can also aid in corrective measures. In-service lubricant performance testing and evaluation generates big quantities of data from various equipment and sensors. Therefore, 'Big Data' concepts can be applied in several ways. One such example can be combining lubricant data with machine data, another can be correlation study of amount of soot produced, change in oil viscosity, friction, wear of an engine. The benefits of Industry 4.0 include shorter lubricant development time, reduced number of trials, lubricant performance prediction through chemical and physical modeling and simulations. This will transform product development process from being empirically driven to using data and simulation driven approach.

6. Conclusion

Lubricants are vital for the tribological life of the machine elements. As modern machine elements are required to perform in heavy-duty applications in a wide range of environments, newer, better and environmentally sustainable lubricants are required to be designed. Lubricant additives are therefore considered as lubrication engineering design components. Lubricant additives are designed and optimized to meet the performance requirements of the equipment or engine. Various types of lubricant additives and their functional properties were discussed. They are designed to provide oxidation resistance, high temperature viscosity, energy and fuel efficiency among others. Lubricant or grease therefore are a complex mixture of several components blended carefully together to meet the performance requirements. The different components can have synergistic or antagonistic effects due to chemical interactions or competition at the metal surface or among themselves. Therefore, formulation of lubricants requires considerable expertise and expensive performance testing. Green Tribology and Industry 4.0 era will steer the lubricant development, use and disposal.

Acknowledgements

This chapter could not be published without the exceptional support of Prof. Paul B. Davies (University of Cambridge, UK) and substantial industry inputs and scrutiny by Mr. Tony D. Smith (Castrol, UK).

Author details

Debashis Puhan
Department of Chemistry, University of Cambridge, Cambridge, United Kingdom

*Address all correspondence to: dp617@cam.ac.uk

References

[1] Bell JC, Delargy KM, Seeney AM. The Removal of Substrate Material through Thick Zinc Dithiophosphate Anti-Wear Films. In: Dowson D, Godet M, Taylor CM, Childs THC, Dalmaz G, editors. Tribol. Ser., vol. 21, Elsevier; 1992, p. 387-96. https://doi.org/10.1016/S0167-8922(08)70546-7.

[2] Pirro D, Webster M, Daschner E. Lubrication Fundamentals, Third Edition, Revised and Expanded. CRC Press; 2016. https://doi.org/10.1201/b19217.

[3] Dowson D. Men of tribology. J Tribol 1978;100:455-7. https://doi.org/1-0.1115/1.3453250.

[4] Lubrecht AA, Venner CH, Colin F. Film thickness calculation in elasto-hydrodynamic lubricated line and elliptical contacts: The Dowson, Higginson, Hamrock contribution. Proc Inst Mech Eng Part J J Eng Tribol 2009;223:511-5. https://doi.org/10.1243/13506501JET508.

[5] Dowson D. Men of Tribology: William Bate Hardy (1864-1934) and Arnold Sommerfeld (1868-1951). J Lubr Technol 1979;101:393-7. https://doi.org/10.1115/1.3453381.

[6] Mortier RM, Fox MF, Orszulik ST. Chemistry and technology of lubricants: Third edition. Springer Netherlands; 2010. https://doi.org/10.1007/978-1-4020-8662-5.

[7] Rudnick LR. Lubricant Additives: Chemistry and Applications. CRC Press; 2003.

[8] Lauterwasser F, Bartels T, Smolenski D, Seemann M. Megatrend Fuel Economy: How to Optimize Viscosity with VI Improvers. SAE Tech. Pap., vol. 2016- February, SAE International; 2016. https://doi.org/10.4271/2016-28-0030.

[9] Spikes H. The History and Mechanisms of ZDDP. Tribol Lett 2004;17:469-89. https://doi.org/10.1023/B:TRIL.0000044495.26882.b5.

[10] Tang Z, Li S. A review of recent developments of friction modifiers for liquid lubricants (2007-present). Curr Opin Solid State Mater Sci 2014;18: 119-39. https://doi.org/10.1016/j.cossms.2014.02.002.

[11] Zhou Y, Qu J. Ionic liquids as lubricant additives: A review. ACS Appl Mater Interfaces 2017;9:3209-22. https://doi.org/10.1021/acsami.6b12489.

[12] Seddon EJ, Friend CL, Roski JP. Detergents and dispersants. Chem. Technol. Lubr. Third Ed., Springer Netherlands; 2010, p. 213-36. https://doi.org/10.1023/b105569_7.

[13] Lugt PM. A Review on Grease Lubrication in Rolling Bearings. Tribol Trans 2009;52:470-80. https://doi.org/10.1080/10402000802687940.

[14] Scharf TW, Prasad S V. Solid lubricants: A review. J Mater Sci 2013;48:511-31. https://doi.org/10.1007/s10853-012-7038-2.

[15] Berman D, Erdemir A, Sumant A V. Approaches for Achieving Superlubricity in Two-Dimensional Materials. ACS Nano 2018;12:2122-37. https://doi.org/10.1021/acsnano.7b09046

[16] Vazirisereshk MR, Martini A, Strubbe DA, Baykara MZ. Solid lubrication with MoS_2: A review. Lubricants 2019;7:57. https://doi.org/10.3390/LUBRICANTS7070057

[17] Enhancing the mobility of the future: Speciality lubricants for electric vehicles. Munchen: n.d.

Chapter 10

Friction, Lubrication, and Wear

Natarajan Jeyaprakash and Che-Hua Yang

Abstract

The surface properties of a bulk material are not accepted totally and independently. The tribology is the most important field that comprises the component design with static and dynamic relations for a reliability and performance. Hence, it is believed that the surface contacts, atmosphere, and lubrication significantly change the wear resistance of the material surface. However, the wear process is more complicated, in that a surface wear properties based on many tribological factors namely sliding type, mode of loading and working atmosphere. In this chapter, will explore the tribology fundamental, friction, various lubrication, wear types and mechanism on the wear process.

Keywords: tribology, friction, lubrication, wear, mechanism, wear resistance

1. Introduction

Tribology is defined as the science of sliding two surfaces in relative motion. It comprises the investigation and application of the wear, friction and lubrication. The term called tribology has been realized for thousands of years. In the year of 1966, the well-known 'Jost Report' was released to the government of British. Since, the term called 'tribology' has been broadly used and the investigation on this topic has been highly explored. This is an interdisciplinary area and related with materials, wear, friction, surface analysis, wear mechanism and working atmosphere [1]. Also, the surface properties of the base material cannot be measured totally as independent. It involves the component shape with contact types i.e., static or dynamic for the essential performance and reliability. The wear resistance of the material is mainly depending on the loading type, sliding and working environment. However, the phenomena are more complicated, in that a surface wear properties based on many tribological factors namely sliding type, mode of loading and working atmosphere [2]. Further, the wear behavior examination includes wear rate, coefficient of friction, volume loss, worn-out morphologies examination and studies of the mechanical properties and microstructure. Jeyaprakash et al. performed the laser coating and studied the wear resistance, friction behavior and various wear mechanisms [3]. Jin et al. performed high entropy alloy coating and studied the mechanical and wear properties [4]. Si et al. produced the Fe–Mo–Cr–Co coating and studied the wear resistance and corrosion resistance properties [5]. From the above literature, it has been recognized the importance of tribological properties. The main objective of this chapter is to study the tribology fundamental, friction, lubrication, wear, mechanism.

2. Fundamentals of tribology

2.1 Surfaces in contact

The friction and wear are mainly dependent on the characteristics of two sliding surfaces. The difficulty to predict and to clarify with more accuracy such phenomena reveals the complex nature of the surfaces, which can be evaluated through material properties such as microstructure, presence of organic molecules and oxides, water vapor, geometrical irregularities and other impurities which can be adsorbed from the atmosphere. Hence, while the two bodies are coming in to closer contact, the significant features of their sliding surfaces define the nature of the interaction, which includes mechanical character, with the development of a stress-strain on the sliding area, with the strong establishment of physical or chemical bonds [6]. To calculate the contact stresses, the smooth surface concept can be introduced, i.e., the surfaces are free from geometrical irregularities. Generally, the formation of smooth surfaces is difficult at a molecular level. The relation for contact stresses and deformations can be obtained through theoretical analysis which is developed by Hertz for linear elastic bodies. This can be employed while the two bodies are in frictional or elastic contact, with the assumption that the contact body radius is higher while compared with contact zone size.

2.1.1 Elastic contact

The viewpoint of geometrical, the contact between two solid bodies can be classified in to conformal or nonconformal as shown in **Figure 1**. From the **Figure 1 (a)**, it can be observed that the conformal contact happens while mating surfaces fit closely together. This kind of contact can be seen while bearing sliding on shafts and between wire and tool in drawing processes. The **Figure 1 (b)** shows the contact between two bodies which is nonconformal and this can be theoretically occurred. For example, with the presence of point contact in rolling bearing (between seat and ball), whereas a line contact happens in gears (between tooth and tooth). In another case, the contact area has a limited extension and it can be easily determined.

2.1.2 Viscoelastic contact

In the case of polymers, the deformation behavior can be occurred that is affected by plastic, elastic and viscoelastic processes. For example, the polypropylene (PP) sphere hard pressed on the transparent plane with the function of contact deformation displacement (δ) and time. In polymer plate the viscosity influence is

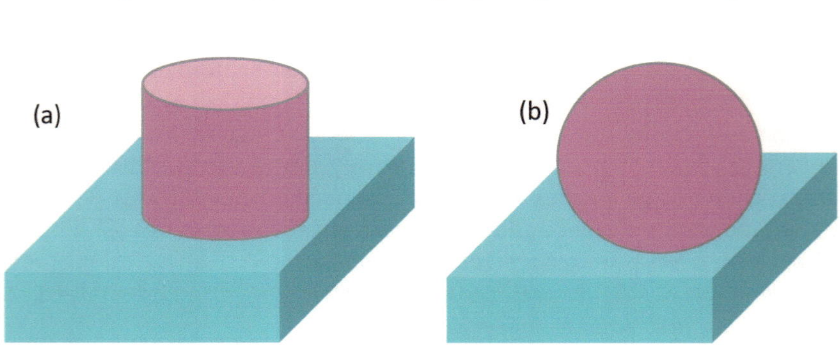

Figure 1.
The conformal contact between a plane and the base of a cylinder (a) and nonconformal contact between a plane and sphere (b).

DOI: http://dx.doi.org/10.5772/intechopen.93796

mainly noticeable while holding the amorphous phase and the heat may be higher than glass transition temperature. Further, the loss of energy is related to the viscoelastic loading and unloading processes. This energy dissipation may create the temperature on the material due to lesser thermal conductivity of material. However, the creation of plastic deformation is stable. Besides, the plastic processes and viscoelastic are based on the temperature and while increasing the temperature their intensity also increased [7]. **Figure 2** shows the various loads displacement and time while polypropylene sliding on a plane.

2.1.3 Elastic and plastic contacts

The material behaves with ductile way; the provided contact load can be induced the plastic deformation. At the same time, the equivalent stress at the critical point influence the material uniaxial yield stress. In this case, the material is not as elastic stage; however, it must be an elastic-plastic condition [8].

Figure 3 illuminates the elastic-plastic contact in detail with sphere and plane contact. In this case, the sphere has a higher hardness while compared with plane. If the applied load is increases, the plastic zone size can be increased. The applied load is taken out while the contact pressure with the below specific limit, then with the same magnitude of additional load, which are applied possibly, results the increase in elastic deformation only.

2.2 Friction

The frictional forces can be recognized as good or bad, without this friction, there is no possibility to use vehicle tires on a road, walking on the road or pickup objects. In some cases, such as machine application like clutches, vehicle brakes and transmission of power (belt drives), friction is increased. But, in many cases like rotating and sliding components such as seals and bearings, friction is unwanted. The higher friction makes more material loss (i.e., wear rate) and energy loss. In these kinds of working atmosphere, the friction is reduced [9].

The term friction is called as the force resisting the relative motion of two mating surfaces in contact with a fluid. The two sliding surfaces move relative to each

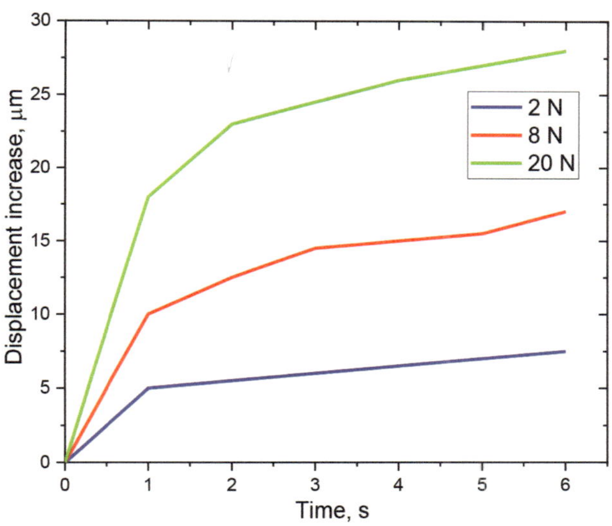

Figure 2.
Polypropylene sphere sliding on a plane with respect to load displacement and time.

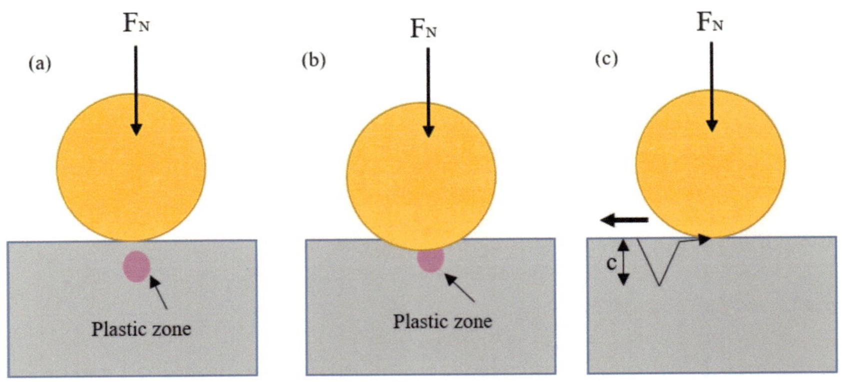

Figure 3.
Elasto-plastic (a) complete plastic, (b) brittle-type contact, and (c) between plane and a sphere.

other, the friction between the mating surfaces converts the kinetic energy in to heat or thermal energy. Generally, the term used to describe the friction is the coefficient of friction. It is denoted with dimensionless scalar value (μ) and explaining the ratio of the friction between two mating surfaces (F) and the applied force on them (F_n). It is described in Eq. (1):

$$F = \mu(F_n) \tag{1}$$

In a sliding condition, it is normally required to monitor or calculate the frictional behavior during the experiment. The changes in friction with wear data normally offer the beneficial data regarding modeling and mechanisms. The friction is generally classified as two categories: (a) static friction and (b) dynamic friction. In the case of two objects is not sliding each other is called as static friction. In another hand, both the mating objects sliding relatively to each other is called as dynamic friction [10]. **Table 1** shows the required parameters which need to be considered while performing wear experiments.

2.2.1 Metal surfaces

Generally, the asperities contact in metal surfaces is normally plastic. The metals such as titanium, cobalt, magnesium with hcp (hexagonal closed packed) crystal lattice provides the coefficient of friction nearly 0.5 while sliding against themselves. In addition to that the hcp metals possess a reduced ability to deform

Parameter/condition	Example
Material	Nonmetal: composites, polymers, ceramics and bio glass Metal: Nickel alloy, iron, stainless steel, cobalt alloy and titanium alloy
Lubrication	Oil, saliva, synovial fluid and blood
Geometry	ball-on-socket, plane-on-plane and point-on-plane
Motion type	Rotation motion, unidirectional, freeing, multidirectional
Loading type	Cyclic loading, constant loading, etc.
Atmosphere	Room temperature, high temperature and corrosive medium

Table 1.
Necessary input parameters to be considered for wear experiment.

plastically with relatively lower temperature. Hence, while sliding the asperity junctions deform with lesser intensity and the adhesion forces is not developed at the junctions [11]. While sliding between two metal alloys, the coefficient of friction is to be lesser than in the corresponding pure metals. In the case of bronze material (Cu-8% Sn), while sliding in dry condition, the coefficient of friction is around 0.6 while the typical value of 1 was recorded for the Cu/Cu pair. In the case of steel, the coefficient of friction value is around 0.6–0.8 while using two steel alloys, i.e., which is lesser than the pure steel-steel pairs.

2.2.2 Ceramic surfaces

In ceramics material the contact at the asperities is normally mixed. This may be fully in elastic while the surface roughness shows with less [6]. Else, if the surface roughness is high, it may be shows with plastic stage. Generally, the coefficient of friction must be independent for the normal applied load in elastic contacts. For example, the alumina balls sliding on the alumina surface and the friction showed around 0.4. It can be noted that the friction coefficient in ceramics material with dry atmosphere is lesser around 0.3–0.7 while applying the minimum load with less than 200°C temperature. The obtained frictional values are relatively similar to metal alloys and this may seem to be quite surprise. In case, the ceramics material is characterized with higher hardness and elastic modulus, and lesser values of the surface energy [12].

Besides, the surface energy of ceramics material is reduced through the surface reactions with water vapor and the presence of other substances on the working atmosphere. Hence, the lesser frictional force can be expected from the sliding wear experiment. Though, while continuous sliding with real contact area is notably increases with increase in friction. Further, the applied load is considered as major input parameter in the ceramic material. If the applied load is increased, the brittle contact may establish, and this will increase the coefficient of friction as much as high around 0.8. Specifically, the brittle contact can be occurred while the tangential stresses owing to higher friction due to the occurrence of critical microcracks on the surfaces. This type of microcracks can be seen on the ceramics surfaces normally and producing the defects such as porosity, flaws and inclusions. The cracks can be initiated from the development of asperity because of continuous applied load during the tribological study. **Figure 4** represents the coefficient of friction and applied load for alumina sliding on the alumina.

2.2.3 Polymer surfaces

The polymer like polytetrafluoroethilene (PTFE) is produces very low friction around below 0.1 while sliding on the same material or other metals. Thus, this material behaves as solid lubricant while sliding with counterpart [13]. Generally, most of the polymer material friction coefficient ranges from 0.2 to 1 while sliding in dry condition. In the case of the work of adhesion in polymer was lesser than in ceramics and metals. However, their stiffness and hardness of the material is lesser, and these two effects are nearly proportional.

Figure 5 indicates the correlation between adhesion and coefficient of friction for various polymers sliding against PA. These experiments were conducted on flat to flat surface contacts with lower sliding speed of 0.24 μm/s. So that the thermal effects on the polymers can be avoided [14]. It can be observed that the coefficient of friction was increased with the work of adhesion. In this kind of working condition, the adhesion was considered as a most important factor in friction determination. In the case of point or line interactions, the produced deformations may be higher and therefore the effect of viscoelastic can play a major role.

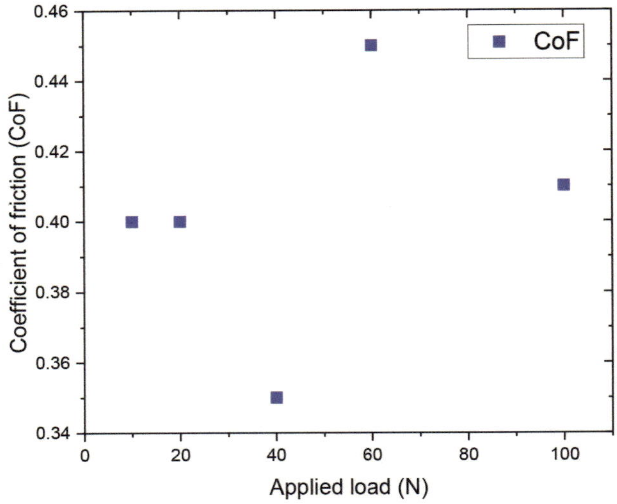

Figure 4.
Coefficient of friction and applied load for alumina sliding on the alumina.

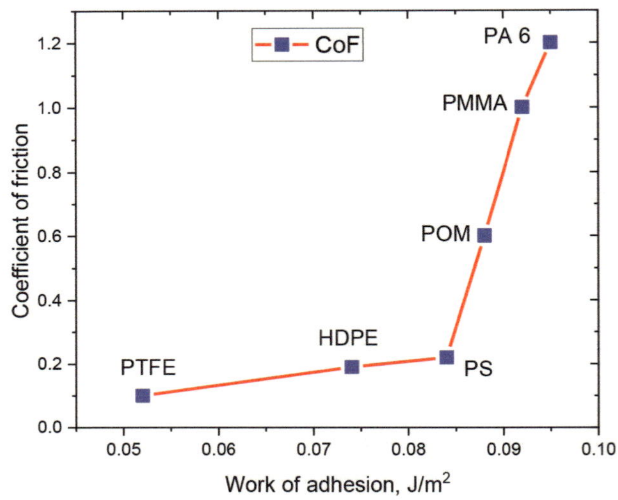

Figure 5.
Work of adhesion and friction coefficient for some polymers sliding against PA 6.

2.3 Lubrication

Two solid components/parts are sliding between surfaces is normally considered through a maximum coefficient of friction and higher wear rate because of the surface properties such as reactivity, lesser hardness, mutual solubility and higher surface energy. The clean surfaces without any rust is freely adsorb traces of other substances from the atmosphere. In addition, the newly manufactured surfaces normally produce the lesser wear and coefficient of friction while compared with clean surfaces. However, there may be chances of external material on the interface of bulk material which can increase the coefficient of friction during continuous sliding process. Hence, the lubricants can be applied to reduce the wear rate and coefficient of friction [8]. The term called lubrication can be applied to two various conditions: namely solid lubrication and fluid film lubrication (liquid or gaseous).

In any kind of material, the solid lubricant such as solid film and powder was used to protect the sliding surface from the unexpected damages during the sliding process and reduce the wear rate and coefficient of friction. The solid lubricants were used in sliding applications. For example, the bearing was operated with low speeds and higher loads, and the hydrodynamically lubricated bearings demanding the start and stop processes. The solid lubricant holds the higher variety of material which can produce the lower wear rate and coefficient of friction [15]. In addition to that the hard materials also were used as lubricant to reduce the friction and wear in extreme working atmosphere.

2.3.1 Fluid film lubrication regimes

A thick film of lubrication was maintained in the region between two solid surfaces with no relative motion or lesser motion through an external pumping agency is called as hydrostatic lubrication [16]. The lubrication regimes which are noticed in fluid lubrication with self-acting can be recognized in the Stribeck curve as shown in **Figure 6**. This graph provides the hypothetical fluid-lubricated bearing structure with coefficient of friction as a function of the rotational speed (N) and absolute viscosity (η) divided by the applied load (P). This curve has a lowest with providing the recommendation that higher than one lubrication system is presented. In some cases, the lubrication regimes can be recognized through lubricant film parameter [17].

2.3.1.1 Hydrostatic lubrication

In hydrostatic bearings, the supporting load on the thicker film provided from an external source, a pump, which can induce the fluid pressure toward the film. Based on this reason, those kinds of bearings are called externally pressurized. Generally, the hydrostatic bearings are considered for usage in compressible and incompressible fluids. Subsequently, the hydrostatic bearings are no need of any relative motion on the surface of bearings to create the load supporting pressures as it essential in the hydrostatic bearings [18]. Further, this type of hydrostatic bearings is used in application with no relative motion or lesser motion between the sliding surfaces. Besides, the hydrostatic bearings offer great stiffness though, this type of lubrication needed the high-pressure equipment and pumps for the cleaning of fluid, which is occupying more space with higher cost.

2.3.1.2 Hydrodynamic lubrication

The hydrodynamic lubrication is called as thick film or fluid film lubrication. The convergent type of bearings starts to move in the direction of longitudinal from the initial position, a less thickness of layer is pulled due to viscous entrainment. Further, it is compressed between the two bearing surfaces and generating the necessary pressure to support the load with no other external devices as indicates in **Figure 6**. This hydrodynamic lubrication mechanism is necessary for the effective working of the hydrodynamic journal and the thrust bearings are extensively used in the modern manufacturing industry [19, 20].

In addition, the hydrodynamic lubrication films thickness is ranging from 5 to 500 μm and referred as the ideal lubricated contact condition. Also, the friction coefficient of the hydrodynamic contacts is as lesser as 0.001 which represents the **Figure 6**. Sometimes the frictional force can be increased slightly while increase in sliding speed due to the viscous drag. Generally, the physical interaction can be

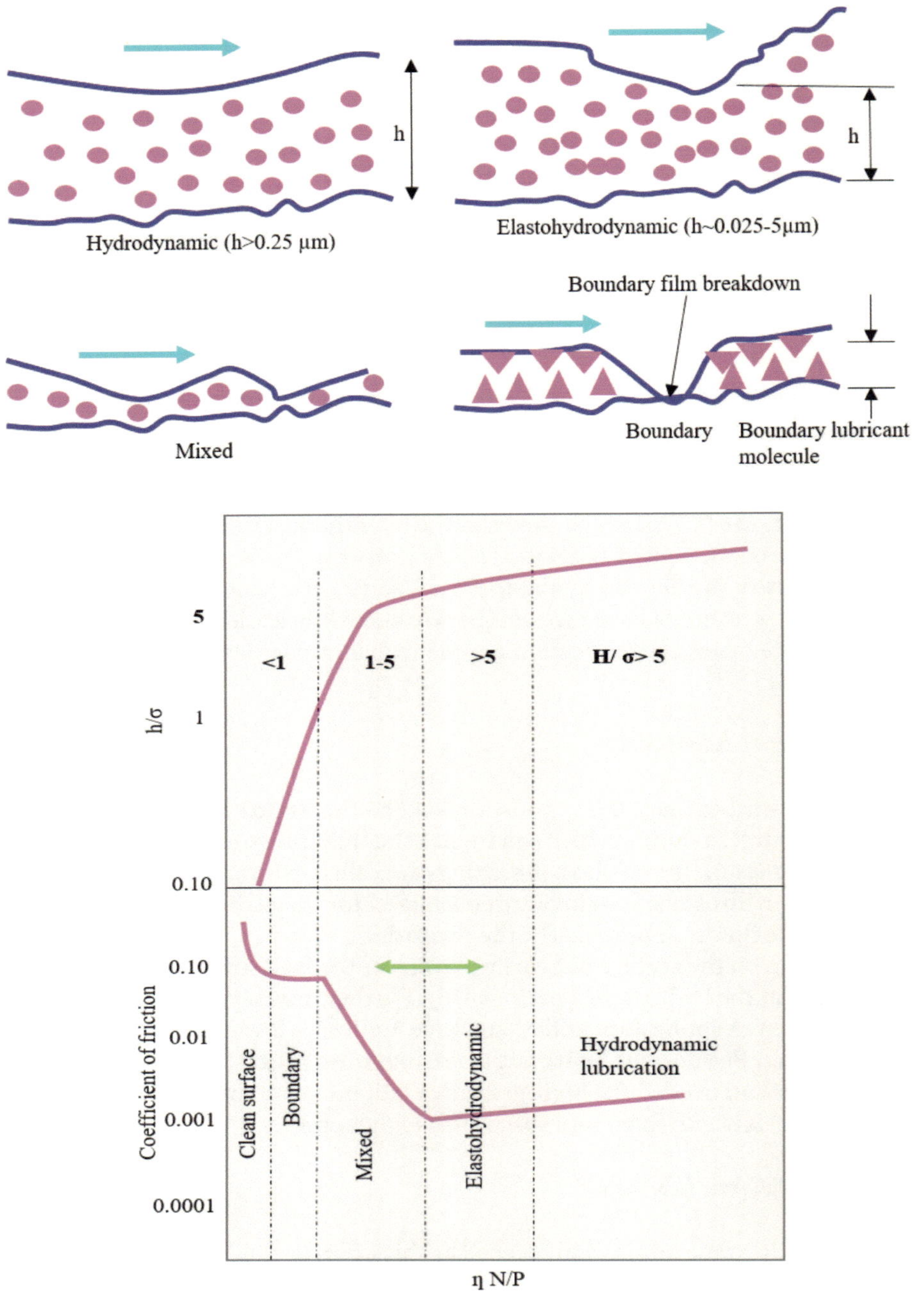

Figure 6.
Coefficient of friction and lubricant film parameter as a function of Stribeck curve showing various lubrication regimes observed in fluid lubrication without an external pumping agency.

occurred while starting and ending up with lesser sliding speeds. The behavior of the interaction is directed through the lubricant bulk properties i.e., viscosity, and the frictional force can be developed based on the shearing of the viscous lubricant. Besides, the corrosive wear can be occurred on the bearing surfaces due to the presence of lubrication. The adhesive wear also was possible while initial and ending up the processes. The corrosive wear can be reduced through lubricant precipitation and formation of film on the bearing surface.

2.3.2 Boundary lubrication

While increasing the load, the fluid viscosity or speed is decreased in the Stribeck graph as shown in **Figure 6**. In this situation, the friction coefficient can be increased as high as 0.1 or higher than this. This situation may also arise in a starved contact. When the solid surfaces are nearby that surface contact between the multi-molecular or monomolecular films of gases or liquids and the dense solid asperities may rule the contact. For more understanding, the cross section of the films and the asperity area contacts is shown in **Figure 6** [21]. Further, with the absence of gases and boundary lubricants (without oxide films), the produced frictional forces may be higher than one. In case with failure in boundary lubrication, causes corrosive and adhesive wear. Generally, the boundary lubricants can easily form the sheared film on the bearing surfaces. Therefore, this formed shear film can be reduced the corrosive and adhesive wear. The major physical properties of the films are hardness, melting point and shear strength. The other properties are cohesion, tenacity or adhesion and formation rates. Also, the viscosity of the lubricant can show a minor impact on the wear and friction behavior [22, 23].

2.3.3 Mixed lubrication

The transition between the boundary lubrication and the hydrodynamic regimes is a gray zone known as mixed lubrication. In this regime, the both mechanisms such as boundary lubrication and hydrodynamic lubrication may be in operational condition. There might be possible for the higher solid contacts, however, a minimum portion of the bearing surface leftover through partial hydrodynamic film. The hard solid interaction between the new metal surfaces can cause to a wear debris formation, adhesion of particle with counterpart, metal transfer from bulk to counterpart and eventual seizure. But, in the case of liquid condition, the chemically formed films protect the surfaces from adhesion during the sliding experiment. This mixed regime is called as thin film lubrication, partial fluid and quasi-hydrodynamic lubrication [21].

2.4 Wear

2.4.1 Introduction

Generally, the term wear is defined as material removal or surface damage on the one or two surfaces while rolling, sliding or impact motion relative to one another. Particularly, the wear happens through surface interactions at asperities. While two objects/components in relative motion, the material can be displaced from the interacting surfaces. Consequently, the properties of the material may be changed at least or interface region. But, there is a possibility for less or no material losses. Then, the displaced material can be removed from the interacting surfaces and may cause the material transfer to the counterpart surface or may break as small wear debris. When material transfer from bulk to counterpart, the net mass or volume loss of the interacting surface is zero while the bulk material surface is worn. The wear loss leads the real material loss, and this may occur sometimes independently.

Generally, the wear is a system output and it is not a material property. In addition to that the working atmosphere affect the interface wear. In some cases, mistakenly assumed that the higher frictional force displays the increase in wear rate. For example, the polymers and solid lubricant interfaces showed with higher wear and lesser friction, whereas ceramic material showed the lower wear but moderate frictional force. In all the dynamic machine components such as cams, bearings and seals, the wear is almost undesirable one. Those components or machines need to be replaced after a small damage or material loss or if the surface showed with higher roughness. If the system is well defined in tribology, the material removal will be very slow and at the same time it must be continuous with steady process.

2.4.2 Classification of wear mechanism

The wear occurs chemically or mechanically means and is normally induced through frictional heat. Mainly, the wear contains six principal quite distinct phenomena that have only one thing in common; the removal of material from the rubbing interface [24, 25]. The wear can be classified as follows: (1) abrasive; (2) adhesive; (3) fatigue; (4) impact by erosion; (5) corrosive; (6) electrical-arc-induced wear. The other commonly raised wear is fretting corrosion and fretting. However, the wear such as abrasive, adhesive and corrosive are the major combinations of wear. Based on the previously encountered issues, the abrasive and adhesives are the major wear mechanisms in the industry.

2.4.2.1 Abrasive wear

The abrasive wear happens while asperities of a rough and hard particles slides on the softer surface and remove the softer material and finally damages the surface through fracture or plastic deformation. Most of the ceramic and metallic materials with high toughness and hard particles result in the plastic stage of the soft material. Even the metal interfaces will deform plastically while applying higher loads. The abrasive wear can be occurred generally in two different situations as shown in **Figure 7**. In the beginning, the hard surface is the harder of two rubbing surfaces like two body abrasion. For example, in cutting, grinding and machining. In another case, the hard surface is the third body. Particularly, this must be small abrasive particle, identified in between the other two surfaces and this may be sufficiently harder. This may be able to scratch either one or both the sliding surfaces.

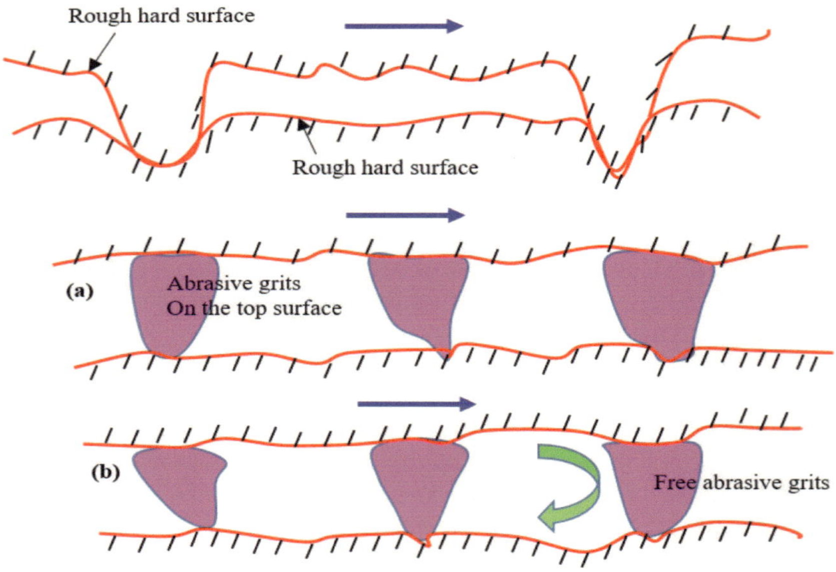

Figure 7.
(a) Schematics of hard rough surface mounted with abrasive grits sliding on a softer surface and (b) free abrasive grits caught between the two surfaces.

In most of the cases, the adhesive is the begging mechanism of wear on the sliding surface and this wear debris get trapped into interfaces of mating surfaces, subsequently resulting the three-body abrasive wear [26]. Further, the scratching can be observed in many cases as a continuous groove which is parallel to the sliding direction. The worn-out surface of steel against hardened steel disc with dry condition is shown in **Figure 8**. From the scanning electron micrograph (SEM), the continuous scratching is visible parallel to the sliding direction.

2.4.2.2 Adhesive wear

Generally, the adhesive wear can be occurred while two solid body sliding on the flat surfaces with dry or lubrication. The bonding of metals is happening at the asperity contacts at the interface, and these contacts is sheared through sliding process, which may produce the plowing of material from softer surface to harder surface. With continuous sliding process, the transferred wear debris is stop on the transferred surface and the same debris may return to original surface or else this may be loose as wear debris particle. In some cases, the debris may have fractured through fatigue process with continuous loading and unloading process and causes with loose particle formation.

There is some mechanism was explained for the fragment detachment of a material. However, still the well identified mechanism for adhesion is shearing of the two solid bodies or the weakest surface from one of the two surface [27]. The schematic representing the two possible way of metal breaking through shearing process as shown in **Figure 9**. Normally, the interface adhesion strength is assumed to be lesser while compared with breaking strength of nearby regions; hence, the break through shearing arise at the interface regions (path 1) in many cases and there will be no material loss happen during these sliding process. In another case, with lesser fraction of contacts, break may happen in any one of the two solid bodies (path 2) and a minor piece of material (the blue dotted semi-circle) can be attached to the harder other surface.

The SEM picture (**Figure 10**) shows the steel worn out surface with adhesion wear. From the **Figure 10**, it can be seen clearly that the adhesive debris pullout from the softer surface. During the sliding process, the surface asperities severely suffer from fracture or plastic deformation. Further, the subsurface also underwent strain hardening and plastic deformation. The SEM micrograph of worn-out surface shows the severe pull out material with plastic deformation. In the picture, the yellow

Figure 8.
SEM picture of steel surface after abrasive wear with dry condition.

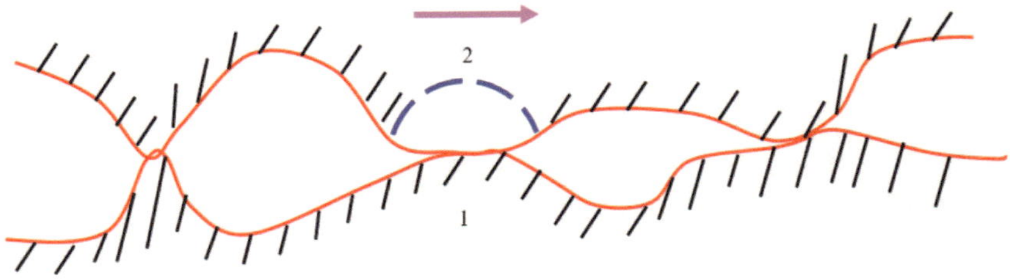

Figure 9.
Schematic representing the two possible ways of metal breaking through the shearing process.

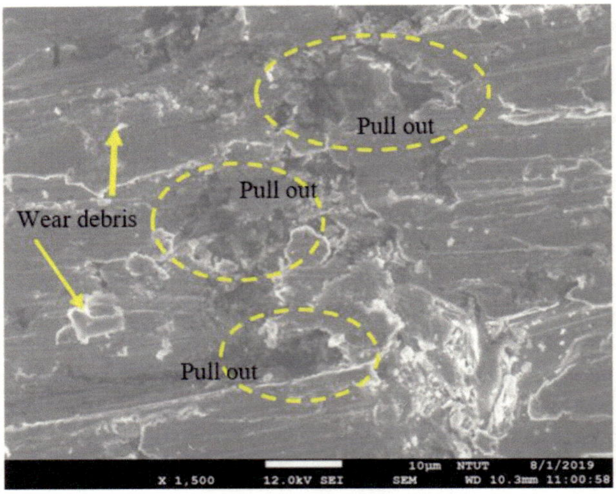

Figure 10.
SEM picture of steel surface after adhesive wear under dry condition.

dotted semi-circle indicates the extensive structural changes by adhesive wear. In addition to that severe type of adhesion wear is called as smearing, galling and scuffing and also this terms are used to describe other type of wear in sometimes.

2.4.2.3 Fatigue wear

Generally, the fatigue wear on surface and subsurface level can be noticed through continuous sliding and rolling atmosphere. The continuous loading and unloading processes, which will induce the surface and subsurface to form the cracks after critical repeated cycles. Then, the surface of the material will breakup into lager fragments and producing the larger pits on the softer surface. However, the material removed through fatigue wear is not considered as major parameter. There are much relevant is the beneficial life in terms of time or revolutions prior occur the fatigue wear.

2.4.2.4 Erosive wear

The erosive wear can be occurred due to impingement of solid hard particle with high velocity on the specimen surface. **Figure 11 (a, b)** shows that the hard particle impinging toward the specimen surface and removes the material from the top surface of the specimen. The contact stresses produce from the particle kinetic energy in liquid or air stream as it meets the surface. In erosive wear, the impingement

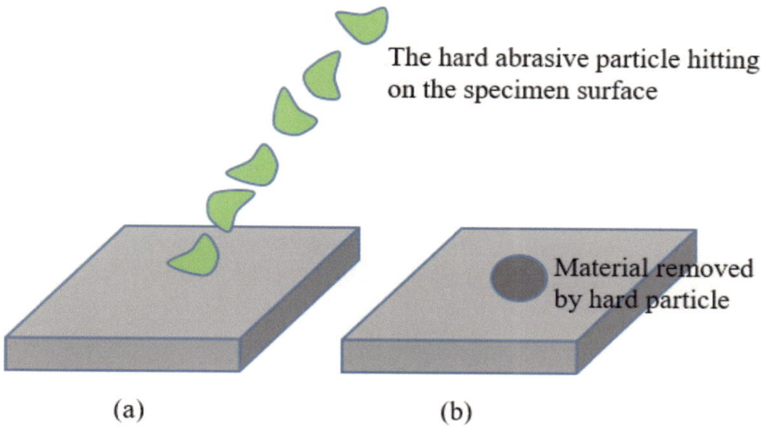

Figure 11.
Schematic of erosive wear with hard particle hitting on the surface; (a) hard abrasive particle hitting the surface, (b) material removed by hard particle.

angle and particle velocity combined with abrasive sizes and provide the measure of kinetic energy for the impinging particle. The wear particles/debris are formed in erosion arises as an outcome of repetitive impacts [28].

2.4.2.5 Corrosive wear

The corrosive or chemical wear happens while sliding takes place in a chemical atmosphere. The oxygen is considered as a most dominant corrosive medium in air atmosphere. So that the corrosive wear in air atmosphere is normally called as oxidative wear. The corrosive wear is significant in many factories such as slurry handling, chemical processing, mining and mineral processing. The chemical wear can arise due to the electrochemical or chemical interaction of the surfaces with the atmosphere. However, the chemical corrosive wear occurs in extremely high corrosive atmosphere and in high humidity and high temperature atmosphere. The electrochemical corrosive wear can occur with chemical reaction accompanied of an electric current. The potential variations can be observed between those two regions. The high potential region and low potential region is known as cathode and anode, respectively. There will be a current flow between the cathode and anode over an electrolyte conductive medium, the metal dissolve at the anode side in the form of liberates electrons and ions [29].

While conducting the experiment, the electron transfers via metal to the cathode and minimize the oxygen or ions. After corrosion test, these surfaces changes to some other appearance with corroded region. Further, the electrochemical corrosion is influenced through the electro potential. The aqueous is a most common liquid environment in corrosion atmosphere. In this working atmosphere, the less amount of gases may dissolve, normally carbon dioxide or oxygen may influence the corrosion. **Figure 12** indicates the electrochemical corrosion testing setup and corroded micrograph with layer formation.

2.4.2.6 Electrical arc-induced wear

While the presence of higher potential on the thin air film during the sliding condition, a dielectric breakdown gives that leads to arcing. The higher power density can be produced with short time during the arcing period. The produced

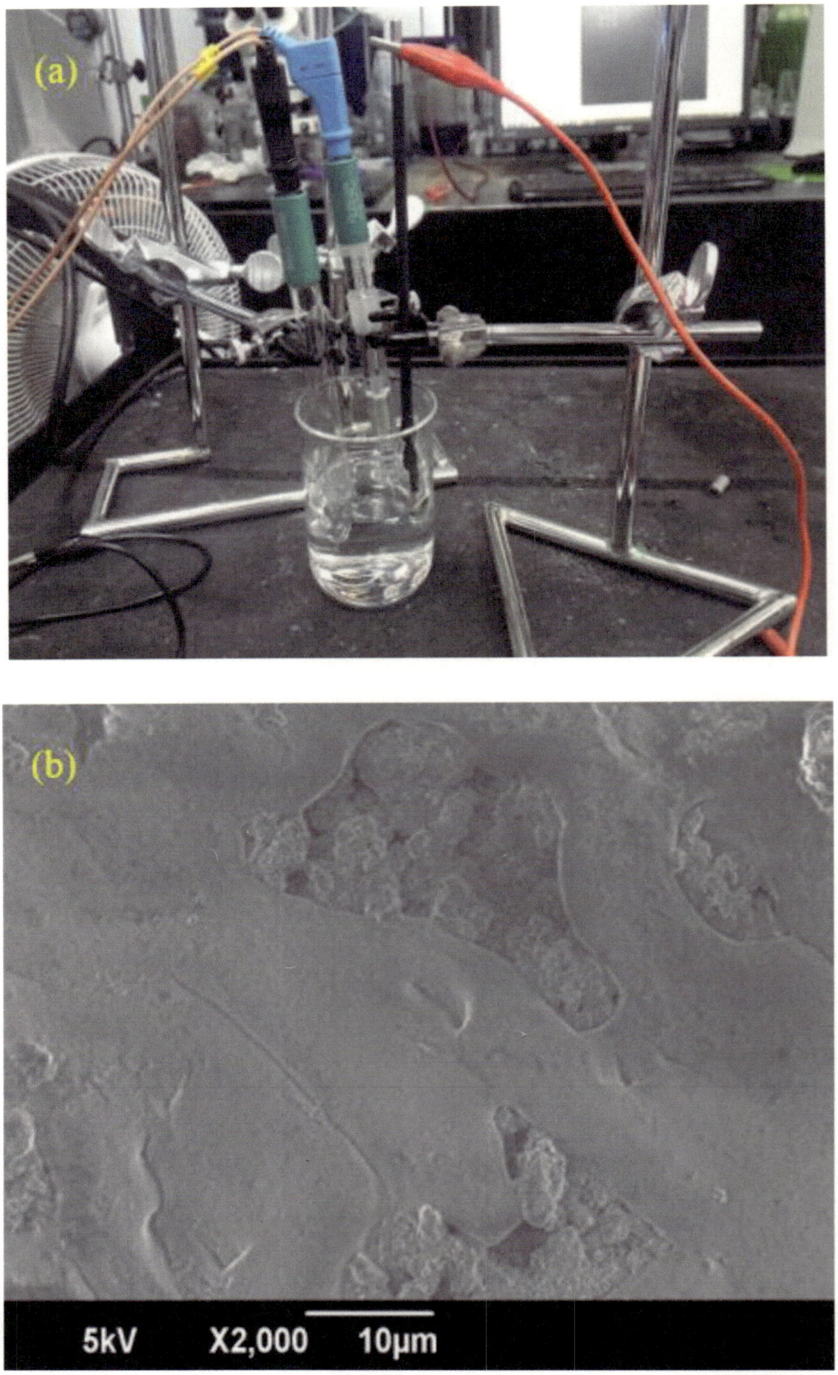

Figure 12.
Macropicture of electrochemical corrosion testing setup (a) and corroded micrograph with layer formation (b).

heating results in extensive melting, solidification, surface corrosion, phase changes and hardness changes, and sometimes ablation of metal can be occurred. This arcing creates the higher craters and after sliding either fracture or shears the material lip, causing the three-body abrasion, surface fatigue, corrosion and fretting. Arcing may create the many type of wear mode and resulting catastrophic failure in the electrical equipment's [30].

3. Conclusion

In this chapter, the fundamentals of tribology, friction, lubrication, and types of wear and mechanism are briefly described. The friction and wear are mainly dependent on the characteristics of two sliding surfaces. The metals such as titanium, cobalt, and magnesium with an hcp (hexagonal closed packed) crystal lattice provide the coefficient of friction of nearly 0.5, while sliding against themselves. In the case of steel, the coefficient of friction value is around 0.6–0.8 while using two steel alloys, i.e., which is lesser than that of the pure steel-steel pairs. Further, the alumina balls sliding on the alumina surface and the friction showed around 0.4. The friction coefficient in ceramics material with dry atmosphere is lesser around 0.3 to 0.7 while applying the minimum load with less than 200°C temperature. The polymer material friction coefficient ranges from 0.2 to 1 while sliding in dry condition. Besides, the solid lubricant, such as solid film and powder, was used to protect the sliding surface from the unexpected damages during the sliding process and reduce the wear rate and coefficient of friction. Wear is defined as material removal or surface damage on the one or two surfaces while rolling, sliding or impact motion relative to one another. Particularly, the wear happens through surface interactions at asperities. Therefore, the wear characteristic must be taken into account while performing the sliding wear processes in mechanical components.

Acknowledgements

The authors wish to thank the Ministry of Science and Technology, Taiwan ROC for the financial support to carry out this work.

Conflict of interest

The authors declare that they have no known competing financial interests or personal relationships that could have appeared to influence the work reported in this paper.

Author details

Natarajan Jeyaprakash[1*] and Che-Hua Yang[1,2]

1 Additive Manufacturing Center for Mass Customization Production, National Taipei University of Technology, Taipei, Taiwan, ROC

2 Institute of Manufacturing Technology, National Taipei University of Technology, Taipei, Taiwan, ROC

*Address all correspondence to: prakash84gct@gmail.com; prakash@ntut.edu.tw

References

[1] Yan Y. University of Leeds, UK, Tribology and Tribo-Corrosion Testing and Analysis of Metallic Biomaterials. United Kingdom: Woodhead Publishing Limited; 2010

[2] Meng Y, Xu J, Jin Z, Prakash B, Hu Y. A review of recent advances in tribology. Friction. 2020;**8**(2):221-300

[3] Jeyaprakash N, Yang C-H, Tseng S-P. Wear Tribo-performances of laser cladding Colmonoy-6 and Stellite-6 Micron layers on stainless steel 304 using Yb: YAG disk laser. Metals and Materials International. 2019:1-14. DOI: 10.1007/s12540-019-00526-6

[4] Jin B, Zhang N, Yu H, et al. AlxCoCrFeNiSi high entropy alloy coatings with high microhardness and improved wear resistance. Surface and Coatings Technology. 2018;**402**:126328. DOI: 10.1016/j.surfcoat.2020.126328

[5] Si C, Duan B, Zhang Q, Cai J, Wu W. Microstructure, corrosion-resistance, and wear-resistance properties of subsonic flame sprayed amorphous Fe–Mo–Cr–Co coating with extremely high amorphous rate. Journal of Materials Research and Technology. 2020;**9**(3):3292-3303

[6] Straffelini G. Friction and wear methodologies for design and control. Springer Tracts in Mechanical Engineering. Switzerland; Springer International Publisher; 2015. pp. 1-158

[7] Czichos H. In: Lee LH, editor. Polymer Wear and Its Control. Washington: American Chemical Society; 1985. pp. 3-26

[8] Khonsari MM, Booser ER, Tribology A. Bearing Design and Lubrication. Wiley: New York; 2008

[9] Bharat Bhushan. Introduction to Tribology. 2nd ed. Publication. USA: A John Wiley & Sons, Ltd.; 2013

[10] Loomis WR. New Directions in Lubrication, Materials, Wear and Surface Interaction. United States: Noyes Publications; 1985

[11] Società italiana SKF, I cuscinetti volventi, Catalogo generale (versione ridotta), Società italiana SKF; 1989:1-512

[12] Jahanmir S, editor. Friction and Wear of Ceramics. New York: Marcel Dekker; 1994

[13] Blanchet T, Kennedy F. Sliding wear mechanism of polytetrafluoroethylene (PTFE) and PTFE composites. Wear. 1992;**153**:229-243

[14] Czichos KH, Habig KH. Tribologie Handbuch, Reibung und Verlschleiss. Wiesbaden, Germany: Springer Vieweg, Springer Fachmedien; 1992

[15] Clauss FJ. Solid Lubricants and Self-Lubricating Solids. New York: Academic Press; 1972

[16] Stribeck R. Characteristics of plain and roller bearings. Zeit. Ver. Deut. Ing. 1902;**46**:1341-1348, 1432-1438, 1463-1470

[17] Szeri AZ. Fluid Film Lubrication – Theory and Design, Second Edition. Cambridge, UK: Cambridge University Press; 2010

[18] Bhushan B. Tribology and Mechanics of Magnetic Storage Devices. United States: Springer Science and Business Media LLC; 1990

[19] Gross WA, Matsch LA, Castelli V, Eshel A, Vohr JH, Wildmann M. Fluid Film Lubrication. New York: Wiley; 1980

[20] Fuller DD. Theory and Practice of Lubrication for Engineers. 2nd ed. New York: Wiley; 1984

[21] Bruce RW. Handbook of Lubrication and Tribology, Vol. II: Theory and Design. 2nd ed. Boca Raton, Florida: CRC Press; 2012

[22] Ku PM. Interdisciplinary Approach to Friction and Wear. SP-181. Washington, DC: NASA; 1970. pp. 335-379

[23] Beerbower A. Boundary Lubrication, AD-747 336, Office of the Chief of Research and Development. Washington, DC: Department of the Army; 1972

[24] Peterson MB, Winer WO, editors. Wear Control Handbook. New York: ASME; 1980

[25] Loomis WR. New Directions in Lubrication, Materials, Wear, and Surface Interactions: Tribology in the 80s. Park Ridge, New Jersey: Noyes Publications; 1985

[26] Bhushan B, Davis RE, Kolar HR. Metallurgical Re-examination of Wear modes II: Adhesive and abrasive. Thin Solid Films. 1985b;**123**:113-126

[27] Archard JF. Contact and rubbing of flat surfaces. Journal of Applied Physics. 1953;**24**:981-988

[28] Finnie I. Erosion of surfaces by solid particles. Wear. 1960;**3**:87-103

[29] Wagner C, Traud W. Interpretation of corrosion phenomena by superimposition of electrochemical partial reaction and the formation of potentials of mixed electrodes. Zeitscrift für Elektrochemie. 1938;**44**:391-402

[30] Guile AE, Juttner B. Basic erosion process of oxidized and clean metal cathodes by electric arcs. IEEE Transactions on Components, Hybrids, Manufacturing and Technology. 1980;**PS-8**:259-269

Chapter 11

Biolubricant from Pongamia Oil

Sabarinath Sankarannair, Avinash Ajith Nair, Benji Varghese Bijo, Hareesh Kuttuvelil Das and Harigovind Sureshkumar

Abstract

Recent researches focus on the development of lubricants from non-edible vegetable oil which are environment friendly and renewable. In the current work, an industrial lubricant is formulated from a non-edible vegetable oil viz. pongamia oil (PO) by blending it suitable additives. The additives such as silicon dioxide (SiO_2) nanoparticles, tert-butylhydroquinone (TBHQ) and styrene butadiene rubber (SBR) were selected as antiwear, antioxidant and viscosity improver additives respectively for the study. Various lubricant properties of the formulated oil (FO) are studied and comparisons were made against neat PO and popularly available mineral oil lubricant viz. SAE 20W40. It is found that the FO possesses superior viscosity index, and lower coefficient of friction than the commercial SAE 20W40. Moreover, the viscosity range, oxidative stability and the wear scar diameter of the FO is also in the range of SAE 20W40. This work is done with an aim of promoting Pongamia agriculture and reducing soil pollution.

Keywords: biolubricant, pongamia oil, agriculture, nanoparticle, viscosity improver, tribology

1. Introduction

The technique or process of using a material to reduce friction and wear between contact surfaces, which are in relative motion, is known as lubrication. It also helps in force transmission, foreign particle transportation and heat transfer. Lubricity is the property of a lubricant to reduce the friction. Lubricants are classified into solid lubricants (e.g.: graphite), semi-solid lubricants (e.g.: grease), liquid lubricants (e.g.: mineral oils based) and gaseous lubricants (e.g.: air). Liquid lubricants are further classified into fixed oil based, mineral oil based and synthetic oil based lubricants on basis of lubricant base stock [1].

Evidences for usage of lubricants for thousands of years have been found. Several methods were adopted by human race from time to time to solve the issues regarding friction and wear. Egyptians used animal fats in ball bearings for lubrication back in 1000 B.C. Oil-impregnated lumber were used to slide building stones in the time of the pyramids and on the axles of chariots dated to 1400 BC, calcium soaps have been found [2]. Romans used thrust bearing and lubricants having rapeseed and olive oil as well as animal fat as base back in 40 A.D. Mineral oil based lubricants were widely used since 1850, but they are non-biodegradable, fast depleting and also have adverse effects on the environment [3, 4]. Nowadays, synthetic oil is considered as a better alternative to these mineral oil based lubricants.

However, these synthetic oil based lubricants are much expensive than the mineral oil based lubricants [5]. The improper after-use disposals of the available lubricants are creating severe environmental issues by polluting the water bodies [6]. Developing an efficient lubricant from a non-edible plant oil base stock is an effective solution to the above issues, as they are biodegradable, renewable and environment friendly [7]. Vegetable oil base stocks also possess high thermal stability, low volatility, good biodegradability, non-toxicity and good lubrication properties in comparison to mineral oil base stocks [8].

The present article aims at developing a lubricant from a non-edible vegetable oil. In the current work, pongamia oil (PO) is selected as the base stock due to its high oleic acid content and non-edible nature [9–14]. Formulated oil (FO) is developed by blending suitable additives in PO. Rheological, oxidative and tribological properties of PO and FO are evaluated and compared against the properties of a commercially available lubricant SAE20W40. Hence the primary objectives of the current work are to:

- To develop a suitable bio-lubricant from a non-edible vegetable oil viz. PO.
- To add suitable additives to PO and enhance its various properties.
- To compare the FO with a commercially available lubricant viz. SAE20W40.

2. Methodology

Rheological properties, oxidative stability and tribological properties of the PO with and without the addition of suitable additives are studied. Various PO blends with the additives are prepared using a magnetic stirrer. Rheological properties (dynamic viscosity, kinematic viscosity and viscosity index) of the oil blends are evaluated by using a rheometer (Anton Par MCR 102) having parallel plate geometry and redwood viscometer. Oxidation stability of the sample is determined using hot oil oxidation test (HOOT) in a dark oven. Tribological properties viz. wear scar diameter (WSD) and coefficient of friction (COF) are acquired with the help of a four-ball tester apparatus. Worn-out portions of the ball specimens are examined initially using an optical microscope and later by a scanning electron microscope (SEM). Chemical properties of the PO such as total acid number, total base number, iodine value, saponification value are also analyzed as per ASTM Standards (**Table 1**).

2.1 Materials and methods

2.2 Experimental procedure

a. BLENDING OF POLYMERIC ADDITIVES TO PO: In the present study SBR and EVA were selected as the viscosity improver additives. The PO is blended with these polymer additives to enhance the rheological properties. The samples of 0.5, 1.5 and 2.5 SBR and EVA in PO are prepared separately with the help of a magnetic stirrer having hot plate.

b. BLENDING OF ANTIOXIDANT ADDITION TO PO: The PO is blended with TBHQ to enhance the oxidative stability. 0.5, 1.5 and 2.5 wt% of antioxidant TBHQ is blended with PO respectively. The blends are prepared using a magnetic stirrer.

Materials	
Category	**Item**
Base oil	Pongamia oil
Commercial oil	SAE20W40
Viscosity improver additive	Styrene butadiene rubber (SBR) & ethylene-vinyl acetate (EVA)
Antioxidant additive	Tert-butylhydroquinone (TBHQ)
Antiwear additive	SiO_2 nanoparticle
Instruments	
Device	**Description**
Magnetic Stirrer with hot plate	Used to blend additives in oil sample
Rheometer (Anton Par MCR 102)	It is a device used to measure the dynamic viscosity of oil samples
Redwood viscometer	It is a device used to measure the kinematic viscosity of oil samples
Hot air oven	Used for hot oil oxidation test of oil samples
Four ball tester	Device used to study the tribological properties of oil samples
Thermo gravimetric analyzer (TGA)	It is used to continuously measure mass while the temperature of a sample is changed over time
Differential scanning calorimetry (DSC)	It is used to measure the difference in the amount of heat required to increase the temperature of a sample and reference is measured as a function of temperature

Table 1.
Materials and instruments used for the analysis.

c. BLENDING OF ANTIWEAR ADDITIVES TO PO: The PO is blended with nanoparticles of SiO_2 (Particle Size: 15–40 nm) to improve its tribological properties viz. wear scar diameter and coefficient of friction. The blend is prepared by adding SiO_2 in 0.6, 0.8 and 1.0 wt% respectively to PO using a magnetic stirrer.

2.3 Tests conducted

a. *TOTAL ACID NUMBER (TAN):* The TAN of the PO is obtained by titration method. It is a measure of acidity of the oil that is done by dissolving the PO in toluene and then titrating it against potassium hydroxide (KOH) using phenolphthalein as indicator. ASTM D664 standard was used for calculations.

b. *TOTAL BASE NUMBER:* The TBN of the PO is obtained by titration method. It is a measure of basicity of the oil done by dissolving PO in chlorobenzene and then titrating it against hydrochloric acid (HCl) using phenolphthalein as indicator. ASTM D2896 standard was used for calculations.

c. *SAPONIFICATION VALUE:* Saponification value denotes the number of milligrams of potassium hydroxide needed to saponify 1 gof fat according to the conditions specified. It is a calculation of the average molecular weight (or chain length) of all the fatty acids present. ASTM D5558-95 standard was used for calculations.

d. *IODINE VALUE:* The iodine value (or iodine adsorption value or iodine number or iodine index) is the mass of iodine in grams that is consumed by 100 g of a chemical substance. Iodine numbers are used to find the quantity of unsaturation in fatty acids. ASTM D1959 standard was used for calculations.

e. *DYNAMIC VISCOSITY ANALYSIS:* The dynamic viscosity is measured using Anton Par Rheometer MCR 102 in rotation mode having parallel plate geometry. Dynamic (absolute) viscosity is the tangential force per unit area required to move one horizontal plane with respect to another plane, at a unit velocity, while maintaining a unit distance apart in the fluid.

f. *KINEMATIC VISCOSITY ANALYSIS:* Kinematic viscosity is the ratio of absolute (or dynamic) viscosity to density. Force is not involved in this quantity. Kinematic viscosity can be found out by dividing the absolute viscosity of a fluid with the fluid mass density. The kinematic viscosity is measured using Redwood viscometer for PO from 40–100°C.

g. *WEAR SCAR DIAMETER ANALYSIS:* Four ball testing of the oil is done in a four ball tester apparatus, with the sample, given a load of 40 kg at 75°C. The ball is then analyzed using a scanning electron microscope and wear scar diameter is recorded.

h. *COEFFICIENT OF FRICTION ANALYSIS:* Four ball testing of the oil is done in a four ball tester apparatus, with the sample, given a load of 40 kg at 75°C. Calculations were done as per ASTM D 5183-05 standard to test COF.

i. *HOT OIL OXIDATION TEST:* The quickened aging of vegetable oil and PO added with antioxidant are stimulated with HOOT. This is done to find the oxidation stability.

j. *THERMOGRAVIMETRIC ANALYSIS:* The thermal stability of antioxidant selected and that of the formulated oil (FO) are evaluated using the thermo-analytical curves obtained from TGA, Q50 equipment, TA-Instruments.

3. Results

3.1 Preliminary analysis and enhancement of properties of PO

A series of experiments have been conducted as per standards and results are taken as an average of 3 readings having standard deviation of the sample as the error bar. The experimental data are given in the graphs and tables. **Table 2** represents the physicochemical properties of PO and compared with two widely studied bases stock viz. sesame oil and coconut oil.

The density of PO is found to be 0.92 g/cm^3 using a pycnometer, which is lower than that of water. It is found that the acid number of PO is slightly higher than that of the sesame oil and coconut oil. However, for lubricant TAN value should be low as possible [15]. PO has the least saponification value among the three. Low SV indicates long fatty acid chain, which helps in the formation of thick tribolayer [16]. Iodine value of a triglyceride indirectly shows the amount of unsaturation present in it [17]. From the iodine value, it is clear that unsaturation in PO lies between sesame oil and coconut oil. Thus, it is clear from the evaluation of physicochemical

Sl. no	Analysis	Pongamia oil	Sesame oil [9]	Coconut oil [9]
1	Iodine value (g I2/100 g)	88.18 ± 0.19	105.1	9
2	Saponification value (mg KOH/g)	168.75 ± 3.72	191	261
3	Total acid number (mg KOH/g)	3.94 ± 0.14	3.18	0.56
4	Total base number (mg KOH/g)	0.37 ± 0.01	0.41	0.16
5	Density (g/cm^3)	0.92	0.9216	0.92429

Table 2.
Physiochemical properties of PO, sesame oil, and coconut oil.

properties of PO that most of the chemical properties are well suited to the desirable properties of a base stock, which can lead to the development of an eco-friendly lubricant.

The temperature below which the liquid loses its flow characteristics is known as pour point of a liquid [18]. **Figure 1** represents the pour point of PO, which is evaluated from differential scanning calorimetry (DSC). The pour point of PO was found as 6.29°C. The pour point of PO is found to be lower than that of coconut oil [9] due to the presence of more unsaturated fatty acids. Pour point of the lubricants can be reduced using a suitable pour point depressant.

The thermal stability of PO is studied using TGA [19]. From the TGA results (**Figure 2**), the onset temperature of PO for thermal degradation of 98% is found out as 197.6°C. The weight percentage reduction 2% of PO was done by assuming

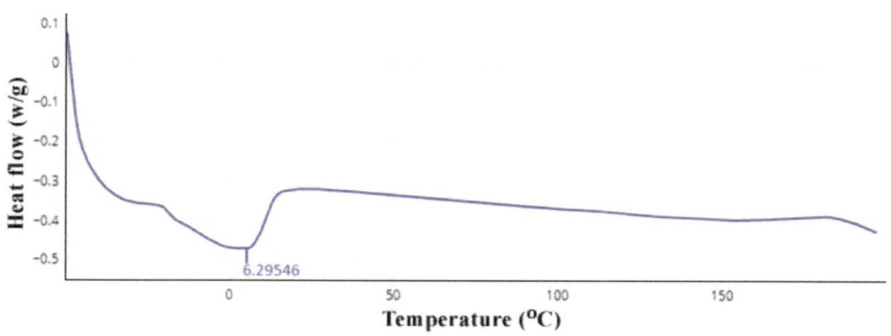

Figure 1.
Differential scanning calorimetry (DSC) curve of PO.

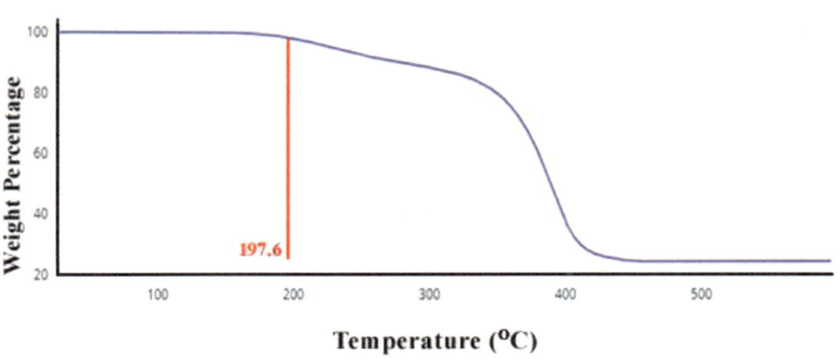

Figure 2.
TGA of PO.

the loss of moisture content and volatile components from it [19]. Thus, by observing the TGA, it is evident that PO is a well suited environment-friendly base stock for a wide range of temperatures. The thermal compared to SAE20W40 which have thermal degradation of at 204.39°C [9].

Table 3 shows the fatty acid profile of PO, compared with that of sesame oil and coconut oil. High amount of oleic acid present in the PO can improve the tribological properties [10]. A larger proportion of saturated fatty acids can adversely affect the pour point of the lubricants.

3.1.1 Analysis of rheological properties

It is clear from **Figures 3** and **4** that the dynamic viscosities of PO at various temperature range is inferior w.r.t commercially available SAE 20W40 (**Figures 3** and **4**). Hence, the dynamic viscosity of PO is enhanced using different weight percentages of EVA and SBR polymers. The variation of viscosity is studied from 25–100°C [20–22].

From **Figure 3**, it is observed that the plain PO is having the lowest viscosity both at lower as well as higher temperatures. The viscosity is found to be improving on addition of EVA in different weight percentages to PO. PO with 2.5% w.t. EVA has shown the highest change in viscosity both at lower as well as higher temperatures.

From **Figure 4**, it is found that SBR tends to show better improvements in viscosity than EVA at different weight percentages. The curve of SBR at 2.5% w.t. in PO is found to be close enough to reference oil SAE20W40. However, from the results obtained from **Figures 3** and **4**, at equal weight percentages of the additives PO + SBR combinations have shown higher viscosity enhancements. Hence, SBR has been selected as the viscosity enhancer in the current study.

Constituent	Pongamia oil [10]	Sesame oil [9]	Coconut oil [9]
Oleic acid	62.98%	42%	5%
Linoleic acid	16.84%	38%	1%
Palmetic acid	9.1%	13%	7.5%

Table 3.
Fatty acid profile of PO and sesame oil.

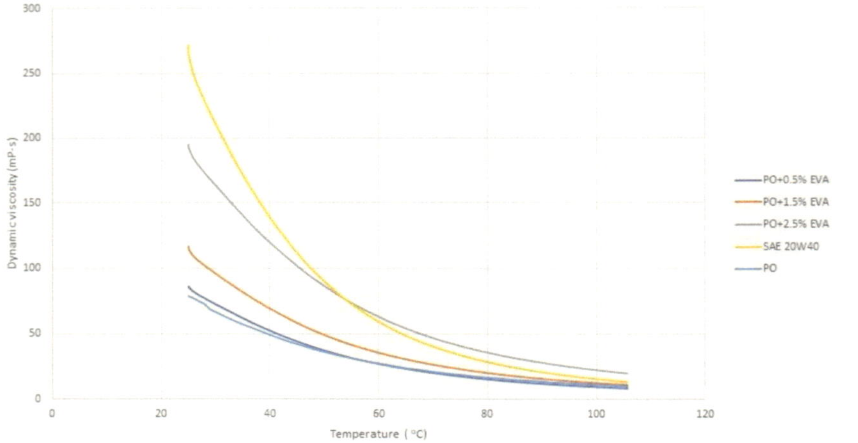

Figure 3.
Dynamic viscosity v/s temperature curve for PO with EVA compared against SAE20W40.

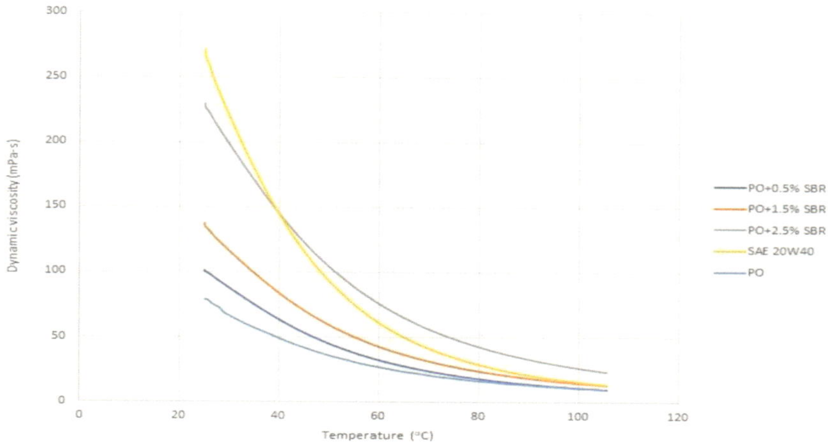

Figure 4.
Dynamic viscosity v/s temperature curve for PO with SBR compared against SAE20W40.

3.1.2 Analysis of oxidation stability

High oxidation stability of TBHQ is because of the longer alkyl chains present in it [23]. Oxidation stability of the oil samples are estimated by measuring the change in viscosity at 40°C after hot oil oxidation test (HOOT) from the fresh sample without subjected to HOOT [24]. Change in dynamic viscosity of neat PO, PO oil samples blended with different weight percentages of TBHQ and SAE20W40 at 40°C are calculated and plotted as shown in **Figure 5**. From the analysis of different oil samples blended with different weight percentages of TBHQ, PO + 2.5% TBHQ showed the least change in viscosity. It is observed that for the weight percentage of TBHQ beyond 2.5%, it is difficult to dissolve in PO. The change in viscosity at 40°C of PO + 2.5% TBHQ is found to be very lesser than neat PO comparable with that of SAE20W40. Thus PO + 2.5% TBHQ have excellent oxidation stability.

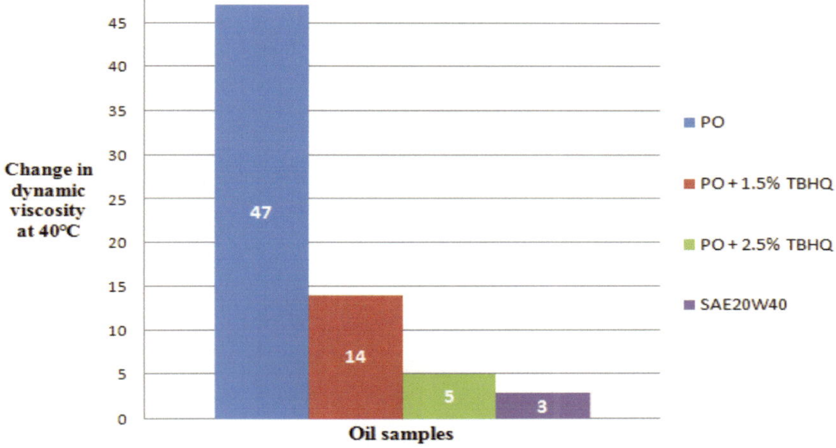

Figure 5.
Change in dynamic viscosities of oil samples at 40°C after HOOT.

3.1.3 Analysis of tribological properties

Tribological properties were evaluated with the help of a four ball tester equipment [25]. Coefficient of friction (COF) and wear scar diameter (WSD) are the parameters used to evaluate tribological properties. From **Figure 6**, it is noted that with the addition of nanoparticles, the WSD is reduced [26–29]. The WSD is found to be decreased by 18.31% with the blending of 0.85% weight percentage of SiO_2. The small size of SiO_2 makes improvement in the tribological properties by reducing friction and wear by rolling, mending, and protective film formation [30].

From **Figure 7**, it is noted that with the addition of nanoparticles, the COF is increasing for PO blends with SiO_2. However, in comparison with SAE20W40, the COF value of PO blends is considerably low.

3.2 Formulated oil results

Based on the tests conducted, the formulated oil (FO) is PO + 2.5 wt% SBR+ 2.5 wt% TBHQ +0.85 wt% SiO_2.

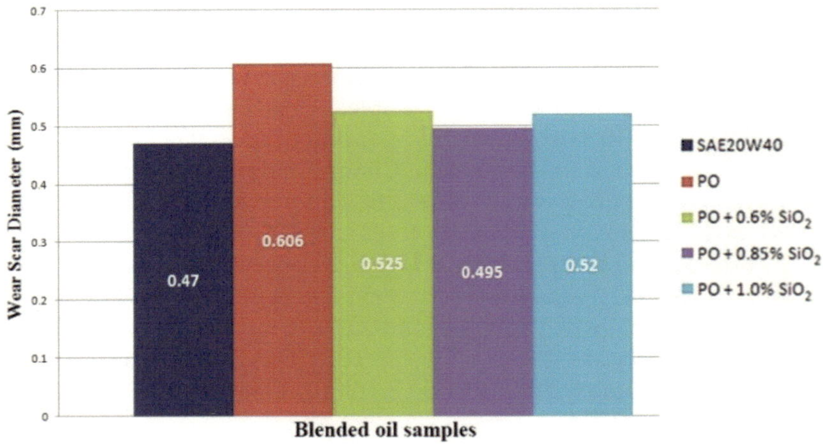

Figure 6.
Variation in WSD with the blending of different weight percentages of SiO_2 with PO.

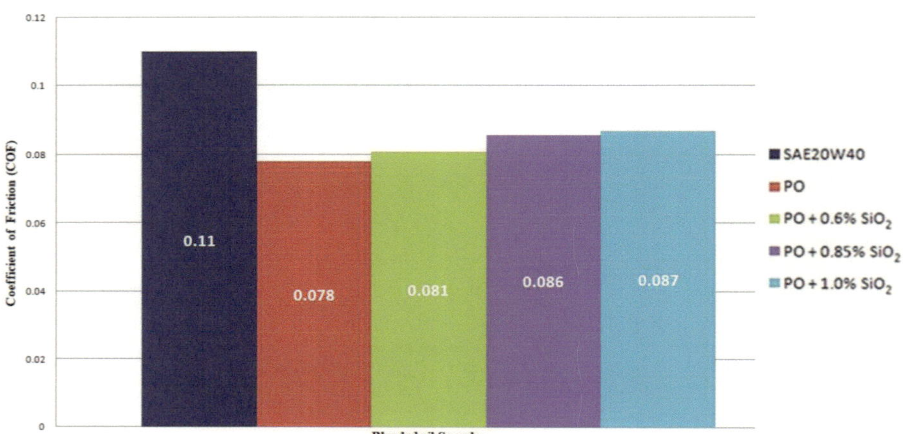

Figure 7.
Variation in COF with the blending of different weight percentages of SiO_2 with PO.

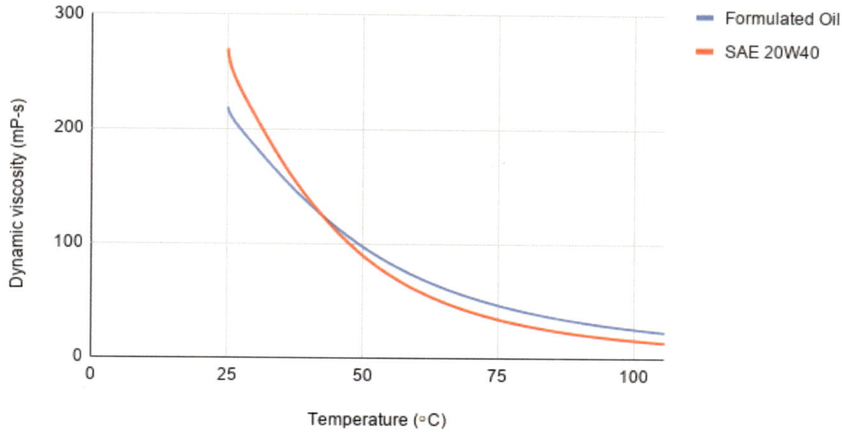

Figure 8.
Dynamic viscosity v/s temperature curve comparison of formulated oil and SAE 20W40.

Properties	FO	SAE20W40 [9]
COF	0.092	0.107
Wear scar diameter	0.48 mm	0.47 mm
Viscosity index	227.4	135.57
Increase in dynamic viscosity after HOOT at 40°C	5.4 cP	3 cP

Table 4.
Evaluated properties and results of the FO compared with SAE20W40.

Figure 8 represents the dynamic viscosity v/s temperature curve of the formulated oil and was found to be comparable with that of reference oil SAE 20W40.

The evaluated properties and results of FO compared with SAE20W40 is as shown in **Table 4**. The COF of FO is found to be comparatively lower than that of SAE20W40. This implies that the frictional effect produced by FO will be much lower than the latter. The viscosity index of the FO shows much improved results than that of SAE20W40. This ensures the use of FO at wide range of temperatures. The WSD of the FO is much closer to that of SAE20W40. Therefore, the FO has almost similar anti-wear effects as that of SAE20W40. Increase in dynamic viscosity after HOOT at 40°C indicates that the oxidation stability of the FO is superior that PO and is comparable with that of SAE20W40.

4. Conclusion

The project was aimed on formulating a bio-lubricant from a non-edible vegetable oil as base stock. In addition to non-edible nature, pongamia oil (PO) was chosen as the base oil due to its high oleic acid content. SBR, TBHQ and SiO_2 nanoparticle additives are added to PO for improving its viscosity, oxidation stability and antiwear properties respectively. The final base oil and additive combination of the formulated oil (FO) is PO + 2.5 wt% SBR + 2.5 wt% TBHQ +0.85 wt% SiO_2. Rheological studies show that the performance of the FO is identical to the dynamic viscosity trend of SAE20W40 and possess a superior viscosity index than SAE 20 W40. HOOT indicates that the oxidation stability of FO is much closer to that of SAE20W40. Tribological studies indicate that FO possess identical WSD and lower COF than SAE20W40.

Pongamia (milletia) is a genus of legume plant family. It is distributed in tropical and subtropical regions of the world. It is easily cultivable [31] and can grow in semi-arid conditions [32]. The cultivation of these trees has increased due to its recent interests in the field of bio-fuels and lubricants. This project is based on the enhancement of pongamia oil (PO) to a bio-lubricant, the cultivation of these trees will be promoted to a great extend thereby improving the agricultural sector.

Author details

Sabarinath Sankarannair[1]*, Avinash Ajith Nair[2], Benji Varghese Bijo[2], Hareesh Kuttuvelil Das[2] and Harigovind Sureshkumar[2]

1 Department of Mechanical Engineering, TKM College of Engineering, Kollam, Kerala, India

2 Department of Mechanical Engineering, Saintgits College of Engineering, Kottayam, Kerala, India

*Address all correspondence to: sabarioch@gmail.com

References

[1] Srivastava SP. Classification of lubricants. In: Srivastava SP, editor. Developments in Lubricant Technology. USA: John Wiley & Sons; 2014. DOI: 10.1002/9781118907948.ch2

[2] Pirro DM, Webster M, Daschner E. Lubrication Fundamentals (Third Edition, Revised and Expanded). Abingdon, UK: CRC Press; 2016

[3] Höök M, Tang X. Depletion of fossil fuels and anthropogenic climate change—A review. Energy Policy. 2013;**52**:797-809

[4] Tilton JE. The Hubbert peak model and assessing the threat of mineral depletion. Resources, Conservation and Recycling. 2018;**139**:280-286

[5] Bart Jan CJ, Gucciardi E, Cavallaro S. Biolubricants: Science and Technology. Amsterdam, Netherlands: Elsevier; 2012

[6] Wu X, Yue B, Su Y, Wang Q, Huang Q, Wang Q, et al. Pollution characteristics of polycyclic aromatic hydrocarbons in common used mineral oils and their transformation during oil regeneration. Journal of Environmental Sciences. 2017;**56**:247-253

[7] Sharma BK, Biresaw G. Environmentally Friendly and Biobased Lubricants. Abingdon, UK: CRC Press; 2016

[8] Chatra KS, Jayadas N, Kailas SV. Natural oil-based lubricants. In: Green Tribology. New York, USA: Springer; 2012. pp. 287-328

[9] Nair S, Nair K, Rajendrakumar P. Evaluation of physicochemical, thermal and tribological properties of sesame oil (Sesamum indicum L.): A potential agricultural crop base stock for eco-friendly industrial lubricants. International Journal of Agricultural Resources, Governance and Ecology. 2017;**13**:77. DOI: 10.1504/IJARGE.2017.084037

[10] Dauenhauer P. Handbook of plant-based biofuels. Edited by Ashok Pandey. ChemSusChem. 2010;**3**:386-387. DOI: 10.1002/cssc.201000009

[11] Srivastava P, Verma M. Methyl ester of karanja oil as an alternative renewable source energy. Fuel. 2008;**87**:1673-1677. DOI: 10.1016/j.fuel.2007.08.018

[12] Bobade SN, Khyade VB. Detail study on the properties of Pongamia pinnata (Karanja) for the production of biofuel. Research Journal of Chemical Sciences. 2012;**2**:16-20

[13] Mahipal D, Krishnanunni P, Mohammed P, Jayadas NH. Analysis of lubrication properties of zinc-dialkyl-dithio-phosphate (ZDDP) additive on Karanja oil (Pongamia pinnatta) as a green lubricant. International Journal of Engine Research. 2014;**3**:494-496. DOI: 10.17950/ijer/v3s8/804

[14] Chuah LF, Suzana Y, Aziz A, Klemeš J, Bokhari A, Abdullah Z. Influence of fatty acids content in non-edible oil for biodiesel properties. Clean Technologies and Environmental Policy. 2016;**18**(2):473-482. DOI: 10.1007/s10098015-1022-x

[15] Hasannuddin AK et al. Performance, emissions and lubricant oil analysis of diesel engine running on emulsion fuel. Energy Conversion and Management. 2016;**117**:548-557

[16] Mamada K et al. Friction properties of poly (vinyl alcohol) hydrogel: Effects of degree of polymerization and saponification value. Tribology Letters. 2011;**42**(2):241-251

[17] Kemp AR, Mueller GS. Iodine value of rubber and gutta-percha hydrocarbons As determined by iodine chloride. Industrial and Engineering

Chemistry, Analytical Edition. 1934;**6**(1):52-56

[18] Erhan SZ, Asadauskas S. Lubricant basestocks from vegetable oils. Industrial Crops and Products. 2000;**11**(2-3):277-282

[19] Jayadas NH, Prabhakaran Nair K. Coconut oil as base oil for industrial lubricants—Evaluation and modification of thermal, oxidative and low temperature properties. Tribology International. 2006;**39**(9):873-878

[20] Sabarinath S, Prabhakaran Nair K, Rajendrakumar P, Parameswaran P. Styrene butadiene rubber as a viscosity improver: Experimental investigations and quantum chemical studies. Proceedings of the Institution of Mechanical Engineers, Part J. 2018;**232**(4):427-436

[21] Topal A, Yilmaz M, Kok BV, Kuloglu N, Sengoz B. Evaluation of rheological and image properties of styrene-butadiene-styrene and ethylene-vinyl acetate polymer modified bitumens. Journal of Applied Polymer Science. 2011;**122**:3122-3132

[22] Varkey JT, Thomas S, Rao SS. Rheological behavior of blends of natural rubber and styrene–butadiene rubber latices. Journal of Applied Polymer Science. 1995;**56**:451-460

[23] Domingos AK et al. The influence of BHA, BHT and TBHQ on the oxidation stability of soybean oil ethyl esters (biodiesel). Journal of the Brazilian Chemical Society. 2007;**18**(2):416-423

[24] Nair SS, Nair KP, Rajendrakumar PK. Experimental and quantum chemical investigations on the oxidative stability of sesame oil base stock with synthetic antioxidant additives. Lubrication Science. 2019;**31**(5):179-193. DOI: 10.1002/ls.1441

[25] Yadav G, Tiwari S, Jain ML. Tribological analysis of extreme pressure and anti-wear properties of engine lubricating oil using four ball tester. Materials Today: Proceedings. 2018;**5**(1):248-253

[26] Nair SS, Nair KP, Rajendrakumar PK. Micro and nano particles blended sesame oil bio-lubricant: Study of its tribological and rheological properties. Micro & Nano Letters. 2018;**13**(12):1743-1746. DOI: 10.1049/mnl.2018.5395

[27] Shafi WK, Charoo MS. Rheological properties of sesame oil mixed with H-Bn nanoparticles as industrial lubricant. Materials Today: Proceedings. 2019;**18**(7):4963-4967

[28] Rajubhai VH et al. Friction and wear behavior of Al-7% Si alloy pin under pongamia oil with copper nanoparticles as additives. Materials Today: Proceedings. 2020;**25**:695-698

[29] Sajeeb A, Rajendrakumar PK. Comparative evaluation of lubricant properties of biodegradable blend of coconut and mustard oil. Journal of Cleaner Production. 2019;**240**:118255

[30] Sabarinath S, Rajendrakumar P, Prabhakaran Nair K. Evaluation of tribological properties of sesame oil as biolubricant with SiO_2 nanoparticles and imidazolium-based ionic liquid as hybrid additives. Proceedings of the Institution of Mechanical Engineers, Part J. 2019;**233**(9):1306-1317

[31] Sajjadi B, Raman AAA, Arandiyan H. A comprehensive review on properties of edible and non-edible vegetable oil-based biodiesel: Composition, specifications and prediction models. Renewable and Sustainable Energy Reviews. 2016;**63**:62-92

[32] Kesari V, Rangan L. Development of Pongamia pinnata as an alternative biofuel crop — Current status and scope of plantations in India. Journal of Crop Science and Biotechnology. 2010;**13**:127-137

Chapter 12

Nonlinear Frictional Dynamics on Rolling Contact

Yasunori Sakai

Abstract

The rolling machine element is indispensable for realizing high-precision and high-speed relative motion. In addition, its positioning accuracy is approaching the nanometer order, and its importance is expected to increase in the future. However, since the rolling elements and the raceways are mechanically in contact, various nonlinear phenomena occur. This complicated phenomenon must be clear by theoretically and experimentally. This chapter describes the nonlinear friction behavior occurred with rolling contact condition and its effect on the dynamics of bearings. First, the characteristics of the non-linear friction caused by rolling machine elements and the nonlinear friction modeling method using the Masing rule are described. From the numerical analysis using the friction model, it is clarified that the motion accuracy decreases due to sudden velocity variation caused by nonlinear friction. Also, the author show that the resonance phenomenon and force dependency of the dynamic characteristics of rolling machine element due to the nonlinear friction. Finally, the author indicates nonlinear friction influences on the dynamic characteristics in the directions other than the feed direction.

Keywords: nonlinear dynamics, friction, rolling machine element, elastic contact, bearing

1. Introduction

Rolling machine elements such as linear rolling bearings and rotational bearings are used in various industrial fields because they realize high precision relative motion at high speed. In addition, since the fluctuation of friction force is small, self-excited vibration called stick-slip is less likely to occur. Though rolling elements such as balls and rollers are always in point or line contact with the raceway, resulting in high surface pressure. The rolling contact condition complicates tribological phenomena such as friction and wear, and dynamic behavior of rolling machine elements.

Linear rolling bearings are widely used in semiconductor manufacturing equipment, ultra-precision positioning systems for machine tool, robots, and other equipment. Conventionally, it was necessary to use non-contact bearings such as hydrostatic linear guideways to achieve positioning with nanometer accuracy. However, it has become possible to achieve nanometer positioning accuracy even in rolling bearings by optimizing the rolling contact condition.

On the other hand, since rolling elements and raceways are in rolling contact with each other, the damping capacity of bearing is much lower than other guideways such as hydrostatic guides and sliding guides. In addition, the non-linear behavior of contact friction produces resonance and displacement-amplitude dependency in the dynamics of bearings.

The resonance phenomenon of the bearing and the amplitude dependency of the dynamics lead to instability of the positioning control system and the deterioration of positioning accuracy [1–6]. In order to develop more accurate positioning control system and damping element for precision positioning system, it is important to clarify and model the effects of the nonlinear friction behavior on the dynamics of bearings.

This chapter describes the nonlinear friction behavior occurred with rolling contact condition and its effect on the dynamics of bearings. Section 2 explains the nonlinear spring behavior (NSB) depending on displacement, as well as the modeling method of the nonlinear frictional behavior. Section 3 describes the influence of nonlinear friction on the dynamics of the bearing with some numerical and experimental results. Section 4 shows the influence of nonlinear friction on the dynamics in directions other than the feed direction of the bearing. Section 5 concludes the chapter and gives the future interest of the research related to rolling friction behavior.

2. Nonlinear friction behavior in microscopic displacement regime

2.1 Displacement dependent nonlinearity

Figure 1 shows the relationship between the friction force F of the sliding object and its translational displacement x [7]. After the mass is starting motion, the friction force increases as the displacement increases. When the friction force is saturated with the steady state friction force F_s, it is constant until the mass reaches the reversal point A. After reversing the direction of motion at point A, the friction force changes along the outer hysteresis curve ABA. If the motion is reversed at arbitrary point a on the outer hysteresis curve, the friction force will change along the inner hysteresis curve aba starting at point a. When the internal hysteresis curve is closed, the frictional force changes along ABA again. This hysteresis rule is known as the non-local memory NSB [8]. The NSB is caused by elastic deformation of the contact surface asperities [9], microslip [10], and elastic hysteresis loss [11]. In this chapter, the curve OA is called a virgin loading curve, the displacement region where the friction depends on the displacement is called the pre-rolling region, and the displacement region where the friction force is constant is called the rolling

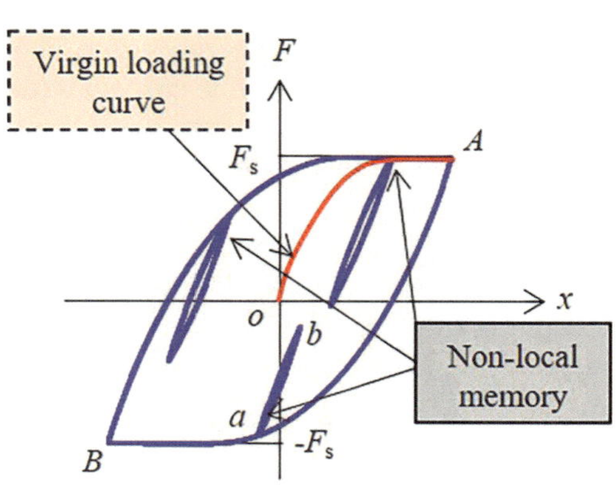

Figure 1.
The relationship between the friction force F *and the displacement* x.

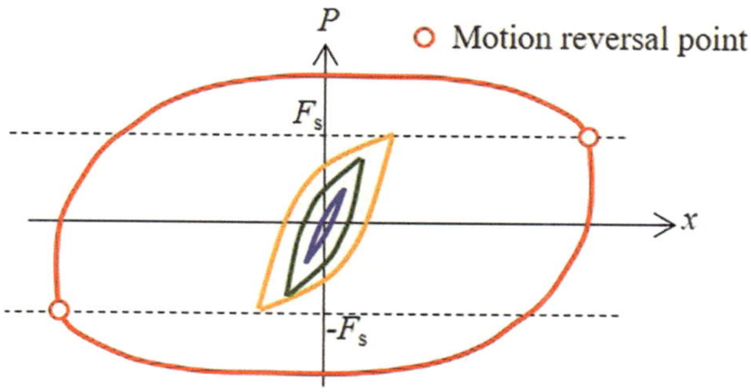

Figure 2.
The relationship between the sinusoidal excitation force P *and the displacement* x.

region. Furthermore, the boundary displacement between the pre-rolling region and the rolling region is called the starting rolling displacement.

Now, the sinusoidal excitation force $P(t) = P_0 \sin\omega t$ acting on the mass where P_0 is the amplitude and ω is the excitation angular frequency. **Figure 2** shows the relationship between sinusoidal excitation force P and displacement x [7]. If the force amplitude P_0 is greater than the static friction force F_s, the displacement will change significantly (such as an arc within $P > F_s$). This suggests that the rolling element initiates a rolling movement. F_s can be determined as P at the motion reversal point [12].

2.2 Nonlinear friction modeling

The friction model proposed based on the Masing rule can describe the NSB and hysteresis behavior of friction in non-local memory. The Masing rule simply describes a hysteresis curve from the virgin loading curve. Therefore, it is widely used in the basic model of seismic response analysis [13] and the hysteresis model [14]. The Masing rule is formulated as follows:

$$F = \begin{cases} f(x) & (x \geq 0) \\ -f(-x) & (x < 0) \end{cases} \tag{1}$$

$$F = \begin{cases} F_r + \lambda f\left(\frac{x - x_r}{\lambda}\right) & \left(\frac{x - x_r}{\lambda} \geq 0\right) \\ F_r - \lambda f\left(-\frac{x - x_r}{\lambda}\right) & \left(\frac{x - x_r}{\lambda} < 0\right) \end{cases} \tag{2}$$

where $f(x)$ is the virgin loading curve. F_r and x_r are the friction force and displacement at the motion reversal point. The hysteresis curve is determined by the geometrically similar curve of the virgin loading curve with a similarity ratio $\lambda = 2$. If $\lambda \neq 2$, the hysteresis curve does not become axisymmetric. In the friction model, $\lambda = 2$.

Figure 3 shows the procedure for calculating the friction force [7]. First, the friction force is calculated by Eq. (1) until reaching the motion reversal point A. After reversing the direction of motion at point A, calculate the friction force using Eq. (2). At this time, the values of x_r and F_r are replaced by the displacement and frictional force at point A. Furthermore, the friction force can be calculated as well when the motion is reversed at the point B or C. The friction force after closing the

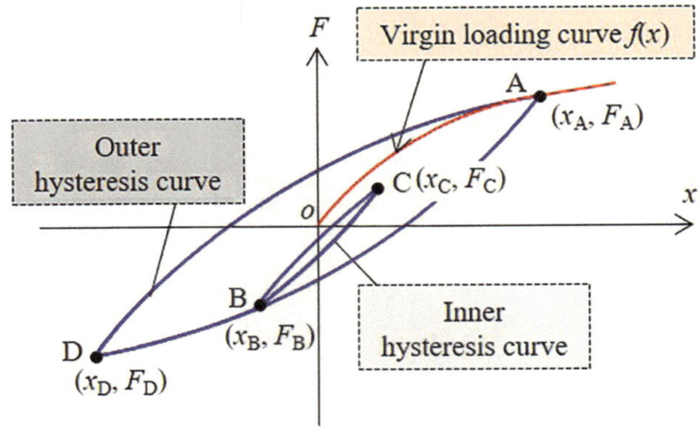

Figure 3.
The relation between friction force F *and displacement* x *calculated by Masing rule.*

internal hysteresis curve aba is calculated with Eq. (2) by replacing the values of x_r and f_r to the displacement and friction at point A.

The hysteresis characteristics of non-local memory can be explained by above mentioned calculations. If x is greater than the maximum displacement value at the previous motion reversal point, the frictional force is calculated by Eq. (1) again.

This model has fewer parameters than previous friction models such as the bristle model [15] and the generalized Maxwell slip (GMS) model [16].

The friction model based on the Masing rule can describe the effect of NSB on the dynamic characteristics by simplifying the friction behavior. Al-Bender uses the exponential and irrational functions as the virgin loading curves to calculate the friction force [17]. However, the rolling region and the starting rolling displacement x_s do not consider. The resonances caused by NSB are depended on x_s [18]. Therefore, the starting rolling displacement and steady-state friction force in the rolling region should be introduced into the friction model.

The virgin loading curve is described by Eq. (3) proposed in this study:

$$f(x) = \begin{cases} A(x + Bx^n) \equiv f_1(x) & (x \le x_s) \\ F_s \equiv f_2(x) & (x > x_s) \end{cases} \tag{3}$$

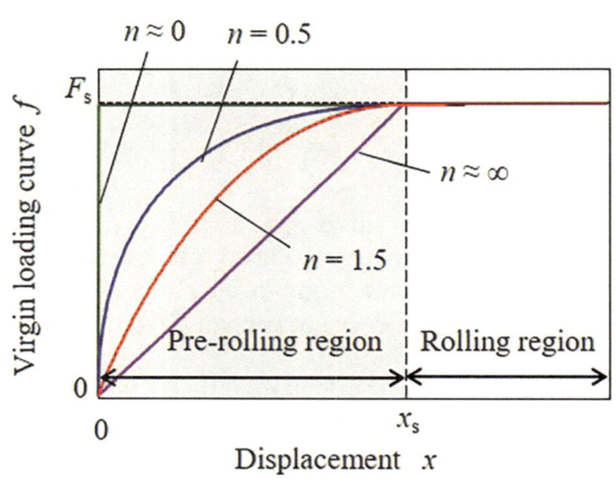

Figure 4.
The proposed virgin loading curve f *for different shape factor* n.

where A and B are constants determined by the continuity condition as below:

$$f_1(x_s) = f_2(x_s) \tag{4}$$

$$\left.\frac{df_1(x)}{dx}\right|_{x=x_s} = \left.\frac{df_2(x)}{dx}\right|_{x=x_s} \tag{5}$$

This virgin loading curve has only three parameters: steady-state friction force F_s, starting rolling displacement x_s, and shape factor n. The shape factor n represents the rate of change in friction of the pre-rolling region and is the most important parameter in this model [19]. These parameters determine the characteristics of the NSB. The proposed virgin loading curves for different n are shown in **Figure 4** [7].

In this model, $n \neq 1$ is always satisfied. Because $n = 1$ means that the friction does not have hysteretic behavior, it is uncommon in the friction characteristics of a rolling bearing.

3. Resonance phenomenon caused by nonlinear frictional behavior

In this section, the frictional effect of NSB on the dynamic characteristics of the sliding object is discussed. The frequency response of the sliding object is an important consideration for developing a highly accurate feed drive system in machine tool and precision machines using friction compensators.

3.1 Numerical analysis of dynamics considering nonlinear friction

Figure 5 shows an analytical model for calculating the dynamics in the feed direction [7]. The equation of motion is as follows:

$$m\ddot{x} = -F(x) + P(t) \tag{6}$$

where m is the mass of sliding object; t is the time; $[\cdot] = d/dt$; $P(t)$ is the exciting force acting on the mass; $F(x)$ is the friction force calculated from Eqs. (1)–(3).

The equation of motion can be expressed by Eq. (7) considering the sinusoidal excitation force $P(t) = P_0 \sin\omega t$ acting on the mass.

$$m\ddot{x} = -F(x) + P_0 \sin \omega t \tag{7}$$

By introducing the dimensionless parameters $K_s = F_s/x_s$, $\omega_s^2 = K_s/m$, $u = x/x_s$, $\gamma_0 = P_0/F_s$, $\tau = \omega t$, $\beta = \omega/\omega_s$ and $[`] = d/d\tau$ into the Eq. (7), the dimensionless equation of motion is described as Eq. (8).

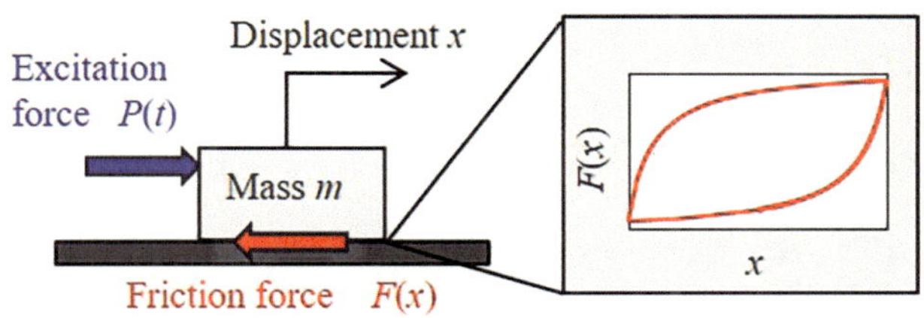

Figure 5.
The analytical model of the rolling guideway for calculating the dynamic characteristics in the feed direction in consideration of the nonlinear spring behavior of friction.

$$\beta^2 u'' = -\frac{F(u)}{F_s} + \gamma_0 \sin \tau \quad (8)$$

As shown in Eq. (8), the motion of the sliding object is governed by the amplitude of the dimensionless excitation force γ_0, the dimensionless excitation frequency β, and the dimensionless friction force F/F_s. F/F_s is a function of shape factor n. Ultimately, the motion of the sliding object is described by determining γ_0, β, and n. Eq. (8) is solved by the fourth-order Runge-Kutta method with nondimensional time derivative $d\tau = \tau/10000$. The frequency response function (FRF) $G(\beta)$ is calculated by Eq. (9).

$$G(\beta) = \frac{u_0}{\gamma_0} = \frac{(u_{\max} - u_{\min})}{2\gamma_0} \quad (9)$$

where u_0 is the displacement amplitude in the steady-state response, $u_{\max}$ and $u_{\min}$ are the maximum and minimum displacements in the steady-state response, respectively.

Figure 6 shows an FRF with dimensionless excitation force amplitude $\gamma_0 = 0.6$ for $n = 0.5$ and 1.5 [7]. As shown in **Figure 6**, the FRF shape and resonant frequency change by n. It means that the NSB affects the dynamic characteristics of the sliding object. There are additional small resonant peaks in the FRF. The small resonance peaks are superharmonic resonances, which are typical phenomenon of nonlinear vibration caused by the nonlinearity of restoring forces [20]. Because the proposed friction model does not take into account the difference between static and dynamic friction, these resonances does not induced by the stick-slip phenomena which is occurred by the negative damping due to the difference between static and dynamic friction forces. In the following, the main resonance peaks are called harmonic resonances and the other resonances are called superharmonic resonances.

By calculating the steady-state response of a specific excitation frequency, the effect of the excitation frequency on the steady-state motion in the feed direction is clarified. **Figures 7–9** show the phase planes of the limit cycles of $n = 0.5$ and 1.5 at frequencies A, B, and C shown in **Figure 6**, respectively [7]. Frequency A is lower than the frequency of the superharmonic resonance of the adjacent harmonic resonance. Frequency B lies between the harmonic resonance and its adjacent superharmonic resonance. Finally, frequency C is higher than the frequency of harmonic resonance.

According to the results as shown in **Figure 7**, when a sliding object is excited by a sinusoidal force of frequency A, velocity reduction and recovery (VRAR) occurs

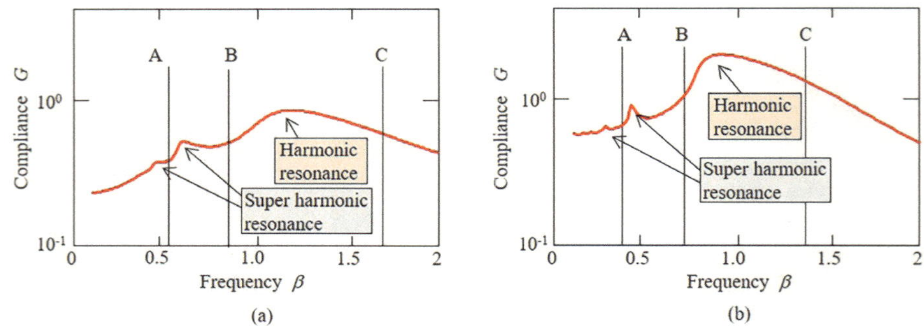

Figure 6.
The frequency response functions in the feed direction with the non-dimensional excitation force amplitude $\gamma_0 = 0.6$ *for different shape factor* n. *(a)* n = 0.5. *(b)* n = 1.5.

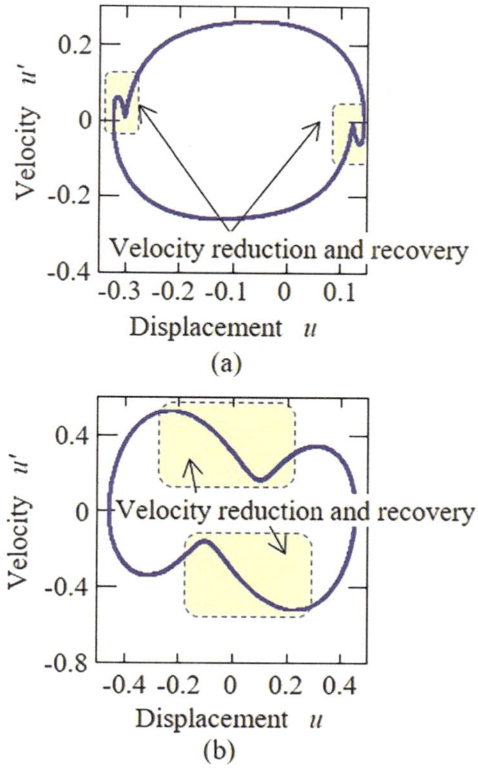

Figure 7.
The phase plane at frequency of A. (a) n = 0.5, β = 0.54. *(b)* n = 1.5, β = 0.45.

after reversing the direction of movement. When the sliding object is excited at frequency B, VRAR does not occur after reversing the direction of movement, as shown in **Figure 8**. However, the shape of the phase plane is distorted. On the other hand, as shown in **Figure 9**, when the sliding object is excited at frequency C, the phase plane becomes circular. If the sliding object is excited by a force with a frequency higher than the harmonic resonance, the non-linearity due to NSB is negligible.

Next, the effect of excitation and friction conditions on the time history of steady-state motion is examined. **Figures 10** and **11** show the dimensionless acceleration u , dimensionless velocity u', dimensionless displacement u, and dimensionless friction force F/F_s at frequencies A and B with n = 0.5 and 1.5, respectively [7].

Comparing **Figures 10** and **11**, the variation of acceleration direction (VOAD) unrelated to sinusoidal displacement motion with the excitation frequency and VRAR are clear when excited with frequency A. Hence, the VOAD causes VRAR. The displacement spike is clearly observed in **Figure 10**, but it does not observe in **Figure 11**. Thus, the displacement spike is caused by the VOAD. Furthermore, the displacement spike sharpens when the VOAD is caused discontinuously. On the other hand, it loosens when the VOAD is caused continuously. The VOAD is affected by NSB of friction and becomes discontinuous when n is small.

Displacement spikes, known as quadrant glitches, are one of the causes of poor feed drive operation accuracy. Previous studies have concluded that quadrant glitches are produced due to motion delays caused by the difference between static and dynamic friction [21]. The results of this study show that displacement spikes

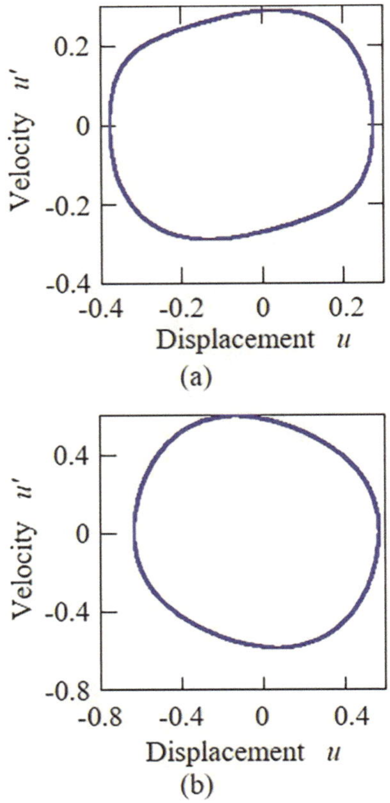

Figure 8.
The phase plane at frequency of B. (a) n = 0.5, β = 0.90. *(b)* n = 1.5, β = 0.71.

(quadrant glitches) are generated without modeling the difference between static and dynamic frictional forces. It means that the quadrant glitches are caused not only by the difference between static and dynamic friction, but also by NSB friction. Also, as shown in **Figure 10**, the quadrant glitches occur even at zero speed. Previous studies have concluded that quadrant glitches are produced by reduced acceleration [22]. These conclusions correspond to our result that quadrant glitches are caused by VOAD.

Figure 12 shows the FRF for n = 0.5 and 1.5. The figure also contains the results of various dimensionless excitation force amplitudes γ_0 [7]. As shown in **Figure 12**, as γ_0 increases, the resonant frequency decreases. Furthermore, resonance does not occur even if γ_0 increases. The compliance at frequencies of superharmonic and harmonic resonance increases with higher values of n. For small n, not only the nonlinearity but also the damping increases. Therefore, when n is small, the compliance at resonance is not high than the case with large n.

Then, the results of the proposed friction model are compared with the experimental results using commercially available roller guideways for verifying the validity.

Figure 13 shows the relationship between the excitation force amplitude and the frequency and compliance at the harmonic resonance [7]. **Figure 13** shows the measurement results of the compliance in the feed direction of the linear rolling bearing obtained in the vibration test conducted under the same conditions as the numerical analysis. The analytical results for n = 1.5 conform well to the experimental results. The comparison results prove that the proposed simple analytical

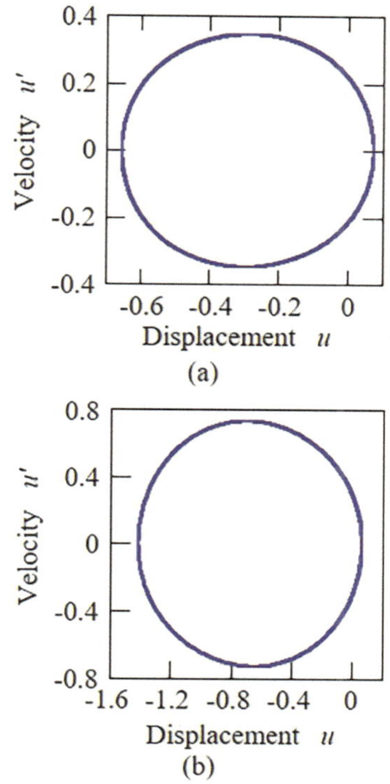

Figure 9.
The phase plane at frequency of C. (a) n = 0.5, β = 1.65. *(b)* n = 1.5, β = 1.40.

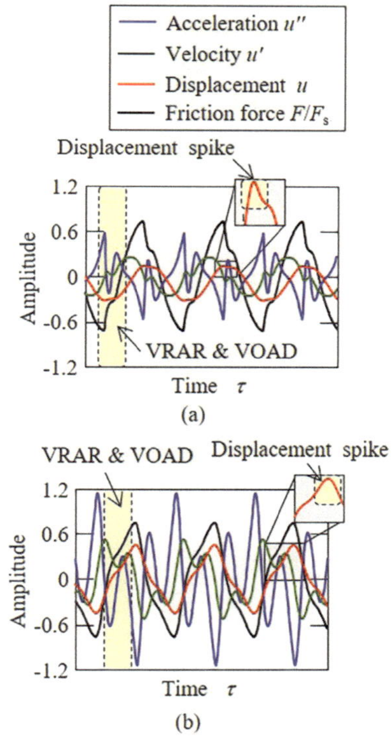

Figure 10.
Time history at the frequency of A. (a) n = 0.5, β = 0.54. *(b)* n = 1.5, β = 0.45.

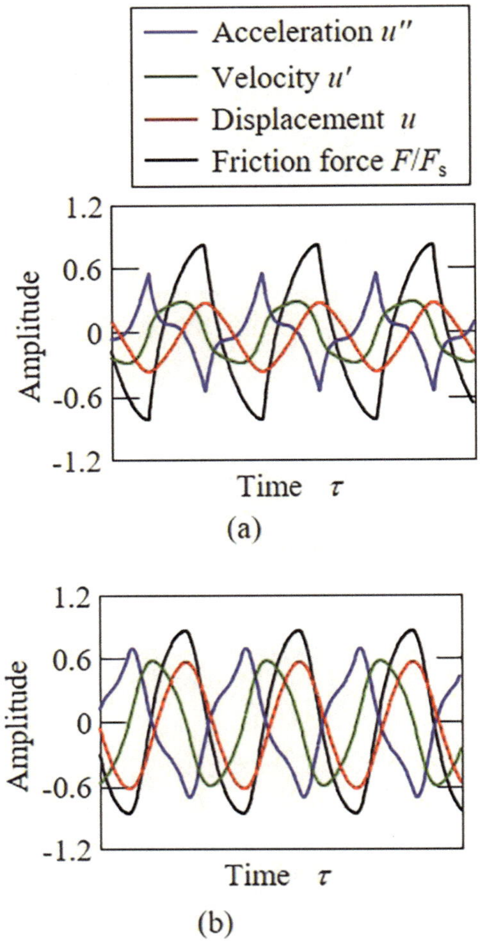

Figure 11.
Time history at the frequency of B. (a) n = 0.5, β = 0.90. *(b)* n = 1.5, β = 0.71.

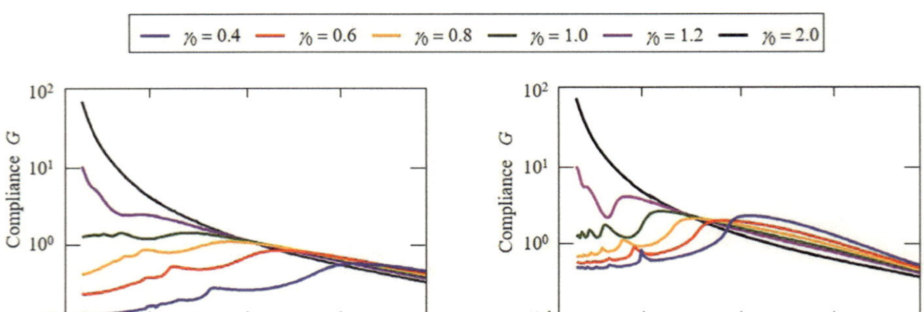

Figure 12.
The force amplitude dependency of the frequency response caused by the nonlinear spring behavior. (a) n = 0.5. *(b)* n = 1.5.

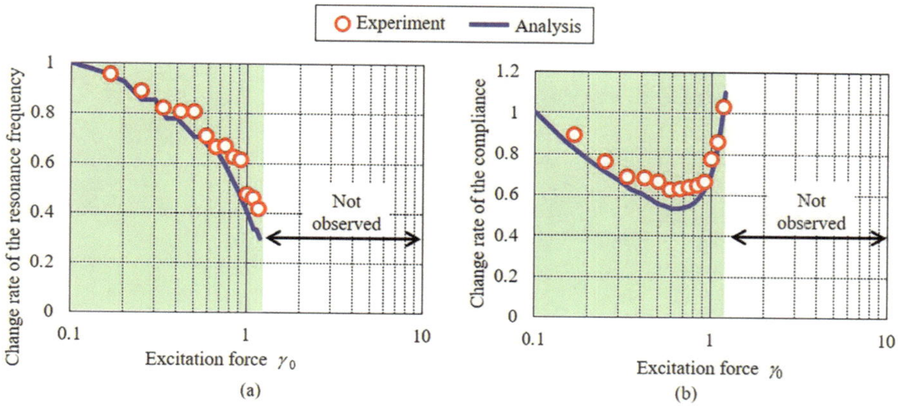

Figure 13.
Relationship between the non-dimensional excitation force amplitude γ_0 and the resonance frequency and compliance at the harmonic resonance. (a) Harmonic resonance frequency. (b) Compliance at the harmonic resonance.

model can qualitatively predict the force dependency of the dynamic characteristics of a rolling bearing in the feed direction.

4. Nonlinear behavior of rotational direction induced by friction

4.1 Experiment for evaluating the dynamics of linear rolling bearing

In a previous study, Yi et al. evaluated the vibration characteristics of a carriage with a steel block (additional mass). Feed, lateral, and vertical response accelerations were measured using a 3-axis accelerometer when the steel block was excited by the shaker [23]. Ota et al. measured the vibration of the carriage moving at a constant speed in the feed direction. They also identified key vibrational components by frequency spectrum analysis of the detected response sounds and accelerations [24]. Through the impulse hammering test, Rahman et al. evaluated the natural frequencies and damping ratios of machine tool tables supported by linear rolling bearings. The natural frequency and damping ratio were identified based on the frequency response function [25].

In contrast, the author measured the acceleration of a carriage carrying a column which is imitated the long workpiece. The vibration measuring system for evaluating the dynamic characteristics of the linear rolling bearing is shown in **Figure 14** [26]. The system comprises a rail, a carriage, a column made of S50C steel, an electrodynamic shaker, and a stinger made of S45C. The column was fixed on the carriage with bolts. The rail was fixed on the stone surface plate through the steel base. The shaker was supported with cloth belt to obtain a free boundary condition. The stinger has a circular cross-sectional shape. One of stinger ends was screwed to the free end of the column through the impedance head, and the other one was screwed to the shaker.

The column was used as a pseudo workpiece to satisfy the condition that the center of gravity is higher than the upper surface of the carriage. This condition is often observed in the general usage of linear rolling bearings, such as in the feed drive mechanism of a machine tool. The excitation force in the feed direction was applied to the free end of the column.

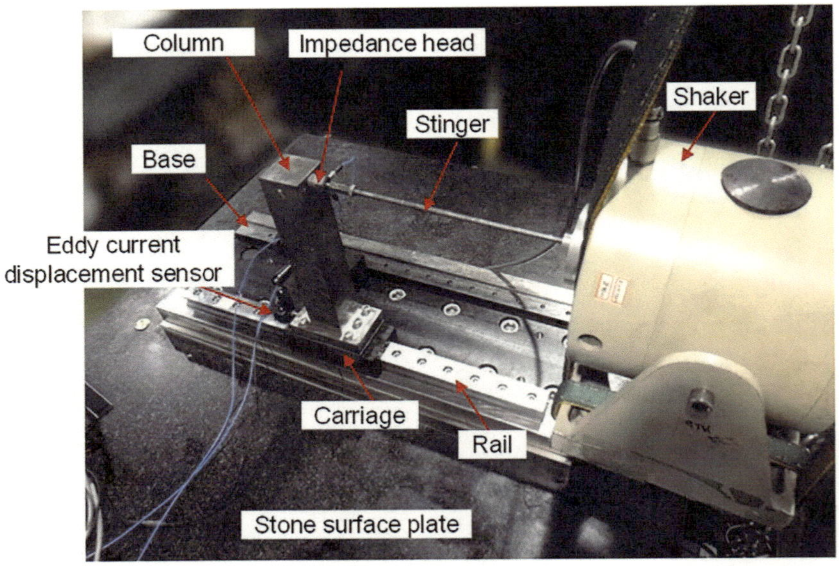

Figure 14.
Experimental setup for evaluating the vibration characteristics.

The sinusoidal excitation signal generated using the function generator was input to the shaker via a power amplifier. The excitation and acceleration signals detected at the free end of the column were amplified by a charge amplifier and analyzed using a digital spectrum analyzer. In addition, at the same time as the excitation, the eddy current displacement sensor was used to detect the response displacement of the carriage in the feed direction.

In the general case, forces with different amplitudes and frequencies act on the linear rolling guide. For example, in the case of a machine tool, cutting force acts on the rolling guide via the workpiece and table. Its frequency and amplitude vary widely depending on machining conditions such as depth of cut and spindle speed. Micro cutting exerts a force of several millinewtons, and heavy cutting exerts a force of hundreds of newtons. Therefore, the exciting force changed from 0.1 N to 100 N in half amplitude and up to 1000 Hz in frequency.

The lubricant (mineral oil) weighed by syringe was manually applied to the raceways (5 ml per raceway). When changing the lubricant oil, the all components of the linear rolling bearing were washed with kerosene and alcohol for degreasing.

4.2 Influence of excitation force in rotational direction

Figure 15 shows the FRF when excited in the feed direction. Two resonance peaks were observed in the FRF [26]. The vibration modes at these resonance frequencies were measured by experimental modal analysis. **Figure 16** shows the mode shapes of the carriage and column at the two resonances [26]. At the lower resonance frequency, the carriage and column vibrate in the feed direction without any elastic deformation. It was caused by the nonlinear spring behavior of friction in the microscopic region discussed in the previous chapters. At the higher resonance frequency, the carriage and column vibrated in the pitch direction. This natural vibration is caused by the elastic deformation of the contact part between the raceways and the rollers. Since the center of gravity is higher than the upper surface of the carriage, not only the pitching motion but also the translational motion in the feed direction is occurred at the same time. In followings, these

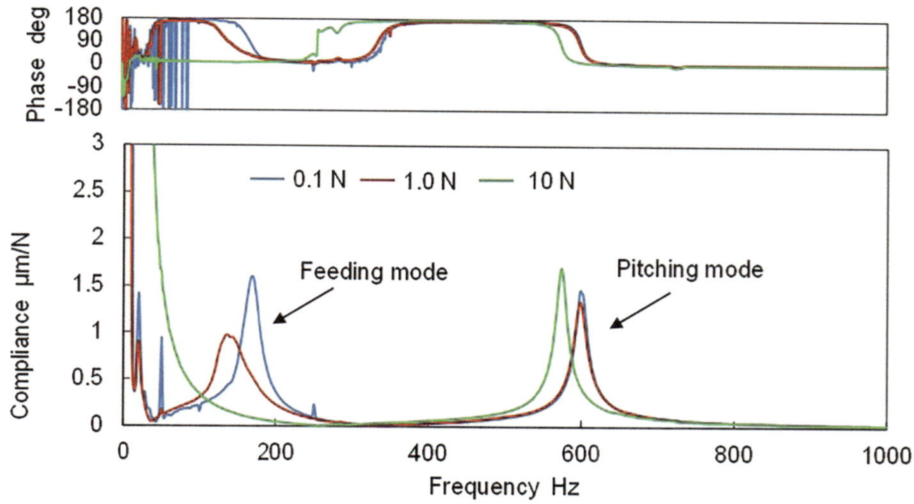

Figure 15.
Force amplitude dependency of dynamic characteristics.

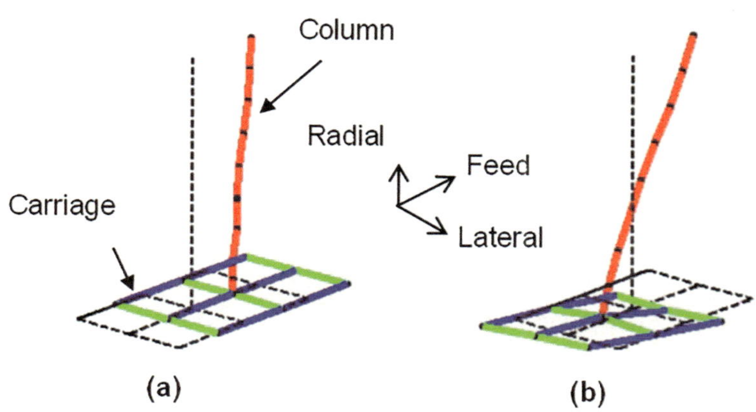

Figure 16.
Measured natural vibration modes of causing each resonance peaks in FRF. (a) Feeding mode. (b) Pitching mode.

vibration modes are called "feed mode" and "pitching mode" in order from low frequency.

Figure 15 also shows the FRFs for different excitation force amplitude. As shown in figure, the dynamic characteristics of the linear rolling bearing change depending on the exciting force. In feed mode, the resonance frequency decreased as the excitation force amplitude increased. Eventually, the feeding mode became unobservable as the excitation force became large. The resonant peak due to the pitching mode has the same tendency as the feed mode for the excitation force amplitude. However, even if the exciting force became large, the resonance peak did not disappear.

Pitching mode is the natural vibration by the rigid body motion of the carriage due to elastic deformation at the contact point between the roller and the raceway.

In the previous study, Ota et al. used a single carriage model to analyze the resonant frequency in pitching mode [27]. In their case, the center of gravity was close to the center of the rail cross section, and the natural vibration characteristics were determined based on the stiffness of the contact parts. However, under actual usage conditions of linear rolling bearings, the center of gravity is higher than the top surface of the carriage. As a result, the natural vibration characteristics are affected by nonlinear friction. The carriage rotates around the center of gravity like a cradle and moves slightly in the feed direction.

In this state, the natural vibration characteristics of the pitching mode depend on the exciting force as shown in **Figure 16**, which is due to the influence of the nonlinear spring behavior.

As shown in the **Figure 17**, the resonance frequency of the pitching mode becomes high when the excitation force is small [26]. This indicates that frictional stiffness affects the pitching mode as "additional spring." Considering the feed mode, the resonance peak disappeared because of an increase in the excitation force amplitude. It indicates that the influence of the nonlinear spring behavior of friction decreases as the response displacement amplitude increases. Thus, the influence of nonlinear friction on the dynamic characteristics of pitching mode becomes small. In addition, the damping ratio was constant in the small excitation force range up to

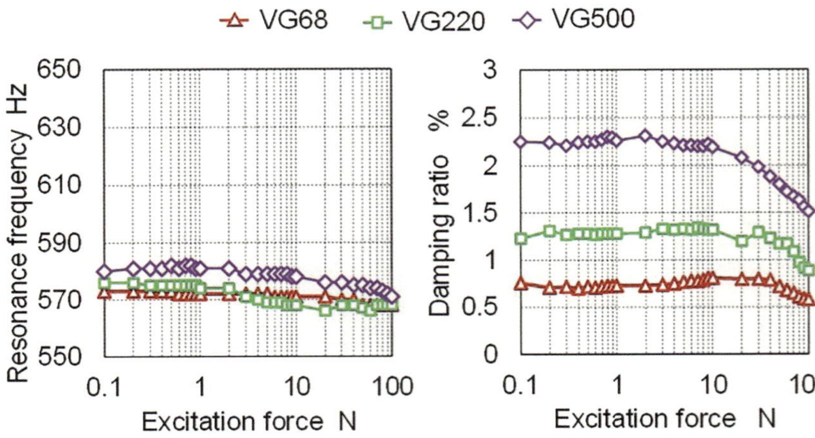

Figure 17.
Influence of oil specification on the nonlinearity of natural vibration characteristics in the pitching mode.

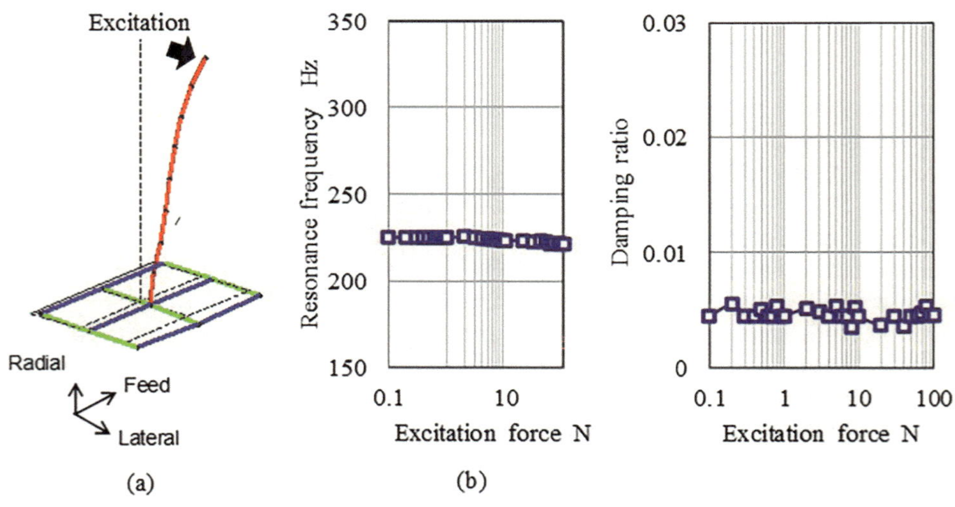

Figure 18.
Influence of excitation force amplitude on the nonlinearity of natural vibration characteristics in the rolling mode. (a) Rolling mode. (b) Dynamic characteristics of rolling mode.

about 10 N, but it decreased in the large excitation force range. According to our previous research, the damping ratio of the linear rolling guideway estimated by the impulse test with 2000 N in the excitation force was about 0.5% with a nonlubricated condition [28]. Thus, the damping ratio converged to about 0.5% when the excitation force increased further.

Figure 17 also shows the results when three types of oils with different kinematic viscosities are used. If the higher kinematic viscosity oil was used, the resonance frequency became higher. The frictional stiffness is high when oil with a high kinematic viscosity is used. These results indicate the difference in the resonance frequency of the pitching mode caused by frictional effect. The damping ratio tends to be higher when oil with higher kinematic viscosity is used. This means that the viscous damping due to the oil film was dominant in the pitching mode.

On the other hand, when the column is excited in the lateral direction, the rolling mode which the carriage and column vibrate around feed axis as shown in **Figure 18** was occurred. However, the excitation force dependence seen in the pitching mode does not occur. This is because the rolling mode is not affected by the additional spring due to non-linear friction.

5. Conclusions

The rolling machine element is indispensable for realizing high-precision and high-speed relative motion. In addition, its positioning accuracy is approaching the nanometer order, and its importance is expected to increase in the future.

However, since the rolling elements and the raceways are mechanically in contact, various non-linear phenomena occur. This complicated phenomenon has to be clear by theoretically and experimentally. In this chapter, the author explained the nonlinear spring behavior of friction that occurs in the rolling contact state and its modeling method. Furthermore, the effect of nonlinear friction on the dynamic characteristics of sliding objects was analyzed numerically. The validity of the model was verified by comparing with the experimental results using the rolling guide.

Finally, it was experimentally shown that the nonlinearity of rolling friction affects the motion in directions other than the feed direction. It is known that rolling friction has velocity dependence and acceleration dependence in addition to the nonlinear spring behavior described in this chapter.

Acknowledgements

This work was supported by Japan Society for Promotion of Science (JSPS) KAKENHI Grant Numbers JP16K21036, JP15J06292.

Nomenclature

A, B	constants
f	virgin loading curve
f_1	proposed virgin loading curve in the pre-rolling region
f_2	proposed virgin loading curve in the rolling region
F	friction force
F_r	friction force at the arbitrary motion reversal point
F_s	steady-state friction force

G nondimensional compliance, u_0/γ_0
K_s stiffness at the starting rolling displacement, F_s/x_s
m mass of a carriage
n change rate of the friction in the pre-rolling region
P excitation force acting on the carriage
P_0 sinusoidal excitation force amplitude
t time
u nondimensional displacement, x/x_s
u_{max} maximum nondimensional displacement in the steady-state response
u_{min} minimum non-dimensional displacement in the steady-state response
u_0 nondimensional displacement amplitude in the steady-state response
x displacement of the carriage in the feed direction
x_r displacement at the arbitrary motion reversal point
x_s starting rolling displacement
β nondimensional frequency, ω/ω_s
dt time differential
$d\tau$ nondimensional time differential
γ_0 nondimensional amplitude of the sinusoidal excitation force
λ similarity ratio
ω sinusoidal excitation force frequency
ω_s natural angular frequency at the starting rolling displacement, K_s/m
τ nondimensional time, ωt
[˙] differential operator with respect to time t, d/dt
[ʻ] differential operator with respect to nondimensional time τ, $d/d\tau$

Author details

Yasunori Sakai
Shibaura Institute of Technology, Saitama, Japan

*Address all correspondence to: sakaiy@shbaura-it.ac.jp

References

[1] Rivin Eugene I. , Author, Balachandran B. , Reviewer. Stiffness and Damping in Mechanical Design. J Mech Des [Internet]. 2000 Mar 1;122 (1):147. Available from: https://doi.org/10.1115/1.533570

[2] Altintas Y, Verl A, Brecher C, Uriarte L, Pritschow G. Machine tool feed drives. CIRP Ann - Manuf Technol [Internet]. 2011;60(2):779–96. Available from: http://dx.doi.org/10.1016/j.cirp.2011.05.010

[3] Majda P. Modeling of geometric errors of linear guideway and their influence on joint kinematic error in machine tools. Precis Eng [Internet]. 2012;36(3):369–78. Available from: http://dx.doi.org/10.1016/j.precisioneng.2012.02.001

[4] Feng G, Wang C. Examining the misalignment of a linear guideway pair on a feed drive system under different ball screw preload levels with a cost-effective MEMS vibration sensing system. Precis Eng [Internet]. 2017;50: 467–81. Available from: http://dx.doi.org/10.1016/j.precisioneng.2017.07.001

[5] Hung J, Lai Y, Lin C, Lo T. Modeling the machining stability of a vertical milling machine under the influence of the preloaded linear guide. Int J Mach Tools Manuf [Internet]. 2011;51(9):731–9. Available from: http://dx.doi.org/10.1016/j.ijmachtools.2011.05.002

[6] Wu JS, Chang J, Tsai G, Lin C, Ou F. The effect of bending loads on the dynamic behaviors of a rolling guide. J Mech Sci Technol. 2012;26(3):671–80.

[7] SAKAI Y, TANAKA T, YOSHIOKA H, ZHU J, TSUTSUMI M, SAITO Y. Influence of nonlinear spring behavior of friction on dynamic characteristics of a rolling guideway. J Adv Mech Des Syst Manuf. 2015;9(1):JAMDSM0004–JAMDSM0004.

[8] Al-bender F. Fundamentals of Friction Modeling. In: Proceedings, ASPE Spring Topical Meeting on Control of Precision Systems,. 2010. p. 117–22.

[9] Kenneth Langstreth Johnson. Surface interaction between elastically loaded bodies under tangential forces. Proc R Soc London Ser A Math Phys Sci [Internet]. 1955 Jul 12 [cited 2020 Sep 11];230(1183):531–48. Available from: https://royalsocietypublishing.org/doi/10.1098/rspa.1955.0149

[10] SHINNO H, KOIZUMI T, ITO Y. Analysis of Damped Vibration of the System with the Contact Surface : 1st Report, Case of the Consideration of Micro-Slip. Trans Japan Soc Mech Eng Ser C. 1983;49(448):2110–5.

[11] YOSHIDA T. Free Rolling Friction and Properties of Material. J Japan Soc Precis Eng. 1961;27(318):468–72.

[12] TANAKA T, OIWA T, OTSUKA J, ONDA H. Study on Linear Ball Guideway for Precision Positioning. J Japan Soc Precis Eng. 2007;73(3):350–4.

[13] Nozomu Y, Sumio S, Yasuto T, Takeko M, Syunichi S. Effect of hysteretic damping characteristics on earthquake response of ground. Proc JSCE Earthq Eng Symp. 2003;27:113–20.

[14] Muravskii G. On description of hysteretic behaviour of materials. Int J Solids Struct. 2005;42(9–10):2625–44.

[15] Olsson H, Åström KJ, Canudas De Wit C, Gäfvert M, Lischinsky P. Friction Models and Friction Compensation. Eur J Control. 1998;4(3):176–95.

[16] Piatkowski T. GMS friction model approximation. Mech Mach Theory [Internet]. 2014;75:1–11. Available from: http://dx.doi.org/10.1016/j.mechmachtheory.2014.01.009

[17] Al-Bender F, Symens W. Dynamic characterization of hysteresis elements in mechanical systems. I. Theoretical analysis. Chaos [Internet]. 2005 Mar 24 [cited 2020 Sep 11];15(1):013105. Available from: http://aip.scitation.org/doi/10.1063/1.1844991

[18] SAKAI Y, TANAKA T, TSUTSUMI M. Influence of the Excitation Force Amplitude on Vibration Characteristics of a Linear Rolling Bearing. J Japan Soc Precis Eng. 2014;80(8):783–91.

[19] Sakai Y, Tanaka T, Yoshioka H. Influence of nonlinear spring behavior of friction on dynamic characteristics of a rolling guideway. J Adv Mech Des Syst Manuf Influ. 2015;9(1):1–15.

[20] Kim TC, Rook TE, Singh R. Super- and sub-harmonic response calculations for a torsional system with clearance nonlinearity using the harmonic balance method. J Sound Vib [Internet]. 2005 Mar 1;281:965–93. Available from: https://ui.adsabs.harvard.edu/abs/2005JSV...281..965K

[21] KAKINO Y. A Study on the motion accuracy of NC machine tools (7th report)-measurement of motion accuracy of 5-axis machine by DBB tests. J Jpn Soc Precis Eng [Internet]. 1994;60(5):718–23. Available from: https://ci.nii.ac.jp/naid/10029041186/

[22] Ohashi T, Shibata H, Futami S, Kishi H, Sato R. Influence of linear ball guide preloads and retainers on the microscopic motions of a feed-drive system. J Adv Mech Des Syst Manuf. 2018;12(5):1–10.

[23] Yi Y, Kim YY, Choi JS, Yoo J, Lee DJ, Lee SW, et al. Dynamic analysis of a linear motion guide having rolling elements for precision positioning devices. J Mech Sci Technol. 2008;22:50–60.

[24] Ohta H, Hayashi E. Vibration of linear rolling guideway type recirculating linear ball bearings. J Sound Vib. 2000;235:847–61.

[25] Rahman M, Mansur MA, Lee LK, Lum JK. Development of a polymer impregnated concrete damping carriage for linear guideways for machine tools. Int J Mach Tools Manuf. 2001;41(3):431–41.

[26] Sakai Y, Tanaka T. Influence of lubricant on nonlinear vibration characteristics of linear rolling guideway. Tribol Int [Internet]. 2020;144(December 2019):106124. Available from: https://doi.org/10.1016/j.triboint.2019.106124

[27] Ohta H. Sound of Linear Guideway Type Recirculating Linear Ball Bearings. Trans Am Soc Mech Eng. 2017;121(October 1999).

[28] SAKAI Y, TSUTSUMI M. Dynamic Characteristics of Linear Rolling Bearings for Machine Tools. J Mach Eng. 2013;13(3).

Section 3

Wear Studies

Chapter 13

Wear Protective Effects of Tribolayer Formation for Copper Based Alloys in Sliding Contacts: Alloy Dependent Sliding Surfaces and Their Effects on Wear and Friction

Ulrike Cihak-Bayr, Robin Jisa and Friedrich Franek

Abstract

High sliding wear resistance is generally attributed to high hardness and high mechanical strength. Novel near net shape process technologies such as metal injection moulding (MIM) or lost foam casting (LF) lack forming processes that typically increase strength. Consequently, the materials exhibit large-grained microstructures with low defect densities. Commercial copper alloys (CuSn8, CuNi9Sn6, CuSn12Ni2) well known for good sliding properties were produced using MIM and LF and characterised in the current study. Their wear and friction behaviour was compared to conventionally produced variants in a lubricated, reciprocating sliding test against steel. The results showed an equal or superior wear resistance and lower friction levels for large-grained microstructures evolving in MIM and LF. SEM, FIB and EBSD studies revealed a tribolayer on the surface and a tribologically transformed layer (TTL), composed of a nano-crystalline zone or partially rotated grains, and selective hardening of grains. The extent of the TTL was different for alloys that were chemically identical but exhibited different initial microstructures. Innovative production routes investigated here showed no tribological drawbacks, but present the potential to increase lifetime, as nano-crystalline zones may render the sample more prone to wear. We present a hypothesis on the cause for these behaviours.

Keywords: Cu-alloys, net-shape forming processes, wear behaviour, tribologically transformed layers, nano-crystalline zone, deformation induced processes

1. Introduction

Innovative technologies that enable production of new materials and complex composites unfortunately do not find their way into engineering application. This is due to a widespread, fundamental lack of trust in new materials or materials produced via non-conventional production processes [1]. With growing awareness for limited energy and environmental resources, paired with cost aspects, innovative near net shape technologies gain interest [2]. The effects of these processing technologies on microstructural details and furthermore on friction and wear are not sufficiently understood to predict material behaviour in a sliding contact to regulate wear rates and or frictional levels. Apart from classical structural mechanical properties, these tribological characteristics have to be known for a proper design with a new material.

In the present chapter two alternative production routes are evaluated for three copper based alloys. The near net shape technologies metal (powder) injection moulding (MIM) and lost foam casting (LF) are described in part 2, both are known for steel [3] and aluminium [4] but have not been commercialised for copper-based alloys and lack substantial basic knowledge in published literature. Yet, near net shape technologies are especially interesting for copper-based alloys due to the high raw metal costs. Here, a special focus is put on the characterisation of wear and friction in a lubricated sliding contact of representative alloys shown in part 3.1. The chosen experimental set-up is depicted in part 3.2 and their tribological and analytical results in part 4. Based on the observations of a formed tribolayer and the nano-crystalline zones forming a tribologically transformed layer (TTL) as described in part 4. Part 5 forms the discussion that links the findings to literature and suggests a hypothesis for the formation of tribolayer and TTL and their effect on wear and friction levels. Finally, the chapter finishes with a conclusion in part 6.

2. Copper-based alloys as tribomaterials

Copper alloys are materials with a good track record for tribological applications comprising pronounced sliding. However, the demand of alloys is limited by high raw material costs and traditional energy intensive production routes; such as melt metallurgy, casting, hot and/or cold forming; the subsequent machining processes lead to large amounts of chips that have to be collected and re-melted to be recycled. As consequence, the production of components made of copper-based alloys demand excessive energy and consequently result in high ecological impact and costs.

Energy-efficient technologies offer economical production methods and additionally large options for complex component shapes. Currently, additive manufacturing and near net shape manufacturing are the main avenues towards achieving these goals. **Figure 1** depicts a schematic of the processes involved: MIM and LF

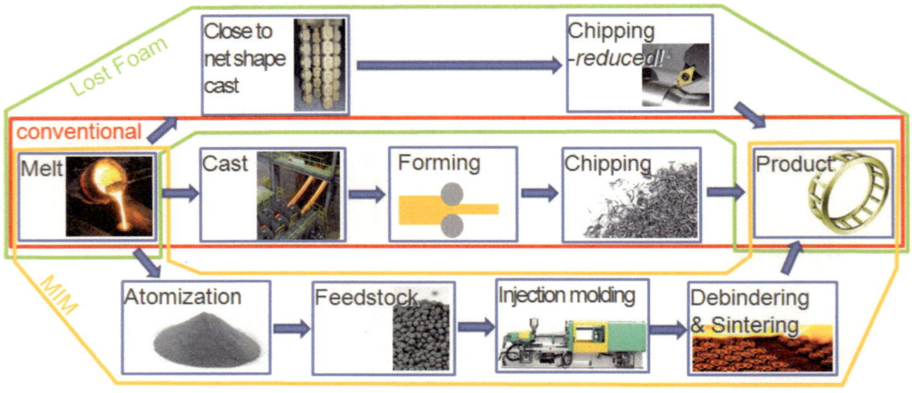

Figure 1.
Technological paths of the analysed materials.

production technologies; the basic principles are described in [5]. Such innovative production technologies usually result in different material properties compared to traditional production methods; the functionality of the component produced by alternative processes are often perceived to be inferior. For a more widespread application of new production technologies, a prediction of the target properties would be necessary. Since a correlation between the material structure and the resulting properties are not trivial, the material selection process is still mainly based on empirical knowledge of engineers. For tribologically loaded components the material selection usually relays on bulk material properties and the assumption that the harder material is the more wear resistant, which is conformed in many cases. This material rating is based on the well-known Archard's wear law [6], which correlates wear volume inversely with hardness.

Tribological studies in the nineties, have revealed that ductile alloys form an altered, fine-grained microstructure in the volume just beneath the sliding contact area as published by Rigney in [7]. Following classical approaches of material hardening, the material increases in strength as the grain size decreases and according to Archard's wear law this results in lower wear.

Most published tribological sliding experiments studying material changes were carried out on pure metals, such as Au or Cu, in a ball on plate configuration in dry contact [8–10]. They show the formation of a microstructure with much finer grain sizes compared to the initial grain size, which is only preserved in the bulk. Some studies can be found on alloys, e.g. steel [11, 12] or Co-based alloys [13, 14], which revealed grain refinement beneath the surface, too. Recent tribological studies focus on the evolution of microstructural changes such as [9, 13], but correlations of grain-size to properties e. g. strength or work-hardening are treated by material science studies: Studies on work-hardening behaviour due to grain size changes, as described for copper by [15], do not deal with grain refinement processes exhibited by tribological contacts. Besides, the strength increasing effect of refinement according to Hall–Petch relations was reported to apply until a gain size of approximately 10 nm is reached for copper [16].

Usually, superior wear resistance is attributed to those newly formed structures [13, 14], but often no explicit correlation to wear or friction is given [9]. Thus, the effect of microstructural changes in subsurface regions in a sliding contact on wear and also friction levels is not really understood and often underrated. Classical material selection strategies follow the idea that material with higher strength exhibit higher wear resistance.

3. Experimental

3.1 Investigated material

The current study compares three copper-based alloys, each produced via two different process technologies – one conventional technology route via casting, forming and machining and a near net shape new technology. The alloys are listed in **Table 1** together with their hardness values and the range of grain sizes observed in the cross sections. The microstructures resulting from different production routes are shown in **Figure 2**. The wrought alloys CuSn8 and CuNi9Sn6 are typically continuously cast, followed by forming (pressing, drawing, rolling) and, for CuNi9Sn6, heat treatment through spinodal decomposition [17]. Apart from the spinodally forming precipitations, which cannot be observed in SEM images, this alloy also forms γ precipitations (in different crystallographic structures as described in [18]) that are several μm large and are visible in SEM pictures. In light

Material	Technological route	Grain size	Hardness
CuSn8 conv.	Continuous cast, drawing, heat treatment, chipping	< 20 µm	238 HV1
CuSn8 MIM	Injection moulding, debindering, sintering	100–150 µm	70 HV10
CuNi9Sn6 conv.	Continuous cast, rolling, heat treatment, chipping	~ 50/150 µm	180 HV1
CuNi9Sn6 MIM	Injection moulding, debindering, sintering	~ 200 µm	171 HV10
CuNiSn6 MIM, heat treated 1 h@450°C	Injection moulding, debindering, sintering, heat treatment	~ 200 µm	266 HV1
CuSn12Ni2 conv.	Continuous cast, chipping	~ 50 µm	115 HB30 2.5
Cuni12Ni2 LF	Lost foam cast, chipping	Dendrite length > 500 µm	90 HB30 2.5

Table 1.
Characterised copper-based alloys and the grain sizes and hardness for different production routes of nominal identical alloys.

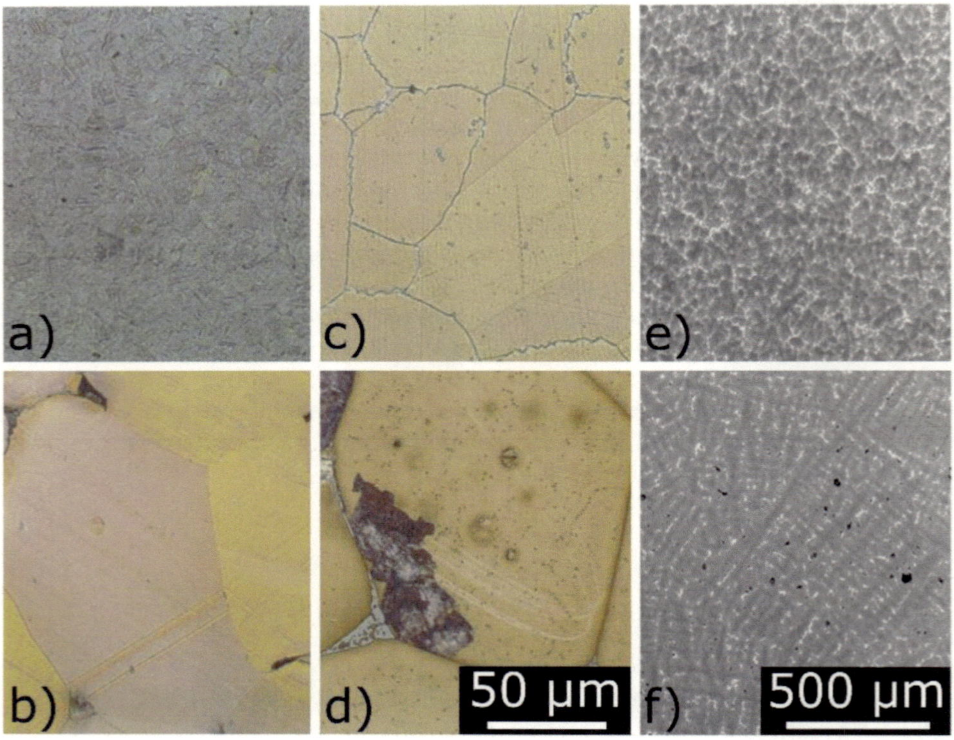

Figure 2.
Microstructures of studied alloys: Conventional: a) CuSn8 c) CuNi9Sn6 e) CuSn12Ni2; MIM: b) CuSn8, d) CuNi9SN6; LF: f) CuSN12Ni2.

microscopy images, their effects on the microstructure can be observed by grain boundary faceting as well as formation of violet lamellar phases during annealing.

Alternatively, these two alloys can be produced by a metal injection moulding (MIM) process from pre-alloyed feedstocks. The MIM production route inherently lacks a forming step and thus any possibility for grain refinement. As a result, the average grain size of the materials produced through MIM is much coarser than the one obtained via conventional routes. **Figure 2** shows representative microstructures of all studied materials. The MIM version of CuNi9SN6 can be heat treated in the same way as the conventional version in order to increase the mechanical strength by precipitation hardening.

Continuous casting (CC) followed by machining is a typical production route for the cast alloy CuSn12Ni2. An alternative, innovative production route, which can help to save energy is a sand casting process called lost foam (LF) casting [4]. The dendritic structure resulting from LF casting is pronouncedly coarser than the CC variant as it is basically a sand-casting process with correspondingly long cooling times and lacking deformation during solidification.

All production variants of the advanced technologies, MIM and LF, lead to larger grain sizes with hardly any defects such as twins. Therefore, these material variants are significantly softer than their counterparts from the conventional production route. As often higher wear resistance is associated with high mechanical strength, the small-grained, conventionally produced alloys are expected to exhibit better wear resistance. This assumption, which is a commonly used basis for design and material choices in mechanical engineering, is verified in the current study for three alloys and two innovative production routes.

3.2 Experimental tests - tribological characterisation

Samples of the different alloy variants were studied in a lubricated reciprocal sliding contact with a modified SRV® test rig. The setup is described in more detail in [19–21]. **Table 2** summarises the main test conditions. Each set of parameters was repeated at least twice. The base oil SN150 was used for CuSn8 and CuNi9Sn6, for CuSn12Ni2 a fully formulated mineral oil, a commercial gear oil, was applied. Thus, one has to be aware of the effect of different viscosities. In order to increase wear of the fully formulated system, CuSn12Ni2 was also examined at a normal load of 240 N. These external conditions differ because the potential applications of the two technologies are different.

The friction coefficients were continuously recorded at a rate of 1 Hz and the data of individual test runs were averaged. Wear scars were characterised after the tests by topographic analysis using confocal microscopy with a Leica DCM3D at 20× magnification, which allowed the measurement of the wear track width. Afterwards, the wear volume was calculated as described in [20].

A wear map is used to illustrate the performance of the individual materials and their variants. It shows the wear volume measured at the end of the test run versus the coefficient of friction at a test time of 90 min, which corresponds to SRV standards [22]. This kind of diagram enables a simple but informative tribological rating, as both friction and wear behaviour are usually relevant. The desired low friction sliding material with a high lifetime can be found in the left lower corner.

3.3 Microstructure analysis

In order to understand modification processes deriving from tribological interactions selected samples were investigated further with light microscopy, nanoindentation, scanning electron microscopy (SEM) and focused ion beam (FIB) cross sections or electron back scatter diffraction (EBSD). EBSD was performed on cross sections normal to the sliding plane with a Zeiss Supra 40VP instrument equipped with an EBSD system from EDAX. The samples were chemically etched prior to the scan. The step sizes (0.20 μm for CuSn8 MIM; 0.10 μm for conventional CuSn8) and the scan areas (100 × 250 μm/19 × 30 μm) were adapted to the grain

Cylinder	Disc	Stroke	Frequency	Normal force	Test duration	Sliding distance	Lubrication	Disc temperature
mm	mm	mm	Hz	N	h	m	—	°C
Ø 4.2 l = 12	Ø 24 l = 7.8	3	30	100 or 240	2	1296	oil drop	50

Table 2.
SRV test parameters.

sizes of the respective samples. Scan sizes of 200 nm were chosen for CuSn8 MIM and 100 nm for conventional CuSn8. The position of the scan along the contact area was pre-selected in SEM images. Three scans were performed on a total area large enough to be representative for the largest grain sizes found in CuSn8 MIM. For microstructures like the conventionally produced CuSn8 a scan directly at the surface was not possible due to the high local deformation. A sufficient Kikuchi pattern quality was only detectable in a depth of 1 µm below the surface. The scans were analyses based on pattern quality images overlaid with the small (SAGB) and large angle grain boundaries (LAGB) as well as on the inverse pole figure (IPF) pictures with the projection direction <100>, which is the surface-normal direction to the plane of the cross section. Numerous precipitations in the MIM version of CuNi9Sn6 complicated the pattern recognition and therefore EBSD scans were omitted. FIB cuts were employed for the deeper analysis of the cast alloy CuSn12Ni2. The FIB cuts were oriented normal to the sliding direction as a cut in the sliding direction would either be located on top or at the bottom of a groove produced by wear debris.

4. Tribological behaviour

Figures 3 and **4** show the frictional behaviour over test time for all studied materials. All friction coefficient curves – except CuSn8 – are at their maximum during running-in, which ranges from 5 to 30 min. All curves end in an almost constant steady level for the chosen test duration and have thus reached steady-state conditions.

The observed friction coefficient values lie between 0.20 and 0.42. These values represent the arithmetic mean of the friction coefficients of different test runs. The data of each run was averaged for each recorded friction value, these mean values are shown as the friction curves in **Figures 3** and **4**. In order to add the information of the scattering of the individual runs for a better interpretation of the friction behaviour, the standard deviation of the friction signal of each run was calculated. The error bars in **Figures 3** and **4** represent the mean standard deviations of individual test runs with the same material, shown only every 5 min for the sake of readability,

The levels of the friction coefficients as well as the friction behaviour differ distinctly for the two materials CuSn8 and CuNi9Sn6. Their sensitivity to variations in the process route seems to be pronouncedly different, as well.

For both materials, MIM manufacturing leads to lower friction coefficients after the run-in phase. For CuSn8 the reduction is roughly 0.08, for CuNi9Sn6 it is less pronounced, namely 0.02–0.03. A heat treatment, which increases the structural strength of the CuNi9Sn6 MIM variant, further reduces the friction coefficient by about 0.07 compared to the conventionally produced CuNi9Sn6.

The MIM variant of CuSn8 shows a high variability throughout the whole test time and after an initial increase to 0.40 smoothly decreases to 0.30 between 30 min and 100 min. The values of the conventional CuSn8 lie within the variability of the MIM variant, but do not show the same characteristics over time. After an initial increase, the friction reaches an almost constant level at 0.40, which is on the upper bound of the scatter of the CuSn8 MIM data. As, the noise of conventional CuSn8 is much smaller than the MIM sample, the different levels of friction of the two CuSn8 variants are regarded as significant.

The error bars largely overlap for the conventional and the MIM version of CuNi9Sn6, indicating just a small tendency for higher average friction levels for the conventional variant. However, the run-in period is distinctly different with

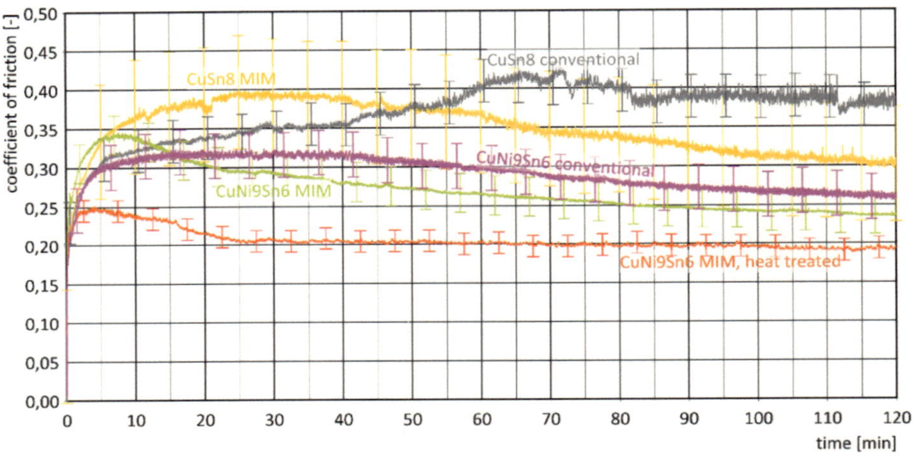

Figure 3.
Friction coefficient of CuSn8 and CuNi9Sn6 both conventionally produced and via the MIM route.

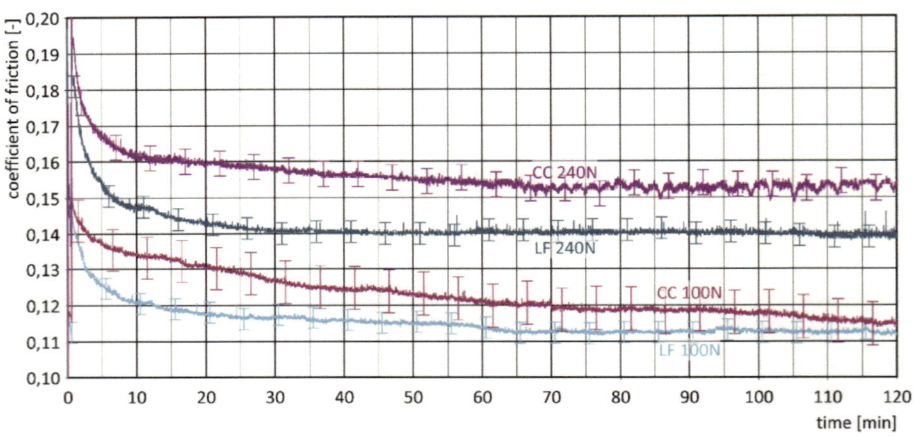

Figure 4.
Friction coefficient of CuSn12Ni2 produced via conventional casting (CC) and via lost foam cast (LF) for 100 N and 240 N normal load.

a pronounced increase for the MIM version, but lasting only 30 min, whereas the conventional material increases and decreases smoothly, reaching a steady state level just before the test ends. The characteristics of the friction curve over time seems to be unaffected by the microstructural changes during heat treatment, as there is again a steep but short increase during run-in. However, the scatter of the friction data is reduced by the heat treatment.

The averaged friction values at a test time of 90 min are depicted in **Figure 5** together with the averaged wear volume at the end of the test in a wear map. The error bars of the coefficient of friction are equivalent to those in the friction curves (**Figure 3**), the uncertainty of the wear volume is the standard deviation of the wear volume at the end of the test. The measured uncertainty is in some cases so small that it is nearly invisible in an appropriately scaled wear map.

The results illustrate a pronounced decrease of wear when the CuSn8 is produced via MIM instead of conventionally. There is also a reduction in wear for CuNi9Sn6 when comparing conventional to MIM in the heat-treated condition, but this decrease is far less pronounced. Between MIM and conventional samples there is no significant change in wear results.

For CuSn12Ni2 the friction coefficient levels are lower, ranging between 0.11 and 0.14 (**Figure 4**). This can mainly be attributed to the fully formulated gear oil used in this study, which was chosen because the tribosystem should be as close to the real application as possible. Therefore, a direct comparison between all the materials discussed before is not permissible. The differences in terms of friction between continuous and lost foam cast variant are little and judged to be irrelevant for applications. Still, the level of coefficient of friction is significantly lower for LF if the contact pressures are increased. Nevertheless, none of the two variants seems to be more sensitive to normal pressure changes than the other. More pronounced differences can be observed in the wear behaviour (**Figure 6**) of CuSn12Ni2, where the LF microstructure shows significantly lower wear volume.

For all investigated materials, the innovative production routes lead to lower wear despite weaker mechanical properties, certainly the degree of improvement depends on the alloy, **Figures 5** and **6**. There are large differences in the final wear volume between the two production routes for CuSn8, which exceeds by far the measurement uncertainty of wear. For CuNi9Sn6 the effect of the production route is much smaller but the heat-treated MIM version shows a distinct wear reduction

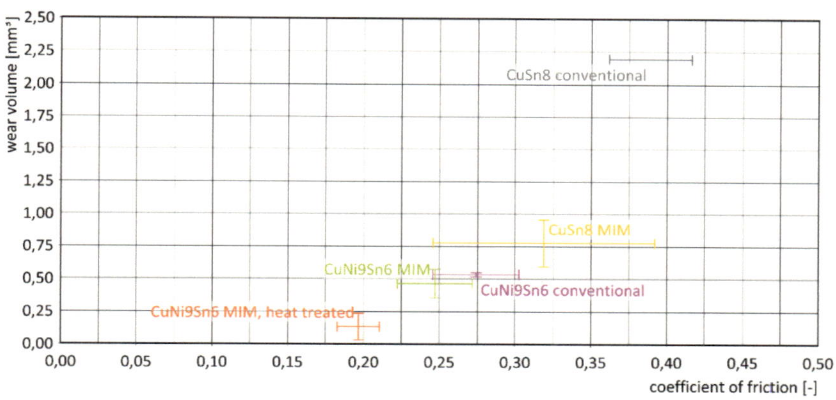

Figure 5.
Wear map of CuSn8 and CuNi9Sn6 in two production variants: Conventional and MIM. The depicted friction coefficient was taken at a test time of 90 min and the error bars are identical to those depicted in ***Figure 3****, the wear volume represents the value at the end of the test and the standard deviation of the measured volume.*

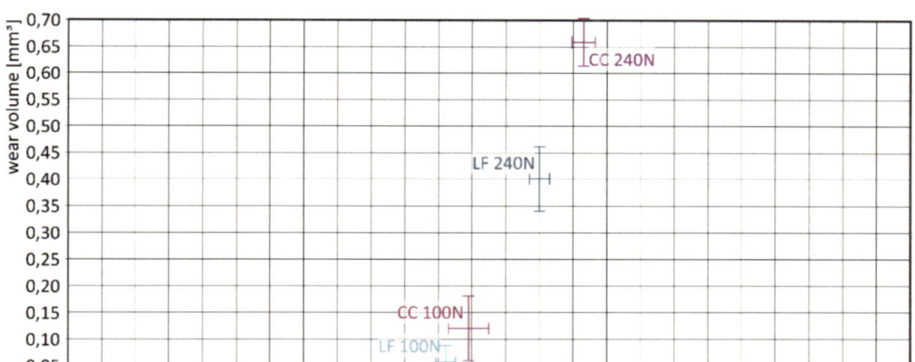

Figure 6.
Wear map of CuSn12Ni2 produced via conventional casting (CC) and via lost foam cast (LF). The depicted friction coefficient was taken at test time of 90 min and the error bars are identical to those depicted in ***Figure 4****, the wear volume represents the value at the end of the test and its standard deviation.*

compared to the conventionally produced sample. Consequently, the increase in mechanical strength due to precipitations does not impair the tribological performance, as it does reduce wear and friction levels. Heat treatment reduced the scatter of the observed friction coefficient level. The results of other samples treated at other temperatures or for different times are not shown here as the wear and friction results nearly coincide for all established heat treatment cycles. In any case, the heat-treated MIM CuNi9Sn6 version offers the lowest wear and friction among the investigated materials lubricated with mineral base oil. The initial microstructure affects the measured wear volume even within systems using additivated oils, which reduces the measured wear volumes due to a tribolayer, forming also in bronze surfaces. Again, the new casting technology (LF) variant of CuSn12Ni2 shows a better performance than the continuous cast one. The effect of different material structures becomes more pronounced for higher normal loads as can be seen in **Figure 6** for 100 N and 240 N. The effects on friction levels are lower than for the systems shown in **Figure 3**, but interestingly the noisiness of the friction coefficient reduces if the structure gets coarse and more dendritic. The latter observation seems to be independent of the applied normal load.

4.1 Tribologically induced changes in the microstructure

The higher wear resistance of softer variants of the respective alloys was initially not expected as hardness is often associated with wear resistance. As the testing conditions were identical, we investigated the microstructure in order to examine the different wear behaviour observed. The idea was that the grain and/or defect structure evolving during the sliding contact is different for these samples. As all other factors were identical, the near surface microstructure behaviour can potentially explain the differences in the macroscopic wear behaviour as well as the observed frictional levels. The light microscopy images in **Figure 7** reveal that no resolvable changes in the microstructure beneath the contact surface occur. Only

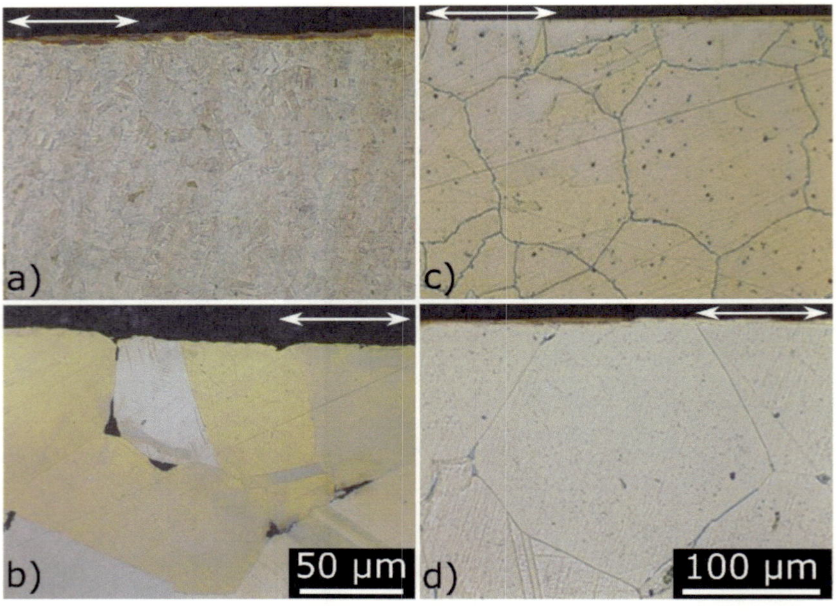

Figure 7.
Light microscopy images of cross sections of the cylindrical samples normal to the sliding direction. a) CuSn8 conventional b) CuSn8 MIM c) CuNi9Sn6 conventional d) CuNi9Sn6 MIM; the 50 µm scale applies to a) and b), the 100 µm scale to c) and d).

the MIM CuSn8 sample exhibits local increase of twins and shear bands in the first layer of grains up to a depth of 20 μm beneath the contact plane.

The LF cast structure does show defects in the light microscopy images (**Figure 8**), such as bent slip lines as well as deformation twins, which are bent additionally. The deformation is very inhomogeneous and many dendrites show no or hardly any changes visible in light microscopy images. The details in **Figure 8b** illustrate pronounced effects in the vicinity of pores. If features of plastic deformation can be observed by light microscopy images, they will reach up to depths of at least 50 μm.

The higher resolution SEM images of the cross sections shown in **Figure 7** are given in **Figure 9a** and **b** for the conventional variants and in **Figure 9c** and **d** for the MIM samples. The CuSn8 is highly twinned and some twins are bent in sliding direction. This occurs not only in the surface grains but also in grains located in deeper zones. Still, the SEM picture is not sufficient to be able to judge if these features are caused by the sliding contact as the initial structure is highly twinned already. In the higher alloyed CuNi9Sn6 the grains are larger compared to CuSn8 conventional and no features indicating deformation close to the surface are visible. The edge is very rounded due to polishing and etching prior to EBSD measurements.

In contrast to the two conventionally produced samples, both MIM alloys revealed the formation of a layer on top of the initial microstructure with a

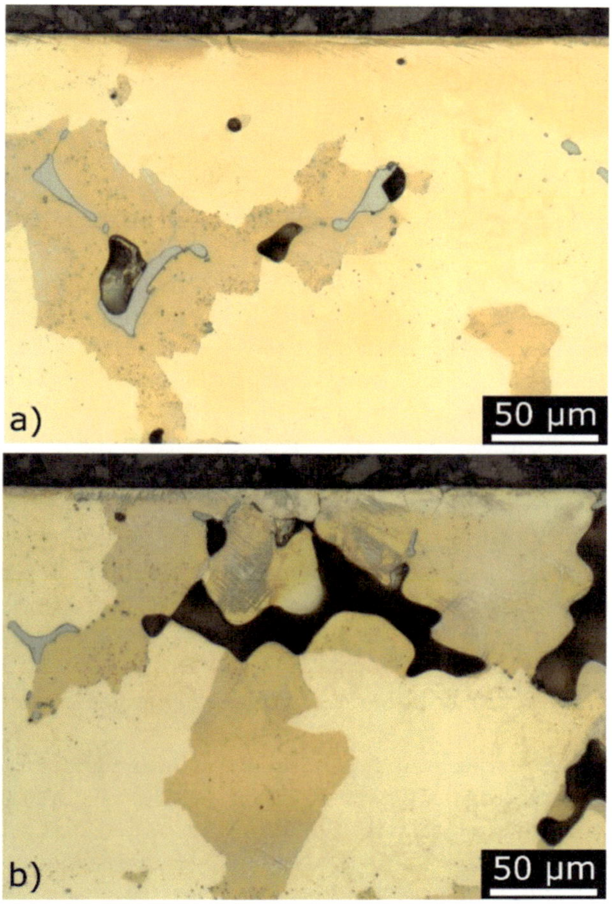

Figure 8.
Light microscopy images of cross sections of the cylindrical CuSn12Ni2 LF samples normal to the sliding direction a) and b) show two position within the same sample, in a) slip lines and in b) deformation twins in the first row of grains are visible.

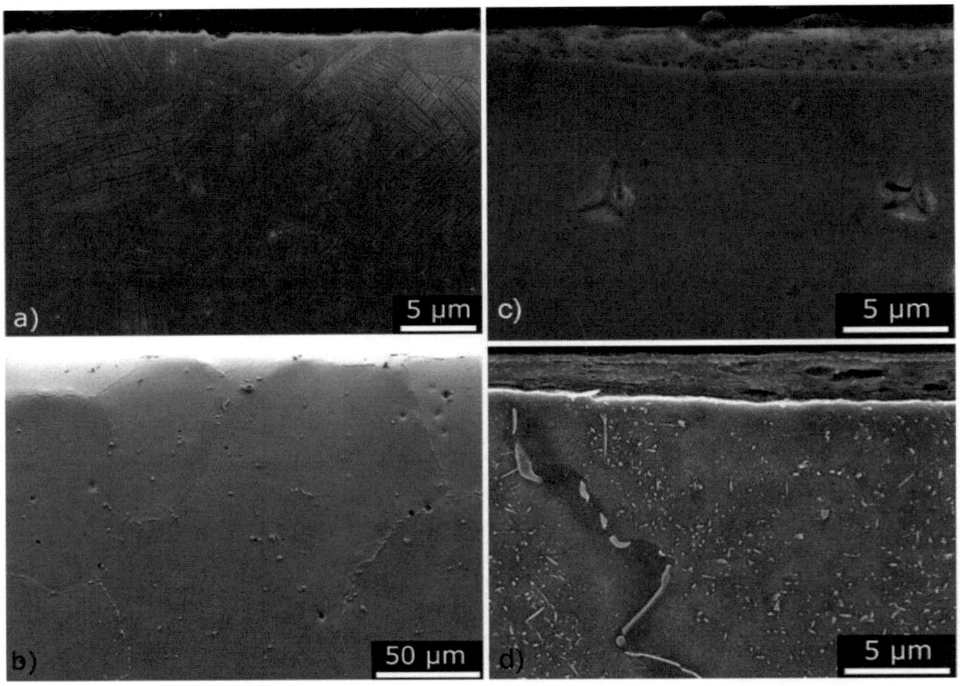

Figure 9.
SEM image of the edge zone on cross-sections oriented parallel to the sliding direction: a) CuSn8 conventional, b) CuNi9Sn6 conventional, c) CuSn8 MIM (mechanically mixed layer, 2 visible nano-indenter marks), CuNi9Sn6MIM (mechanically mixed layer lying and facetted grain boundary with γ precipitations is vertically inclined on the left side of the image).

thickness up to 2–3 µm. This feature was only observed on the MIM microstructures of the studied alloys.

The observed layer seemed to have a patch-like structure as it did not cover the whole length of the cross section. On top views of the contact area these layers were not clearly distinguishable from the rest of the microstructure, thus their size and extent of coverage cannot be given. The EDX measurements (**Table 3**) shows increased contents of oxygen and carbon and some iron. This indicates a mixture with debris from the counter-body, even though the structure appears amorphous or extremely fine-grained. However, no grain structure was resolvable with SEM. What is observable in the SEM images are pores within this layer, which are more pronounced for the layer observed on CuNi9Sn6. In the following, we refer to it as a mechanically mixed layer in order to distinguish it from the TTL, a zone comprising of defects like the slip lines and deformation twins extending over up to 20 µm as seen in **Figure 7c**.

4.2 EBSD as a tool to reveal tribologically induced changes in the microstructure

With the EBSD technique grain orientations and local misorientations can be revealed. Thus, microstructural defects like internal stresses stored in grains or the formation of subgrains and local misorientations can be detected, which cannot be observed with SEM images [23, 24]. As the changes in wear volume were most pronounced for CuSn8, the EBSD analysis focused on this alloy.

For the conventionally produced CuSn8 the forming in the production process resulted in a highly deformed and small-grained structure and any further deformation caused due to the tribological contact is not clearly visible in the SEM images. The pattern quality picture and the IPF picture of an EBSD scan with 1 μm distance to the surface are given in **Figure 10**, with the upper edge of the scan in sliding direction. The quantitative analysis of the grain boundary lengths of large angle grain boundaries (LAGB) and small angle grain boundaries (SAGB) overlaid onto the pattern quality image indicate no increase closer to the surface and so they are not shown. However, the twin density within the first 15–20 μm appears to be higher as numerous micro-twins form. Especially in the first 5 μm, the grains are strongly elongated parallel to the sliding direction. Generally, the grain size is easier to see in IPF images and they reveal high intragranular strains by different shades of colours in nearly all grains, independently of their location with respect to the surface. Hence, internal lattice rotations do not seem to increase close to the contact but are a result of the high deformation during production. Comparing the IPF image of the initial structure to the one at the surface indicates a tendency to slightly smaller grains sizes in the first 15 μm. However, the change in size is far less pronounced than the change in shape.

Thus, the microstructure of conventional CuSn8 can hardly adapt to the stresses accumulated during sliding and has little work-hardening capability. So presumably the microstructure is just worn off soon after the initial contact during the sliding process and cannot form a wear resistant subsurface zone.

The MIM version of CuSn8 was very large grained and therefore investigated by two closely situated EBSD scans situated close together, shown in **Figure 11**.

	C	O	Si	S	Sn	Fe	Ni	Cu
At %	54.58	6.16	1.07	1.90	0.62	0.31	3.33	32.03

Table 3.
EDX analysis of the layer observed on CuNi9Sn6 in the MIM version as shown in **Figure 9c**.

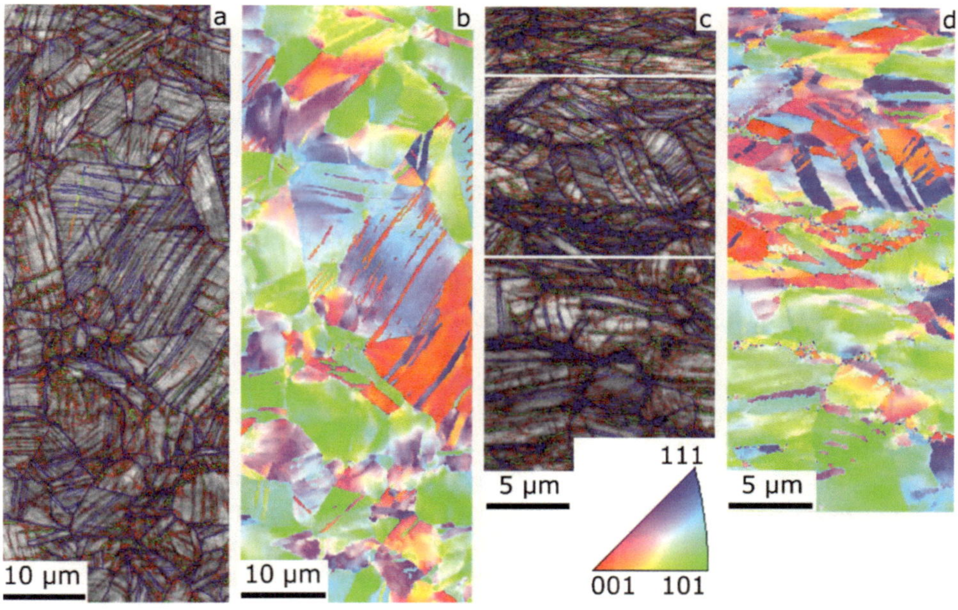

Figure 10.
CuSn8 conventional samples with the sliding direction is parallel to the upper edge. EBSD scan at a depth of 1000 μm beneath the surface: a) pattern quality image with LAGB in blue, SAGB in red and b) IPF image EBSD scan at a depth of 1 μm beneath the surface: c) pattern quality image with LAGB in blue, SAGB in red d) IPF image.

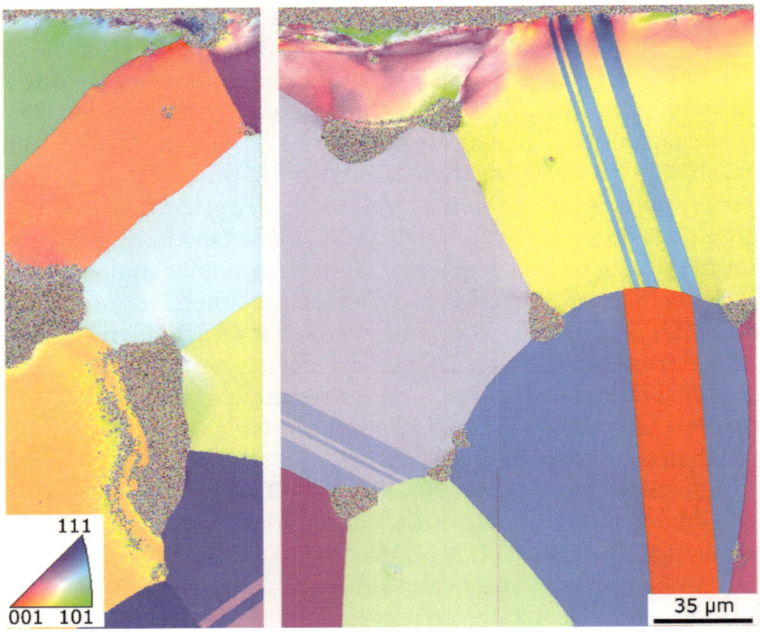

Figure 11.
CuSn8 MIM EBSD scan.

The initially undeformed MIM structure of CuSn8 shows some grains like the green coloured ones the left with no deformation features, but the grain on the right changes its orientation from yellow to pink which indicates a partial lattice rotation and an internal stress field. The respective stress field does not extend over the whole grain but is localised in the first 20 μm beneath the contact and does not affect the twin. In the middle, there is a pore close to the surface and the remaining grain located above this pore seems to be most suitable for storing plastic strain generated by the tribological sliding process. It has various shades of pink and is not as uniformly coloured as non-affected grains. The analysis of small- and large-angle grain boundaries showed no additional information and a figure was therefore omitted. Based on our observations there is no indication for an interrelationship between local lattice rotations and the position of the mechanically mixed layer lying on top of the tribo-contact area.

In addition to SEM and EBSD scans, some apparently affected grains were investigated with nanoindentation, the results of which are given in **Figure 12**. The defect structure shows a high local concentration of defects that lead to an increase in hardness by 1–1.5 GPa over the first 20 μm beneath the surface. However, grains that show no orientation changes in the EBSD scan also showed no hardening in the nanoindenter measurement.

The modulus of elasticity was quantified during nanoindentation and turned out not to vary with distance from the contact surface and is thus not given in **Figure 12**. Altogether, the results of the MIM sample suggest that if grains are modified due to the sliding process they have to be located directly at the surface, grains further away from the contact are unaffected in any case, but even at the surface many grains do not adapt by e.g. lattice rotations.

The cast alloy CuSn12Ni2, though tested with additivated oil, showed smaller but significant differences in the wear volumes of CC and LF cast samples. The tribo-surfaces had different characteristics, which are illustrated in **Figure 13a** and **b**. The surface exhibits grooves, presumably formed by wear debris, but resembles a

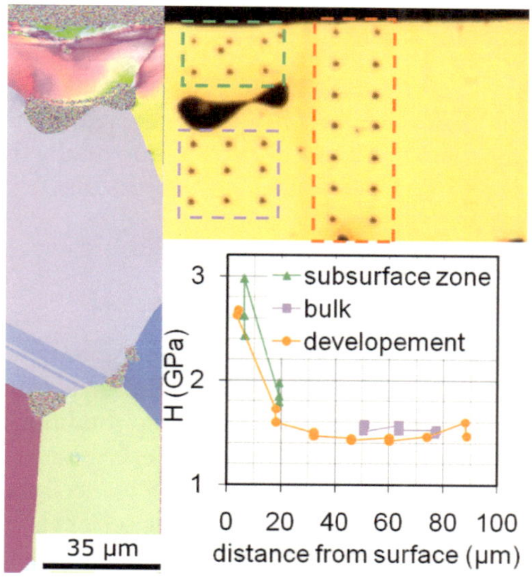

Figure 12.
Location and gradient of the nano-hardness measurements performed on the MIM version of CuSn8.

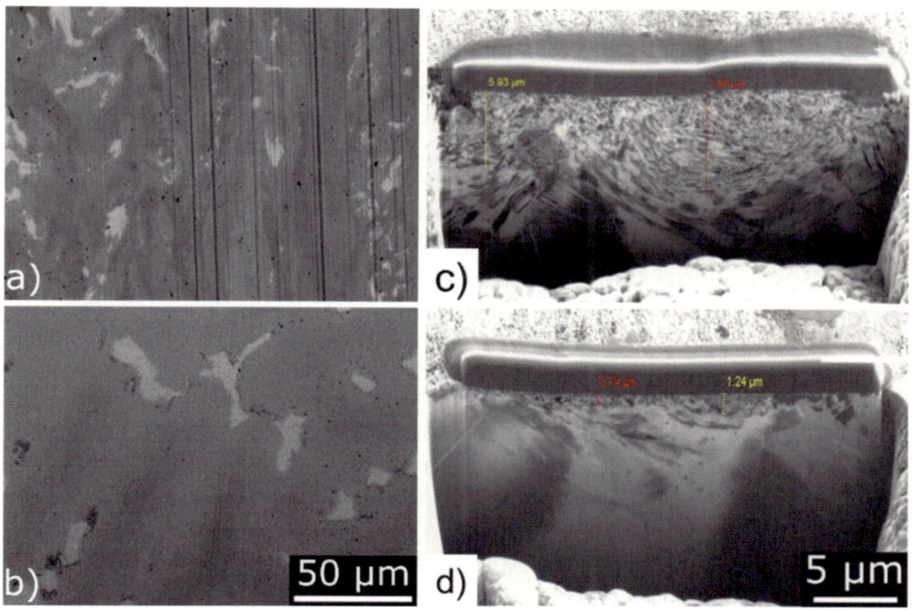

Figure 13.
Wear tracks on CuSn12Ni2: Top view SEM image - sliding direction going parallel to the vertical image edge a) conventional sample, b) LF sample; sub-surface microstructure in field ion image mode - sliding direction normal to the FIB cutting plane, c) conventional sample d) LF sample.

metallographic cross-section as it clearly reveals all phases. The LF surface appears even smoother than the CC surface. As the conventional cross sections indicated some edge effects, but did not enable a proper identification of the structure, only the respective FIB cuts, normal to the sliding direction, are given in **Figure 13c** and **d**. Here samples tested at a normal load of 240 N are examined as the wear volume differences are larger than for 100 N normal load. Both samples form a nanocrystalline layer during the sliding process. However, the thickness of this layer is clearly larger for the CC version with about 8 µm compared to 1.2 µm for LF. As the CC samples had higher wear volumes, this layer does not seem to have wear protective effect, at least under the conditions investigated here. The nanocrystalline layer is not homogeneously thick, but rather shows a wavy interface in both studied cast structures. The average grain size within the nanocrystalline layer with about 250 nm seems to be constant over the thickness and in both samples. Potentially there are even smaller grains, but they cannot properly be resolved in SEM pictures.

5. Discussion

In lubricated reciprocating sliding contacts, all the studied copper alloy variants exhibited different tribological performance, if the microstructure was modified by changes in the production route. For the chosen loading conditions, the coarser and softer microstructures showed superior tribological performance compared to the finer grained and harder materials. Wear was pronouncedly reduced and friction coefficient levels of the run-in friction pairs were lower too for the softer variants of each alloy. This basic finding applies to dendritic cast structures as well as to globular structures.

In this study we compare the production routes MIM versus continuous casting followed by a massive forming step and lost foam casting versus continuous casting. MIM and LF both result in microstructures that are significantly coarser than the ones obtained through conventional processes. Thus these larger grained microstructures exhibit lower mechanical strength. Nevertheless, they showed to be more wear resistant than their smaller grained version from conventional processes. The effect was most pronounced for CuSn8 lubricated with non-additivated base oil SN150. This observation does not follow the conventional assumption that harder materials result in higher wear resistance, which follows the most common wear law of Archard and relates wear volume inversely proportional to macroscopic hardness [6]. As a consequence, small grained microstructures which result in higher tensile strength according to the Hall–Petch relation [16] are usually expected to be more wear resistant. Typically, only the initial or bulk material properties and microstructures are measured and taken into account [1].

In the present sliding systems, the detected wear volume of the initially coarse-grained structure and thus softer version of the alloy was significantly lower for both studied alloys and differed by factor of 2 for CuNi9Sn6 and by a factor of 2.5 for CuSn8. For the dendritic cast structures of the alloy CuSn12Ni2 the same effects on wear could be observed, but reduction was less pronounced. This can be attributed mainly to the use of additivated oil in order to be closer to the real application, yet, additives may contribute to a formation of a chemical wear protective tribofilm.

The analysis of the near-surface zones revealed the formation of a tribologically transformed layer and a mechanical mixed layer lying on top of the contact surface. Such a behaviour is well-known in tribology, but phenomenological descriptions of such microstructural changes often prevail [7, 25]. Few studies treat the effect of these layers on wear [26], or friction [27] More recent studies focus on the internal structure of the subsurface zone like the formation of a fibre texture [27] and/or the development of multiple grain structures, which split up in nano and micrometre sized grains [28]. Kapoor et al. [26] attribute the modified wear rates observed for different loading conditions to the different mechanical properties of these layers, independent of their genesis. Following Hall–Petch and Archard, a formation of a more fine-grained microstructure in the subsurface zone is expected to lead to local hardening and consequently to an increase of wear resistance.

However, the current study showed that the samples forming a thicker and more pronounced fine-grained zone exhibited higher wear than their chemically identical equivalents produced by MIM and LF, respectively. Typically, literature does not cover the role of the material itself or different initial microstructures on wear and friction. As the most prominent difference between the chemically identical samples characterised here can be seen in the initial grain sizes, it was presumed to be a key influencing factor.

For the MIM version of CuSn8 the EBSD scans showed that the large grains react individually to the stress imposed during sliding and some accumulate plastic strain by an extremely localised work hardening processes. The local extension of the hardening is below or equal the single grain size. According to the EBSD results, the lattice orientation of grains exhibiting hardening is substantially different from neighbouring orientations. Thus, the number of their potential gliding planes with low Schmidt factors [29] differs a lot. The number of available slip planes is regarded as the main reason for the inhomogeneity of strain accumulation along the contact surface. Grains showing internal strains in the EBSD scans are assumed to have offered more slip systems with low Schmidt factors, with respect to the sliding direction, than the grains showing no enhanced defect density. If dislocation gliding is possible, the crystallographic lattice can take up strain and will partially rotate. If not, parts of the grain will be abraded directly with hardly any defect formation. The resulting enhanced hardening over a depth of 20 μm below the surface reaching up to 1.5 GPa was proven with nano-indentation measurements (**Figure 12**) for CuSn8. The EBSD scan illustrates that some grains take up plastic strain via local lattice rotations but form no new grain boundary. Quantitative analysis of SAGB or kernel average misorientation (KAM) are not depicted, because the large grains resulting from the MIM process revealed no substructure formation during sliding. The mechanically mixed layer found in patch-like structures on the surface could not be studied by EBSD as the misorientation was too high to identify grain structures. The nature of this layer would have to be verified in detailed TEM analyses in the future.

The TTL and the mechanically mixed patches are considered as two independently forming features (**Figure 14**), which both have the capability to reduce wear. The mechanically mixed layer presumably distributes the normal load and thus reduces the contact stress. The local hardening – though inhomogeneous – may postpone the detachment of wear particles as described in [30]. Surprisingly, we found no mechanical mixed layer on the small-grained structures with high initial twin densities (**Figure 9a**) after identical loading.

In contrast to the MIM version the initial grain structure of the conventional CuSn8 was fine-grained and exhibits very high twin densities. Such a highly deformed microstructure can hardly take up further plastic strain and the local work-hardening potential is low. Consequently, no TTL could be observed in the EBSD scans and the microstructure does not appear to adapt during sliding, but seems to be worn off immediately during sliding.

The dendritic cast alloy CuSn12Ni2 forms a nanocrystalline surface layer, which is similar to the ultra-nanocrystalline layer described for a Ni-alloy by [27].

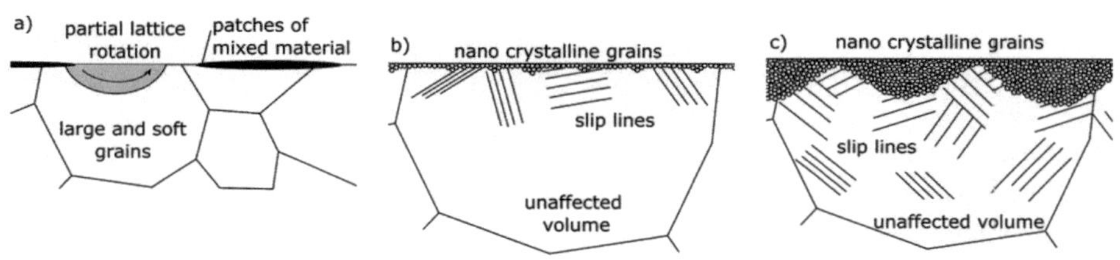

Figure 14.
Principal sketch of tribologically transformed zone beneath the surface a) partial grain rotation and tribolayer patches of mixed material b) formation of a thin layer of nanocrystalline grains accompanied by slip lines and twins, c) thicker nanocrystalline zone in vortex shape together with higher density of slip lines and twins formed also in larger depths.

We again refer to it as TTL. However, under loading conditions as those in the present study, this nanocrystalline layer was thicker for the finer structure resulting from continuous casting and very thin for the softer LF structure with large dendrites ranging over several hundred μm. Therefore, the nanocrystalline layer appears to be not beneficial for higher wear resistance.

Based on these observations we set up the following hypothesis for increased wear resistance of chemically equal, but softer, microstructures. Large grains without or with low initial defect densities can work-harden if the surface grains are appropriately oriented, i.e. so that the Schmidt factor of easy slip planes is low – (111) in the cubic systems of the study. For increased wear resistance, a near-surface layer has to take up plastic strains originating from the external stress state during the sliding process. This layer can form through local rotation of surface grains, as illustrated in **Figure 14**. For large grains, such as in the present study, where the grains have an average equivalent diameter of up to 250 μm, only partial rotation of grains occurs. Although the local grain rotations are rather inhomogeneous over the contact area, as many grains do not participate in local lattice rotations, this zone is relevant for the macroscopically observed wear volume and thus referred to as TTL. If the individual grains are saturated in defect densities, the surface near zone cannot work-harden and, as a consequence, the critical stresses for delamination or detachment of wear particles are much lower, wear occurs earlier and wear rates are higher.

Classical hardening through annealing processes and spinodal precipitations, which are found in CuNi9Sn6 can further reduce the wear volume compared to the non-hardenable CuSn8 as well as compared to the MIM version as shown in **Figure 5**. In the MIM version, the grain boundaries are facetted due to γ precipitation of length up to 2 μm at the grain boundaries, which is accompanied with a matrix depletion in the vicinity of the boundary. During heat treatment these precipitations at the boundaries dissolve and only lamellar phases within the grains and small intragranular precipitates form evenly distributed in the grains. The wear and friction result indicate that the latter structure is superior to adapt to the sliding process. Yet, the precipitation structure and distribution within the grains proved to be irrelevant for wear as well as friction. This indicates that the matrix composition and its element distribution determine the ability of a material to resist abrasion.

In the case of dendritic structures, the TTL comprises a nanocrystalline zone (**Figure 14**) with grain sizes ranging from 100 nm to 300 nm. Again, the thickness of this zone is inhomogeneous with "pockets" of nanocrystalline zones with an extension of about 20 μm below the surface (**Figure 13c** and **d**). As this characteristic is the same for the large grained LF and the finer grained CC sample, the wavy interface could be either a result of the large dendrite thickness and coarse grains or an effect of a periodic pattern of the contact stress state.

Still the differences of the respective maximum thicknesses are significantly different with 8 μm for CC and 1.2 μm for LF. The formation mechanism of such nanocrystalline grains is not entirely clear. [27, 30] compare the sliding process to rolling and torsion processes and claim that the near-surface textures are typical rolling textures. As the matrix of the MIM CuSn8 as well as the CuSn12Ni2 matrix, are both bronze metal matrix lattices, the local rotation found in CuSn8 could be an early status followed by formation of nanocrystalline grains.

The nanocrystalline layer found in LF samples resembles the non-continuous layer described in [28], referred to as nanostructured mixing layer (NML) and dynamic recrystallised layer. They show that these small grains form during the sliding contact and that the kinetics are different for different initial microstructures. The coarse-grained structures show higher wear rates which is not in line with our results on copper-based alloys. We found that the nanocrystalline layer is thicker for smaller-grained initial structures. However, the idea that not the hardness of the initial microstructure determines wear resistance, but the ease with which the nanocrystalline layer can be peeled-off during sliding, seems to be an appropriate explanation for the wear behaviour of the conventional CuSn12Ni2 version forming thick nanocrystalline layers.

6. Conclusions

Innovative material production processes are usually associated with lower mechanical strength and with inferior performance. They lack trust by designers as they suffer comprehensive characterisation, especially tribological performance properties. In the current study, the technology route metal-injection moulding (MIM) and lost foam casting (LF) are applied to well-known commercial bronze alloys. The alternative routes resulted in more ductile and softer materials, but proved to be superior in terms of wear and even showed tendencies for lower friction levels under equal configurations and loadings.

A detailed characterisation executed using SEM, nanointendation as well as EBSD techniques revealed that all samples formed a tribologically transformed layer (TTL) beneath the contact surface, but that the extent of this layer was pronouncedly different. The following observations can be summarised for the current study:

- The thickness of the formed nanocrystalline layer depends on the material production route

- Thicker nanocrystalline layers do not enhance wear, but increase wear.

- Thus, grain refinement as a hardening mechanism does not go along with classical Archard approaches for increasing wear resistance via increasing hardness. The found nanocrystalline layer requires lower shear forces on the grains to be removed and this results in decreased macroscopic friction coefficients.

- The tribolayer, which showed to be a presumable amorphous mechanically mixed layer, forms independently of the TTL on MIM samples and was observed on samples showing lower wear than equivalently loaded conventionally produced variants.

- Large grained initial microstructures, produced by e. g. MIM, exhibit partial lattice grain rotation, which increases hardness very locally and were only observed on samples with lower wear.

Finally, we present a hypothesis for the formation timeline of nanocrystalline zones. Starting with local lattice grain rotation, followed by a monolayer on nanocrystalline grain layer accompanied with slip band and twin formation beneath and finally a thick nanocrystalline layer with a vortex structure. Based on the observation, we postulate that higher alloyed materials are more prone to local lattice rotation and defect formations such as twins and thus form a nanocrystalline zone more easily and quicker under the same loading conditions.

Acknowledgements

Topics presented are based on research projects with support by the Austrian COMET program (project XTribology, no. 849109, project InTribology, no. 872176) and by the respective program management institutions, viz. the Austrian Research Promotion Agency (Österreichische Forschungsförderungs-Gesellschaft mbH – FFG), on behalf of the Federal Government, as well as proportionately by the Governments of Niederösterreich, Vorarlberg, and Wien. We are as well grateful to the involved company partners and scientific partners for their financial support and for the fruitful cooperation.

Author details

Ulrike Cihak-Bayr[1*], Robin Jisa[2] and Friedrich Franek[1]

1 AC2T research GmbH, Wiener Neustadt, Austria

2 Österreichische Tribologische Gesellschaft, Wiener Neustadt, Austria

*Address all correspondence to: ulrike.cihak-bayr@ac2t.at

References

[1] Suffian S, Dzombak R, Mehta K. Future directions for nonconventional and vernacular material research and applications. Nonconv. Vernac. Constr. Mater., Elsevier; 2016, p. 63-80. DOI:10.1016/B978-0-08-100038-0.00003-2.

[2] Marini D, Cunningham D, Corney JR. Near net shape manufacturing of metal: A review of approaches and their evolutions. Proc Inst Mech Eng Part B J Eng Manuf 2018;232:650-69. DOI: 10.1177/0954405417708220.

[3] Khodai M, Parvin N. Pressure measurement and some observation in lost foam casting. J Mater Process Technol 2008;206:1-6. DOI: 10.1016/j.jmatprotec.2007.11.309.

[4] Shivkumar S, Wang L, Apelian D. The lost-foam casting of aluminum alloy components. JOM 1990;42:38-44. DOI: 10.1007/BF03220435.

[5] Li G, Jiang W, Fan Z, Jiang Z, Liu X, Liu F. Effects of pouring temperature on microstructure, mechanical properties, and fracture behavior of Al/Mg bimetallic composites produced by lost foam casting process. Int J Adv Manuf Technol 2017;91:1355-68. DOI: 10.1007/s00170-016-9810-y.

[6] Archard JF. Contact and rubbing of flat surfaces. J Appl Phys 1953;24:981-8. DOI: 0.1063/1.1721448.

[7] Rigney DA. Transfer, mixing and associated chemical and mechanical processes during the sliding of ductile materials. Wear 2000;245:1-9. DOI: 10.1016/S0043-1648(00)00460-9.

[8] Argibay N, Chandross M, Cheng S, Michael JR. Linking microstructural evolution and macro-scale friction behavior in metals. J Mater Sci 2017;52:2780-99. DOI: 10.1007/s10853-016-0569-1.

[9] Greiner C, Liu Z, Strassberger L, Gumbsch P. Sequence of Stages in the Microstructure Evolution in Copper under Mild Reciprocating Tribological Loading. ACS Appl Mater Interfaces 2016;8:15809-19. DOI: 10.1021/acsami.6b04035.

[10] Equey S, Houriet A, Mischler S. Wear and frictional mechanisms of copper-based bearing alloys. Wear 2011;273:9-16. DOI: 10.1016/j.wear.2011.03.030.

[11] Perret J, Boehm-Courjault E, Cantoni M, Mischler S, Beaudouin A, Chitty W, et al. EBSD, SEM and FIB characterisation of subsurface deformation during tribocorrosion of stainless steel in sulphuric acid. Wear 2010;269:383-93. DOI: 10.1016/j.wear.2010.04.023.

[12] Mola J. Considerations in the Design of Formable Austenitic Stainless Steels Based on Deformation-Induced Processes. Austenitic Stainl. Steels - New Asp., InTech; 2017. DOI: 10.5772/intechopen.70939.

[13] Zhao C, Zhou J, Mei Q, Ren F. Microstructure and dry sliding wear behavior of ultrafine-grained Co-30 at% Cr alloy at room and elevated temperatures. J Alloys Compd 2019;770:276-84. DOI: 10.1016/j.jallcom.2018.08.092.

[14] Liu J, Yang S, Xia W, Jiang X, Gui C. Microstructure and wear resistance performance of Cu-Ni-Mn alloy based hardfacing coatings reinforced by WC particles. J Alloys Compd 2016;654:63-70. DOI: 10.1016/j.jallcom.2015.09.130.

[15] Sinclair CW, Poole WJ, Bréchet Y. A model for the grain size dependent work hardening of copper. Scr Mater 2006;55:739-42. DOI: 10.1016/j.scriptamat.2006.05.018.

[16] Schiøtz J, Jacobsen KW. A maximum in the strength of nanocrystalline copper. Science. 2003;301:1357-9. DOI: 10.1126/science.1086636.

[17] Zhao JC, Notis MR. Spinodal decomposition, ordering transformation, and discontinuous precipitation in a Cu-15Ni-8Sn alloy. Acta Mater 1998;46:4203-18. DOI: 10.1016/S1359-6454(98)00095-0.

[18] Virtanen P, Tiainen T, Lepistö T. Precipitation at faceting grain boundaries of Cu-Ni-Sn alloys. Mater Sci Eng A 1998;251:269-75. DOI: 10.1007/s10853-015-9341-1

[19] Jisa R, Ristic A, Brenner J, Lebersorger T, Ilo S, Neumayer H. Effectiveness of lubricant additives for copper-alloy-steel sliding contacts. Lubr Sci 2010. DOI: 10.1002/ls.118.

[20] Jisa R, Laumann S, Cihak-Bayr U, Tomastik C, Eberle R, Keppeler M. Tribological properties of MIM manufactured copper alloys. Tribol - Mater Surfaces Interfaces 2014;8. DOI: 10.1179/1751584X14Y.0000000064.

[21] Cihak-Bayr U, Jisa R, Franek F. Tribological Behaviour of MIM Produced Copper Alloys 2015;273:824187.

[22] Standard Test Method for Determining Extreme Pressure Properties of Lubricating Greases Using A High-Frequency , Linear-Oscillation (SRV) 1985:7-10. DOI: 10.1520/D7421-11.2.

[23] Mason JK, Schuh CA. Representations of Texture. Electron Backscatter Diffr. Mater. Sci., Boston, MA: Springer US; 2009, p. 35-51. DOI: 10.1007/978-0-387-88136-2_3.

[24] Nolze G, Jürgens M, Olbricht J, Winkelmann A. Improving the precision of orientation measurements from technical materials via EBSD pattern matching. Acta Mater 2018;159:408-15. DOI: 10.1016/j.actamat.2018.08.028.

[25] Cai W, Mabon J, Bellon P. Crystallographic textures and texture transitions induced by sliding wear in bronze and nickel. Wear 2009;267:485-94. DOI: 10.1016/j.wear.2008.11.016.

[26] Kapoor A, Franklin FJ. Tribological layers and the wear of ductile materials. Wear 2000;245:204-15. DOI: 10.1016/S0043-1648(00)00480-4.

[27] Padilla HA, Boyce BL, Battaile CC, Prasad S V. Frictional performance and near-surface evolution of nanocrystalline Ni-Fe as governed by contact stress and sliding velocity. Wear 2013;297:860-71. DOI: 10.1016/j.wear.2012.10.018.

[28] Yao B, Han Z, Lu K. Correlation between wear resistance and subsurface recrystallization structure in copper. Wear 2012;294-295:438-45. DOI: 10.1016/j.wear.2012.07.008.

[29] Greiner C, Gagel J, Gumbsch P. Solids Under Extreme Shear: Friction-Mediated Subsurface Structural Transformations. Adv Mater 2019;31:1806705. DOI: 10.1002/adma.201806705.

[30] Rainforth WM. Microstructural evolution at the worn surface: A comparison of metals and ceramics. Wear 2000;245:162-77. DOI: 10.1016/S0043-1648(00)00476-2.

Chapter 14

Characteristic Aspects of Metal Wear: Wear-Induced Wear Transition and Characteristics of Wear Track Profiles

Naofumi Hiraoka

Abstract

This chapter describes two characteristic phenomena of metal wear that are usually not often considered but are related to the basic aspects of wear. The first is a mild-to-severe wear transition caused by the wear itself. Convex sliding pairs are usually accompanied by rolling sliding motion, but rolling sliding motion sometimes produces a peculiar wear profile, leading to high contact pressure. When the contact pressure exceeds a certain value that depends on the material, the wear mode changes to severe wear. This is a common wear transition for convex sliding pairs, but it can also occur for other pairs. The second is the similar appearance of wear tracks on various friction pairs. Rubbing metal under relatively severe conditions creates streaked wear tracks. We found the width and depth of these streaks, that is, wear track profiles are similar regardless of the sliding conditions and the material, which leads to similar appearance of the wear tracks. This suggests the existence of a general mechanism for producing wear tracks.

Keywords: wear track, roughness, profile, appearance, wavelet analysis, power spectrum density

1. Introduction

Two characteristic phenomena of metal wear are described in this chapter. The first is mild-to-severe wear transition [1, 2]. Severe-to-mild wear transition is usually observed in the running-in process. The transition is believed to be due to a smoothing of roughness [3], an increase in morphological conformity of sliding pairs due to wear [4], and/or oxide formation in the wear scars [5–8]. Mild-to-severe wear transition is sometimes observed in friction parts used for a long time. There are various causes of this transition. One of the simple causes is deterioration or depletion of lubricant [9]. Fatigue of the sliding surface or temperature increase [10] is also a possible reason.

We have found that convex sliding pairs, such as those used in some latch mechanisms or sliding electrical contacts, often cause the mild-to-severe wear transitions. This wear transition was found to be caused by an increase in contact pressure between the sliding pairs due to an increase in the wear shape inconformity of the sliding pairs. This wear shape inconformity was caused by the rolling-sliding motion of the convex sliding pair.

The second one is about wear track profiles [11, 12]. When metal sliding pairs are rubbed under relatively severe conditions, streaked wear tracks are generated in many cases. By observing the wear tracks of wear specimens which we have obtained for various test purposes and those of several published papers, we noticed that the appearance of many streaked wear tracks is similar. In other words, the width of the streaks does not change much regardless of the sliding material and wear conditions. We thought this suggests the existence of a general mechanism for producing wear tracks that can suggest the way to prevent severe damage from wear.

Though some researchers seemed to have been aware of this feature, few studies were found that focused on the wear track profile characterization [13], though many studies have been conducted on the wear scar or wear track morphology (e.g., [14]).

These two issues are often overlooked but we think they can be one of the key aspects of wear.

2. Wear-induced mild-to-severe wear transition

2.1 Experimental procedure

Figure 1 shows a schematic of the test rig. Semi-cylindrical upper specimen oscillatorily slides on the cylindrical specimen immersed in mineral oil [1, 2]. The lubrication condition is supposed to be boundary lubrication. This sliding motion is apparently a pure sliding but actually a rolling-sliding motion. **Figure 2** illustrates the specimen motion. The upper specimen rolls an angle of θ, while it slides a distance of $R\,\theta$ on the lower specimen. Note that usually the sliding direction is the same as the rolling direction in the rolling-sliding motion of a gear surface or a traction drive, but this rolling-sliding motion is opposite. Since both contact points of the upper and lower specimens move during sliding, it is more difficult to conform the sliding surface than pure sliding.

Test materials and test conditions are shown in **Tables 1** and **2**, respectively. The lower specimen was silver-plated in order to prevent the initial large contact

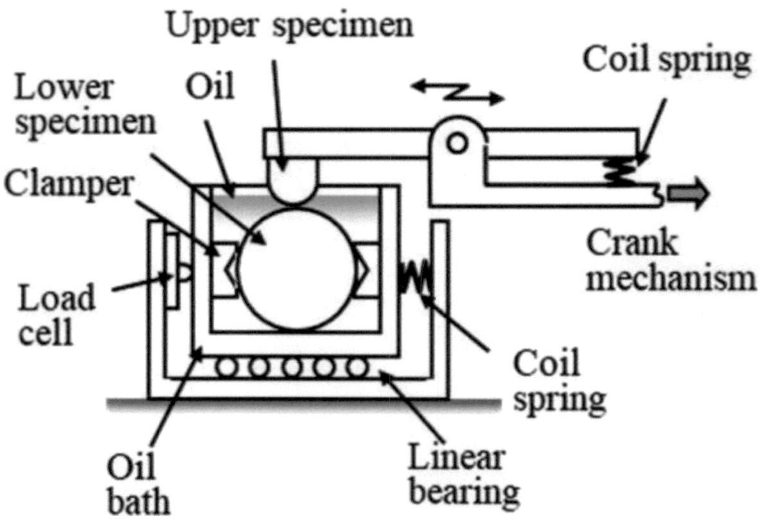

Figure 1.
Schematic of wear test rig [1].

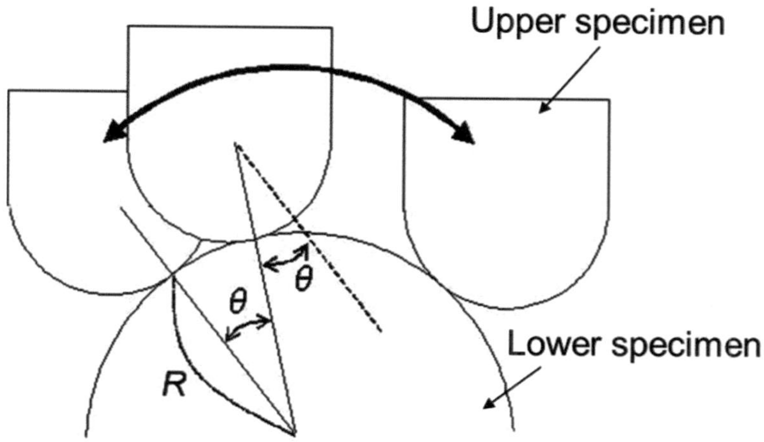

Figure 2.
Rolling-sliding motion of specimens [2].

Upper specimen	Radius; 3 mm, Width; 2 mm Material; 99.9% Cu (JIS C1100H, H_V ~100)
Lower specimen	Radius: 10 mm and width: 10 mm Material: 50 μm silver electroplated 99.9% Cu (JIS C1100H, H_V ~100)
Oil	Mineral and synthesized mixed insulating oil Kinetic viscosity: 5.2 cSt (40°C)

Table 1.
Test materials [1].

Motion	Oscillation
Atmosphere	Laboratory air (20 ~ 25°C, RH 40 ~ 60%)
Load (N)	39, 69, 83, 98
Frequency (Hz)	2
Stroke (mm)	10

Table 2.
Test conditions [1].

pressure due to misalignment. The load was applied by the coil spring. Loads shown in **Table 2** were those when the upper specimen was on top of the lower specimen. Therefore, the load varied during sliding by about 0 ~ −8% due to the vertical movement of the upper specimen. The horizontal force, that was the sum of the horizontal components of the friction and the load, was measured by the load cell to monitor the wear conditions.

2.2 Experimental results

Figure 3 shows the time evolution of horizontal forces. The forces shown were those of the maximum values for one oscillation. All forces indicated sudden increase except for the 39-N load. **Figure 4** indicates the trends of horizontal forces of the 69-N and the 98-N loads obtained from the tests of some oscillation numbers ((a)) and the wear depth at the center of the lower specimen wear scar of them ((b)).

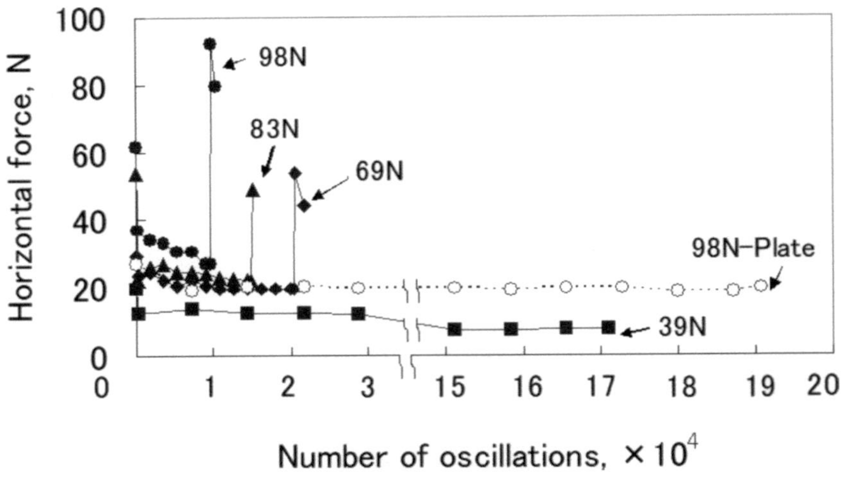

Figure 3.
Time evolution of horizontal forces for the loads tested [1].

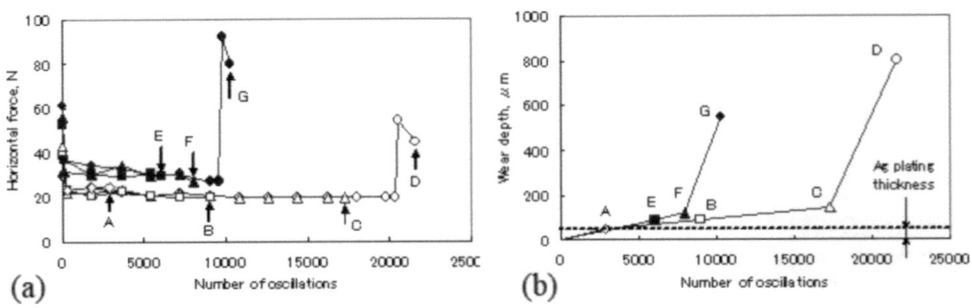

Figure 4.
(a) Trends of horizontal forces of the 69-N (empty symbols) and the 98-N loads (filled symbols) and (b) wear depth of lower specimen wear scar [1].

The wear depth gradually increased and then suddenly and rapidly increased at the oscillation numbers of sudden increase of horizontal forces (points G and D).

Figure 5 shows the microscope images of the wear scars of the lower specimens of B and D in **Figure 4**. From **Figure 5(b)**, the wear scar was rough and indicated the traces of hard adhesion, whereas that appeared relatively smooth in **Figure 5(a)**. From the wear depth progress in **Figure 4(b)** and wear scar appearances in **Figure 5**, the wear can be referred to as mild wear and severe wear before and after the sudden increase of the horizontal force, respectively. Thus, mild-to-severe wear transition occurred above a certain load.

Traces of silver-plating were not observed in **Figure 5**. **Figure 6** indicates the Auger electron energy intensity measured in the wear scars of the lower specimen of the tests stopped before (15,600 oscillations) and after (18,600 oscillations) the point C in **Figure 4** of 69-N load. No silver was detected in **Figure 6** which indicates that the silver-plating was depleted in the early stage of wear and was not related to the wear transition. The intensity of oxygen (O1) was almost the same for two specimens in **Figure 6**. This suggests that surface products of oxide were also not related to the wear transition.

Figure 7 shows the relations between the loads and the oscillation numbers of mild-to-severe wear transition points. It looks like the S-N curve for metal fatigue, indicating the possibility of fatigue as the cause of the wear transition. However,

Figure 5.
Optical microscope images of wears of lower specimens: (a) B and (b) D in **Figure 4** *[1].*

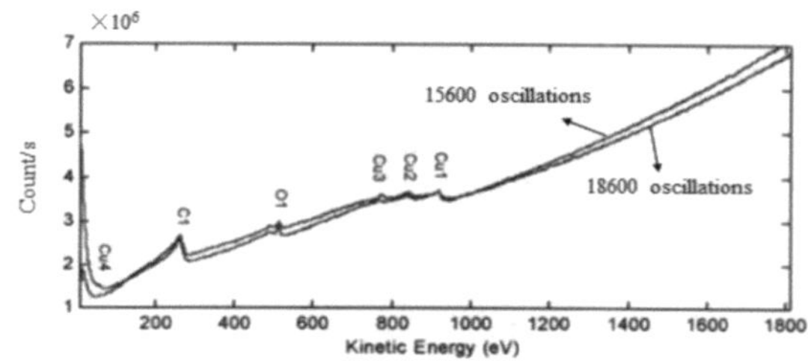

Figure 6.
Auger electron energy intensity in wear scars of lower specimen for 69-N load [1].

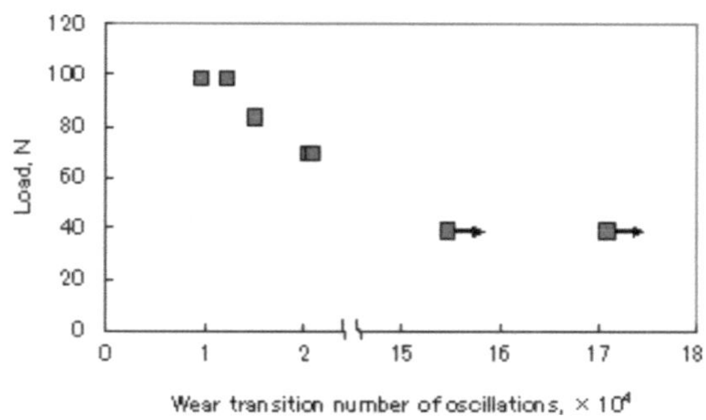

Figure 7.
Relations between loads and oscillation numbers of mild to severe wear transition points [1].

when pure sliding tests of 98-N load were carried out using the same upper specimens and the flat plates, the result is shown in **Figure 3**, no wear transition was observed. The Hertzian contact pressure calculated without considering the silver plate for the flat plate was 529 MPa, which was larger than that for cylindrical lower specimen of 69-N load: 504 MPa. The tests were conducted twice and almost the same results were obtained. This suggests that the fatigue was not the case for cylindrical sliding pairs.

Figure 8 indicates the temperature trend of the upper specimen under 69-N load. The temperature was measured with a thermocouple 1 mm above the contact point. The temperature was almost constant at around 25°C and was not the cause of the wear transition.

2.3 Discussion

The experimental results above indicate that the mild-to-severe wear transition was not due to the lubricant depletion, fatigue, or temperature rise. **Figure 9** shows the wear shapes of the upper and the lower specimens of 69-N load and 18,600 oscillations, measured along the circumferential direction at the center of the wear scar. The trapezoidal wear shape was generated by wear from the original circumferential shape. This peculiar wear shape was probably due to the rolling-sliding motion of the specimens, which made a contact shape inconformity and generated a large contact pressure.

To investigate the generation process of the peculiar wear shape and its effects on the contact pressure, wear simulations were conducted. Calculation model is shown in **Figure 10**. The contact pressure was approximated by the Hertzian contact pressure: $P(x)$. The contact area was divided into discrete slices of Δx width and the upper specimen moved byΔx in one calculation step. The wear depth was calculated by $KP(x)\Delta x$, where K was a virtual specific wear rate, and the new wear shape was obtained by smoothing the calculated discrete wear shape. The detail of the simulation was described in Refs. [1, 15–18].

The results of the simulations are shown in **Figure 11**. Calculation conditions are listed in **Table 3**. Materials for both specimens were supposed to be Cu with Young's modulus of 100 GPa and Poisson's ratio of 0.3. As shown in **Figure 11(a)**, the calculated wear shape reproduced the experimental result of the trapezoidal wear shape showing that the simulation can fairly simulate the wear phenomenon.

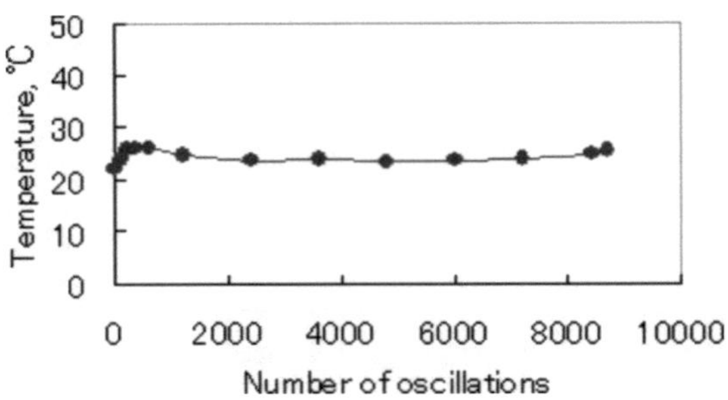

Figure 8.
Temperature trend of upper specimen under 69-N load [1].

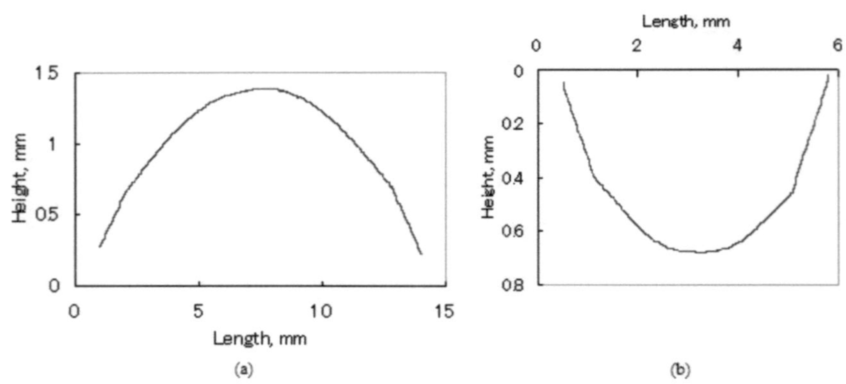

Figure 9.
Wear shapes of upper and lower specimens of 69-N load and 18,600 oscillations: (a) lower specimen and (b) upper specimen [1].

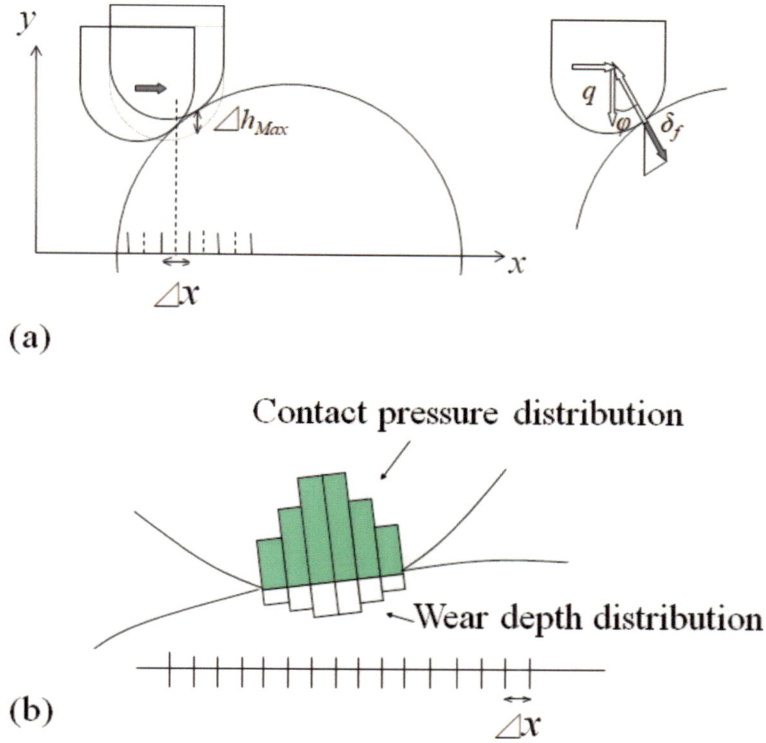

Figure 10.
Calculation model for wear simulation: (a) the whole of the model and (b) schematic of the contact area [1].

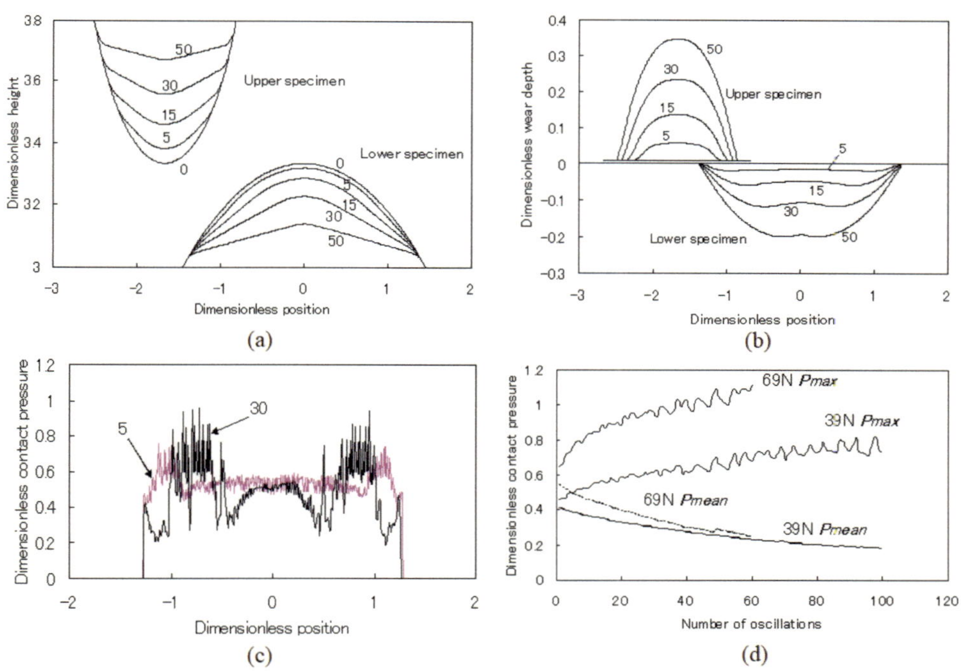

Figure 11.
Results of wear simulation for 69-N and 39-N loads: (a) specimen's shape change, (b) wear depth, (c) contact pressure on the lower specimen, and (d) time trends of the contact pressures (Pmax: maximum contact pressure, Pmean: mean contact pressure, for one oscillation) [1]. The length and the contact pressure were normalized by the upper specimen radius and hardness of the upper specimen material. Numbers correspond to the oscillation numbers.

Note that the numbers shown in the figures correspond to oscillation numbers in the calculation, not the actual ones.

Figure 11(c) shows the calculated largest dimensionless contact pressure for each position on the lower specimen in one oscillation of the upper specimen. The contact pressure was normalized by the specimen material hardness, which means the dimensionless contact pressure indicated 1 when the contact pressure reached the specimen material hardness. The contact pressure was "M"-shaped, and as a result, the wear depth was "W"-shaped as shown in **Figure 11(b)**. The peak contact pressure of "M"-shape increased with the number of oscillations. **Figure 11(d)** shows the time trend of the maximum and mean contact pressure. The maximum dimensionless contact pressure of 69-N load exceeded the value 1 as the oscillation number increased, while that of 39-N load never did. The mean contact pressure of both loads gradually decreased.

Since the excess of the contact pressure over the material hardness or a measure of material strength could cause the wear transition [19], the wear transition of 69-N load occurred when the maximum dimensionless contact pressure exceeded 1.

Figure 12 shows the calculated dimensionless equivalent contact radius, normalized by the upper specimen radius, at each position on the upper specimen in one stroke. The contact point moved from the right to the left in the figure as the upper specimen moved to the right. The contact radius of the rightmost and leftmost parts of the upper specimen decreased as the number of the strokes increased, indicating

Specimen radius	Upper specimen	3 mm
	Lower specimen	10 mm
Specific wear rate	10^{-1} mm^3/Nm	
Line load	34.5 N/mm	
Stroke	10 mm	
Δx	0.02 mm	
Number of points in moving average	17	

Table 3.
Calculation conditions [2].

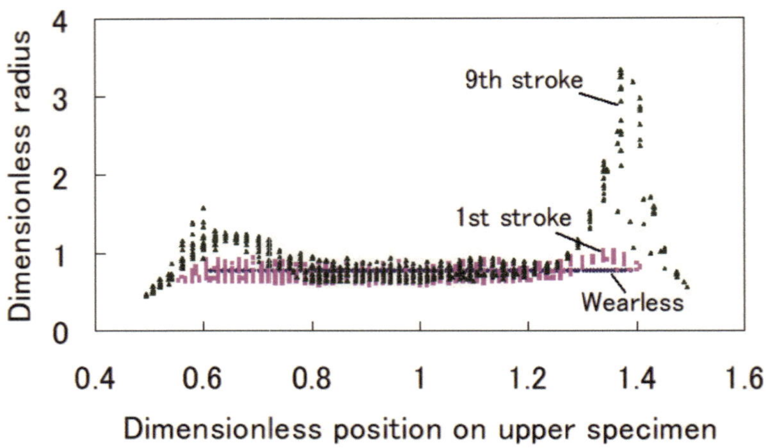

Figure 12.
Calculated dimensionless equivalent contact radius [1].

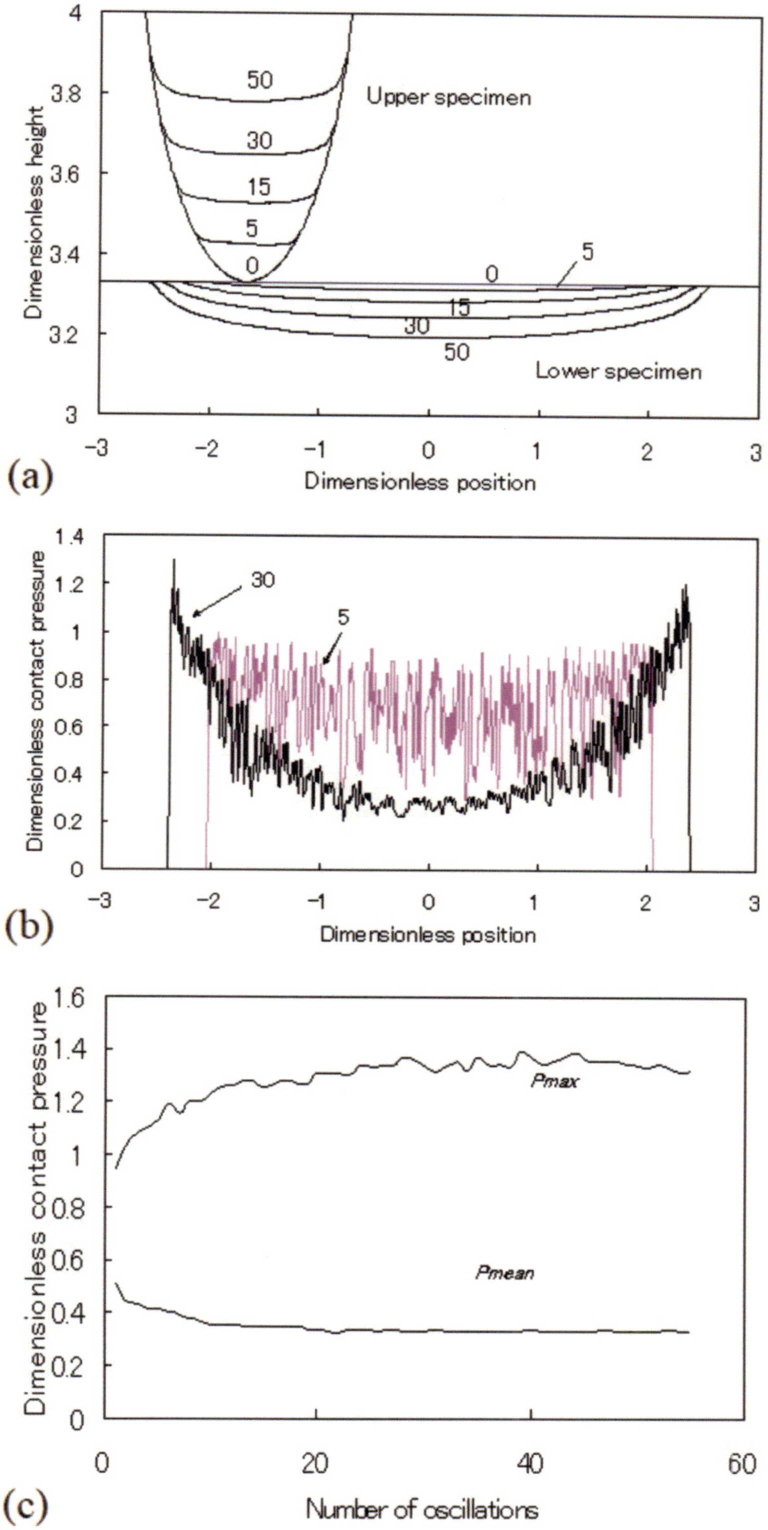

Figure 13.
Simulation results for a pure sliding pair of cylindrical upper specimen and flat plate lower specimen of 98-N load: (a) specimen's shape change, (b) contact pressure on the lower specimen, and (c) time trends of the contact pressure (symbols correspond to ***Figure 11****) [1].*

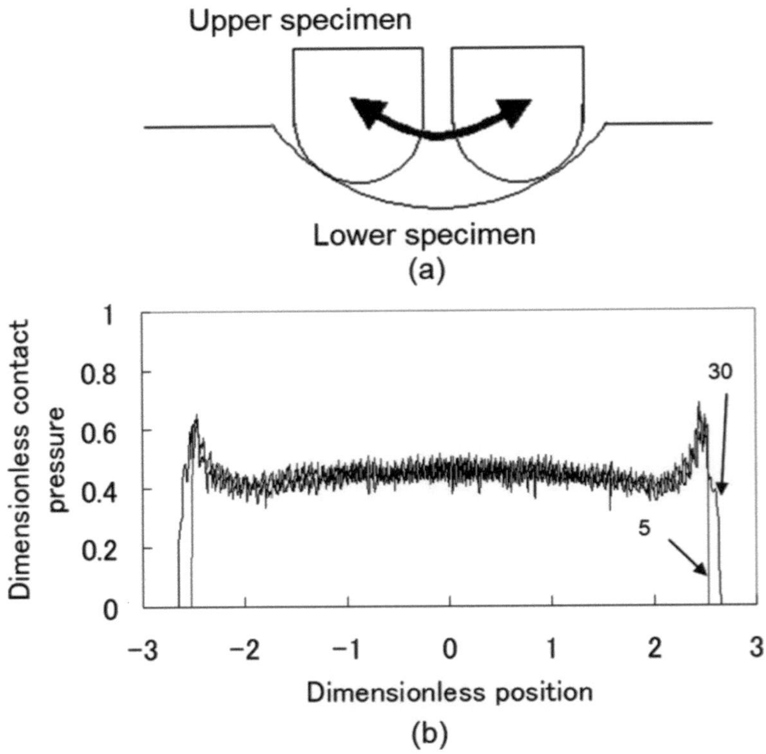

Figure 14.
*Motion and contact pressures for arc-concave lower specimen: (a) motion, (b) contact pressure (with symbols corresponding to **Figure 11**); concave radius: 10 mm (other conditions were the same as listed in **Table 1**.) [2].*

significantly smaller value than 1. This suggests that large contact pressure was generated by increasing inconformity of the contact surface shape.

As described in Section 2.1, the rolling-sliding motion of the sliding pairs in these experiments was the opposite of that often seen on machines, that is, the direction of the rolling and that of the sliding were opposite. In this rolling-sliding motion, a part that has once worn due to sliding may be returned to the contact portion again during one stroke due to the rolling motion. Perhaps this feature causes the inconformity of wear shapes leading to wear transition.

What about other shaped specimens? **Figure 13** shows the simulation results for a pure sliding pair of cylindrical upper specimen and flat plate lower specimen of 98-N load [1]. Dimensionless contact pressure exceeded 1, but its position was the end of the stroke. Therefore, the large contact pressure would have little effect for wear and caused no wear transition as shown in **Figure 2**.

Figure 14 indicates the simulation results for the cylindrical upper specimen and the arc-concave lower specimen of 34.5 N/mm line load (corresponding to 69-N load in **Figure 2**) [2]. As the direction of the rolling and that of the sliding was the same in the rolling-sliding motion for this sliding pair, the recontact of worn portion did not occur leading to lower contact pressure.

3. Characteristics of wear track profiles

3.1 Experimental

Pin-on-disk wear tests were carried out for some materials and under several conditions to obtain wear tracks to be compared and analyzed [11, 12]. The lower part of the vertical set disk was immersed in mineral oil and the contact point with the pin was wetted along with the rotation of the disk. The pin had a flat surface of 8 mm diameter. The pin and the disk surface had a roughness of Ra ~0.3 micrometer. Test conditions and materials are shown in **Table 4**.

Figure 15 shows the wear track photos for the stainless and the brass disks tested in wet condition. Both indicated the streaked wear track and were difficult to discern visually, despite the different materials and test conditions. The wear tracks of the other specimens also showed the similar appearances. **Figure 16** indicates the examples of the wear track profiles. The valleys with 200–500 micrometer width and 20–70 micrometer depth were prominent and these similar dimensions of the valleys may bring the visual similarity of the wear tracks.

3.2 Characterization of wear track profiles

To evaluate the geometric similarity between the wear tracks quantitatively, frequency analysis was applied to the wear track profiles. Since the characteristics of the profile shape curve seems incidental rather than periodic, discrete wavelet transform (DWT) was applied. Details of the analysis were described in Ref. [11].

Figure 17 indicates the power spectral density (PSD) of DWT with regard to the wavelength of the roughness. Although there are some deviations, the PSD curves in the figure had almost the same shapes, that is, they had similar slopes and bended at about the same point, showing the geometrical similarity of the profiles. The profile shape curve of the wavelength at the bending point determines the rms

	Pin material	Disk material	Oil	Sliding velocity (m/s)	Load (N)	Sliding duration (h)
a	JIS S45C steel	JIS SUS304 stainless steel	Wet	0.02	118	1.68, 2, 8, 40, 80
b	JIS C3604 brass	JIS C3604 brass			54, 78, 118	20
c	JIS S45C steel	JIS SUS304 stainless steel	Dry		54, 78	2
	JIS S45C steel	JIS SUS304 stainless steel			118	1, 2, 4, 6

Table 4.
Test conditions and materials [11].

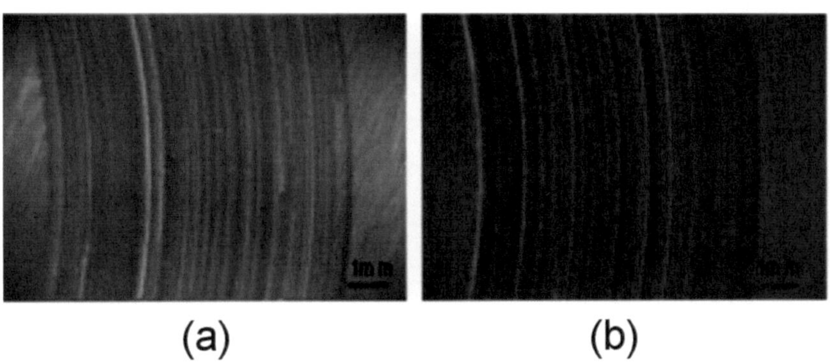

Figure 15.
Photos of typical wear tracks on disks: (a) SUS304 versus S45C, 118-N load, 8-h sliding and (b) C3604 versus C3604, 54 N, 20-h sliding [11].

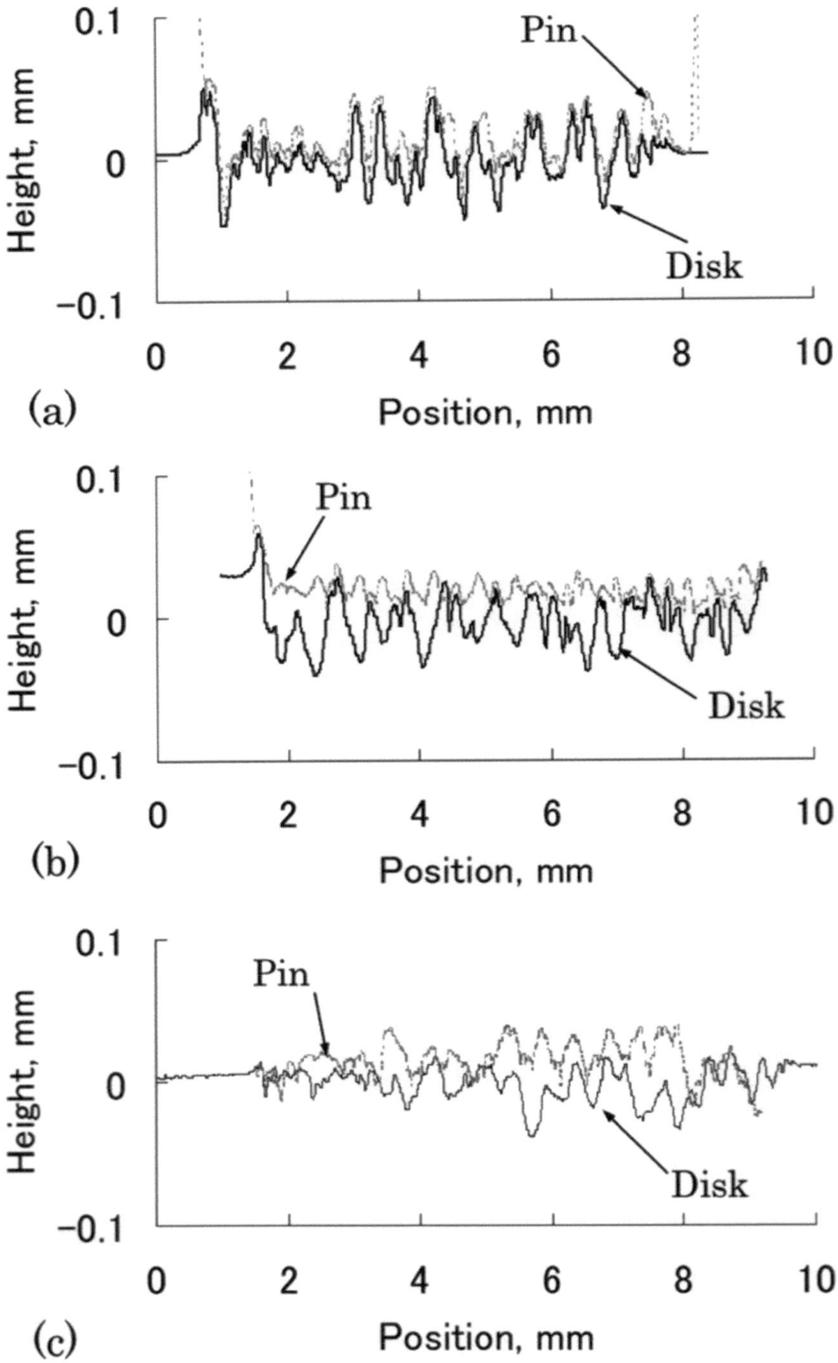

Figure 16.
Examples of combined wear track profiles for disks and pins: (a) S45C versus SUS304, 118-N load, 8 h, wet; (b) C3604 versus C3604, 54-N load, 20 h, wet; and (c) S45C versus SUS304, 118-N load, 4 h, dry [11].

roughness [20] and was visually conspicuous because valleys (or mountains) of the roughness larger than this wavelength had comparatively small depth to their width and were recognized as waviness.

In order to indicate the bending points explicitly, PSD ratios: $\ln\{P_W(\lambda_n)/P_W(\lambda_{n-1})\}$, where $P_W(\lambda_n)$ denotes the PSD at *n*th wavelength in **Figure 17**,

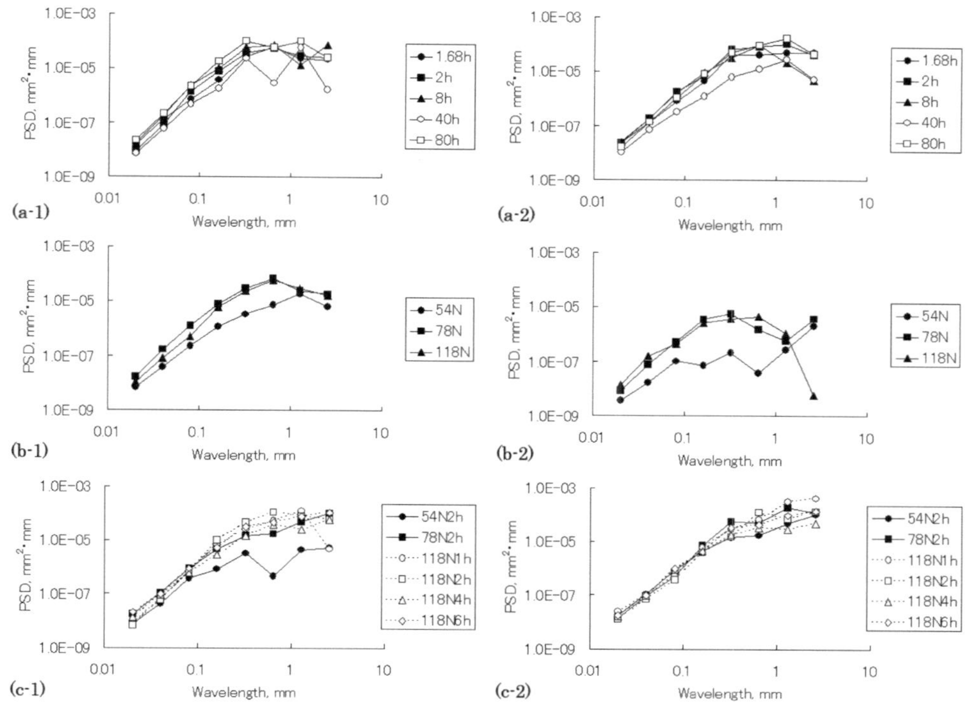

Figure 17.
Power spectral density (PSD) of DWT: (a) S45C versus SUS304, wet; (b) C3604 versus C3604, wet; and (c) S45C versus SUS304, dry, (−1) disks, (−2) pins [11]. Test conditions in (a) ~ (c) correspond to those of a ~ c in ***Table 4****.*

were calculated and shown in **Figure 18**. The PSD ratios showed sharp drops at the wavelength of 0.2–0.5 mm indicating the bending points. The PSD values at these wavelength were 10^{-5}–10^{-4} mm^2 •mm that were approximately equivalent to the valley depth of 30–80 micrometers. These results supported the above observations of valleys of the roughness that characterized the wear tracks.

3.3 Mechanism of generating wear track profiles

Pin-on-disk tests were used to capture the moment of the streak formation. The tester used was as the same as above. JIS S45C steel pins and JIS SUS304 stainless steel disks with surface roughness of Ra ~ 0.3 micrometer are used under 118-N load and wet conditions. The pin-disk assembly is shown in **Figure 19(a)**.

The disk rotation started slowly and then maintained a rotational speed of 12–13 rpm (sliding speed of ~0.03 m/s) to observe the wear track generation process. No signs of wear or damage were observed on the disk on the first few rotations, after which a groove appeared suddenly. As soon as the groove was found, the disk rotation was stopped. Other tests continued rotating after the first groove appeared. In each test, initially only one groove appeared, and after several rotations, the second groove appeared adjacent to the first groove, and so on, and the streak pattern was generated (sometimes, new groove generated at the other place to become another core of the streak).

Figure 19(b) and **(c)** shows the SEM images of the pin surfaces after testing. Transfer particles [21, 22] of about the same size were found on the pin surfaces that matched the number of grooves on the disks. Obviously, these transfer particles plowed the disk surface and generated the grooves of nearly the same size. The

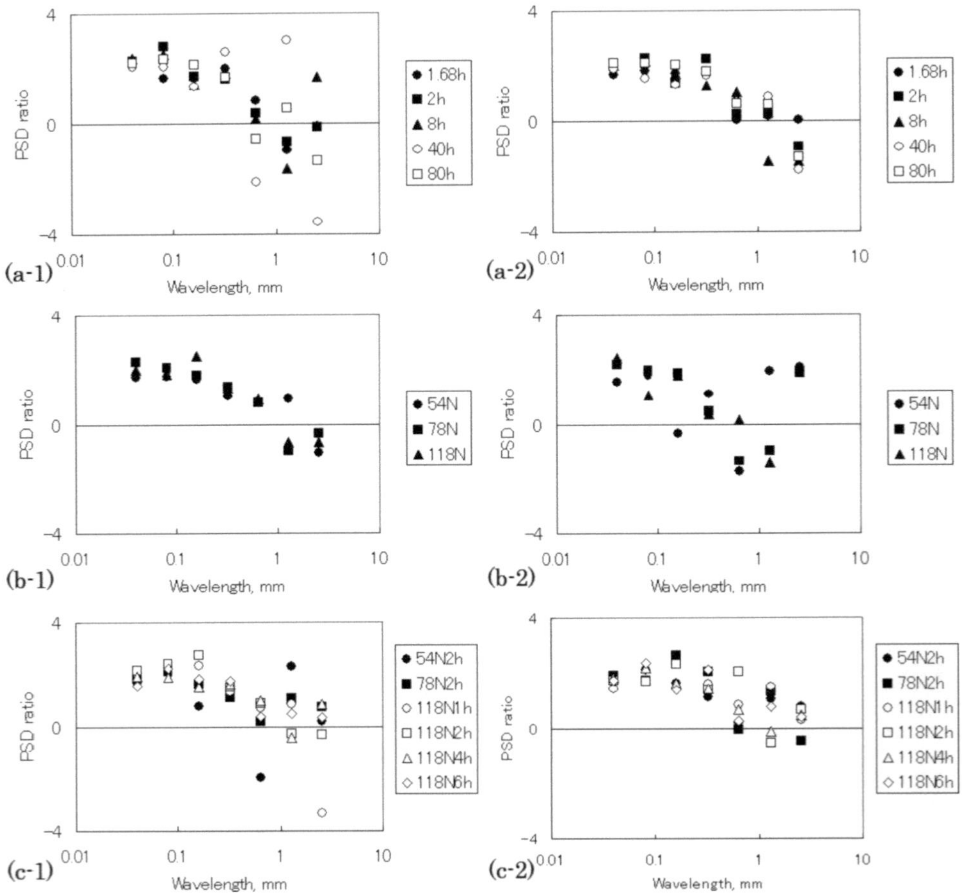

Figure 18.
PSD ratios [11]. Notations (a-1)–(c-2) correspond to those in ***Figure 17****.*

transfer particles included Cr, Ni and Fe according to the energy-dispersive X-ray spectroscopy analysis, suggesting that they originated in a mixture of both the pin and the disk materials and grew to and stopped growing at that size in that place.

Figure 20 shows the surface profiles of the tested pin and the disk. The measuring directions on the pin surfaces were shown in **Figure 19**. As seen in the SEM images, the sizes of transfer particles and naturally those of the grooves were similar and were in the range of the characteristic sizes above. This caused the similar appearances of the wear tracks.

Why were the transfer particles (and their resulting grooves) about the same size? From **Figure 20**, all grooves on the disks were found to have the ridges along their sides that dug into the pin surfaces. These ridges were also found for brass material [11]. These ridges digging into the mating surfaces enclosed the transfer particles and could be assumed to terminate their growth. Also, these ridges digging into the mating disk surface seems to generate the "counter" ridges on the disk and the counter ridges grew to dig into the pin surface generating a new seed of the transfer particle, as shown in **Figure 20(b)**.

3.4 Ridge formation conditions

We assumed that the ridges along the sides of the grooves have a key role to determine the characteristic scales of the wear track streaks and conducted the

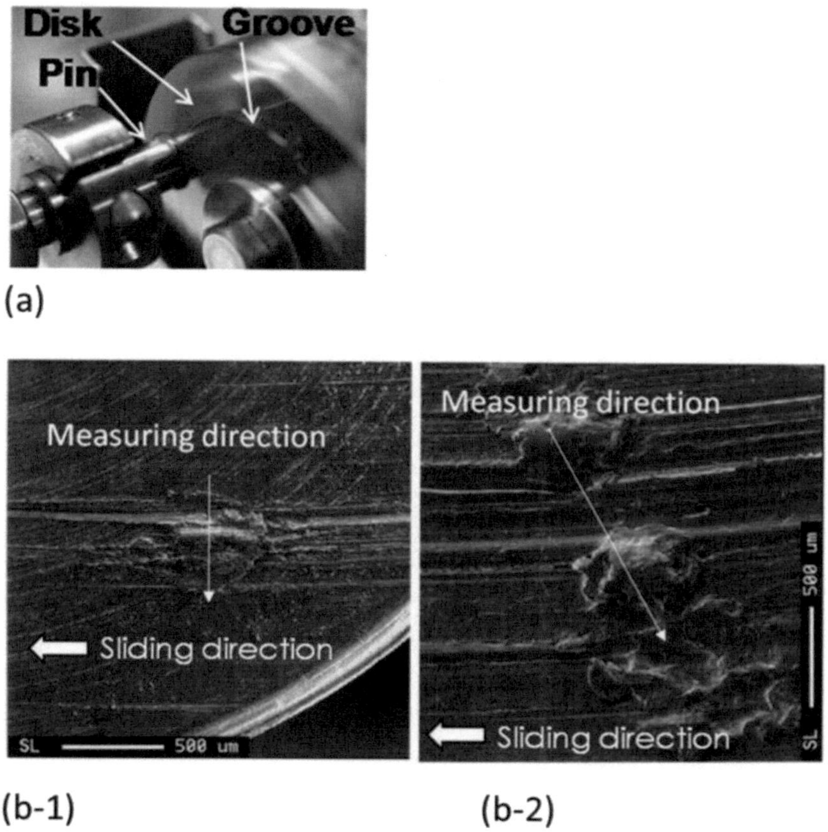

Figure 19.
Perspective of pin-disk assembly and SEM images of pin surfaces after the tests: (a) pin-disk assembly and (b-1) and (b-2) one and 2 transfer particles adhering to pin surfaces, respectively [12].

ridge formation tests. **Figure 21** shows the schematic of the test rig. Two metal plates sandwiched three hard balls and were pressed. Then one plate was rotated at an angle of 90°generating three grooves on the plate. The ball materials and the test conditions are listed in **Table 5**. Plate materials were JIS SUS304 stainless steel and JIS C3604 brass. Some tests were under oil-wetted conditions. The balls were fixed to one plate to prevent rolling in some cases.

Figure 22 shows the examples of groove profiles generated in the tests. You can see the groove with and without the ridge in the figure. **Figure 23** illustrates the groove profile. We designate $2d/\lambda_1$ in **Figure 23** as the "degree of penetration or D_p." The degree of penetration was originally defined as y/a in **Figure 24** for the wear map [23] and is approximately equivalent to $2d/\lambda_1$ in **Figure 23**. **Figure 25** shows the relations between D_p and h/d. In **Figure 25**, h/d is 0, that is, no ridge, for small D_p. h/d increased sharply when D_p exceeded about 0.05–0.1 regardless of the materials. This means that the valleys with higher aspect ratio (larger Dp) had higher ridges on their sides.

From the slopes of PSD curves in **Figure 17**, PSD values were proportional to about the third power of wavelength. This relation is replaced with the relation of $(\text{depth}) \propto (\text{width})^{1.5}$ in the valley profiles. Suppose the grooves grew during sliding maintaining this relation and "depth" and "width" were equivalent to d + h and λ_2, the relation of $d + h = 50 \times (\lambda_2/400)^{1.5}$ holds, when the typical value of d + h = 50 micrometer andλ_2 = 400 micrometer were used. The solid curve in **Figure 26** indicates this relation. The groove can be considered to grow along this curve.

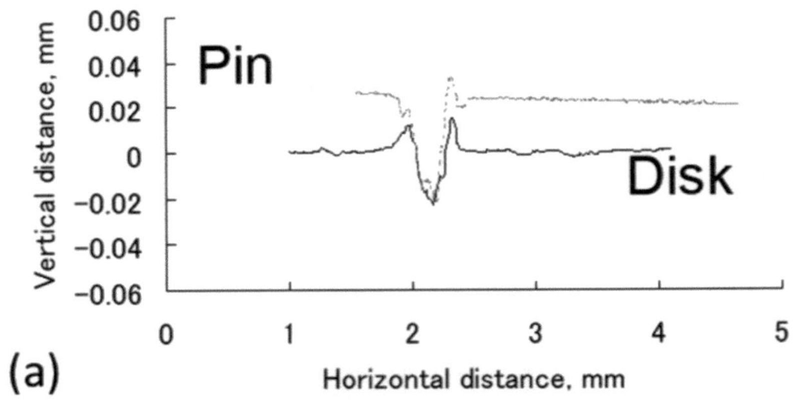

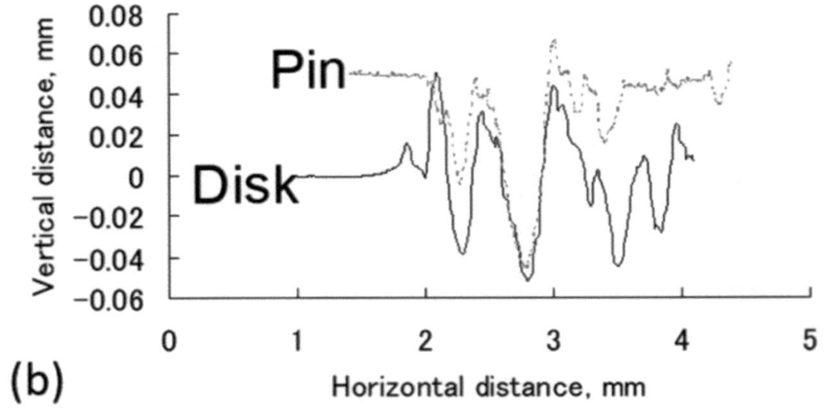

Figure 20.
Surface profiles of pins and disks [12]. Measuring directions of (a) and (b) correspond to those in **Figure 1** *(b-1) and (b-2), respectively.*

If the ridges began to form at $D_p = 0.1$, then $d = 0.05\lambda_1$. The dashed line in **Figure 26** indicates this relation. The intersecting point of the two lines indicates the time when the ridge began to form, the groove width at that time could be read as about 70 micrometers. The time when the ridge grew and dug into the mating surface was after this. Probably the groove width grew to a few 100 micrometers at that time, which was the characteristic scale of the grooves in the streaked wear track. We, therefore, consider this mechanism causes the wear track appearance similarity.

4. Conclusion

In this chapter, we have described two characteristic wear phenomena that are usually less noticeable, but we believe they characterize some aspects of wear and can serve as a background for examining various wear phenomena.

Mild-to-severe wear transition could occur for convex-shaped sliding pair due to their rolling-sliding motion which generates peculiar wear shapes leading to large contact pressure. Streaked wear track were shown to have a similar profile consisting of grooves with 200–500 micrometer width and 20–70 micrometer regardless of friction materials and conditions. This is because the transfer particles that plow the surface and produce grooves grow with sliding, but stop at about the same size, regardless of friction material and conditions.

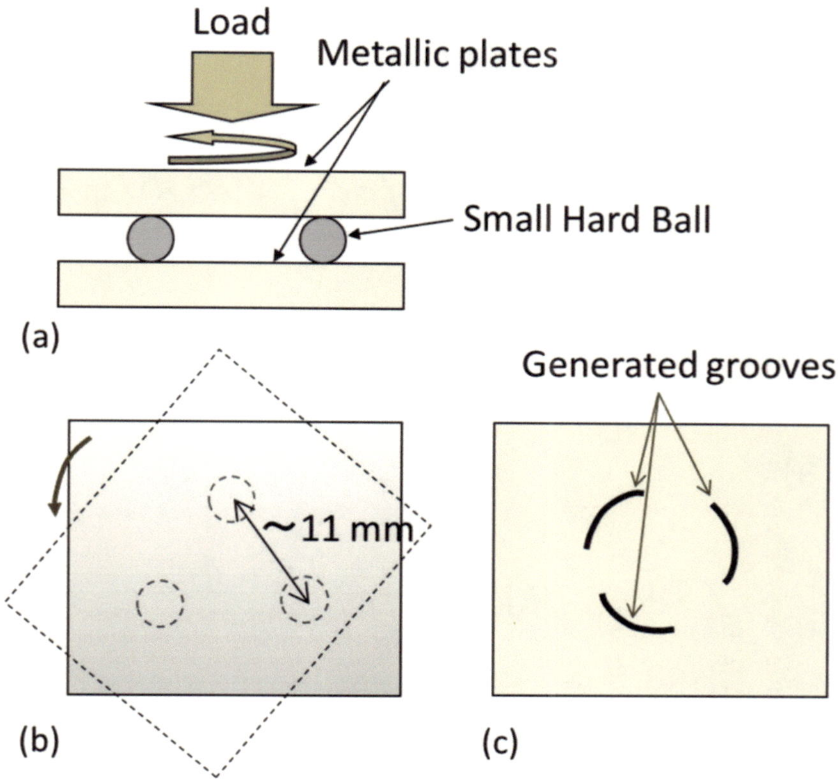

Figure 21.
***Figure 3** Schematic of ridge generating rig: (a) side view, (b) top view, and (c) generated groove trajectories [12].*

Ball material	Ball diameter (mm)	Load for 3 balls (N)
ZrO_2	0.1	15
	0.2	8, 30, 50
	0.5	100
	1	200
JIS SUS440C stainless steel	0.3	4, 15, 60, 100
JIS SUJ2 bearing steel	0.3	15, 60, 100
	0.5	60, 200, 300, 500
	1	20, 60, 200
	1.5	500
	2	500

Table 5.
Ball materials and test conditions [12].

Though we have not actually verified, we can consider the methods to prevent these wear damages. One of the methods to prevent the wear-induced mild-to-severe wear transition is, for example, to reduce the curvature of the surface to reduce the rolling-sliding motion and the contact pressure. One method to prevent the wear track generation is to prevent the transfer particle generation by effective lubrication

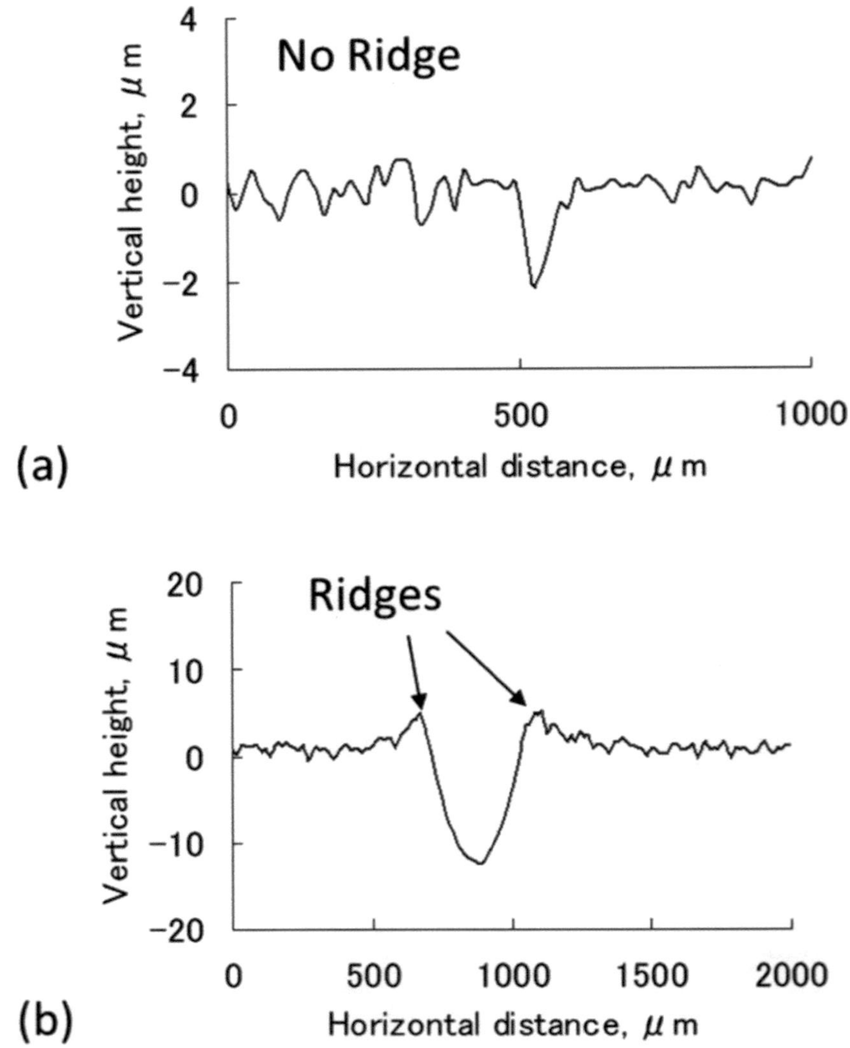

Figure 22.
Examples of measured groove profiles generated with ((a)) and without ((b)) ridges: (a) 0.3 mm in dia. SUJ-2 ball, 5 N/3 balls and (b) 2 mm in dia. SUJ-2 ball, 500 N/3 balls [12].

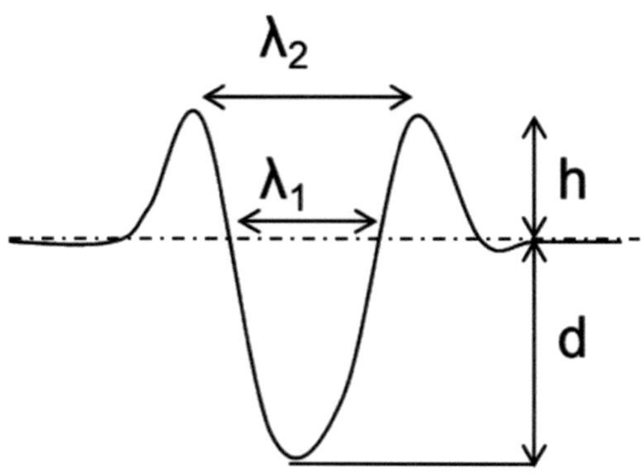

Figure 23.
Illustrated groove profile and notations of dimensions of groove and ridges [12].

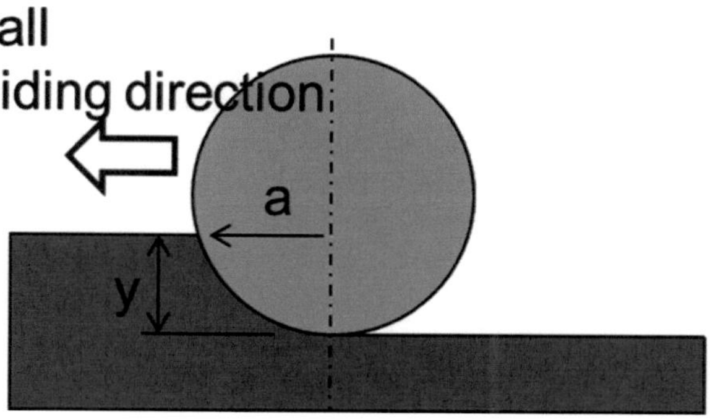

Figure 24.
Degree of penetration: y/a [12].

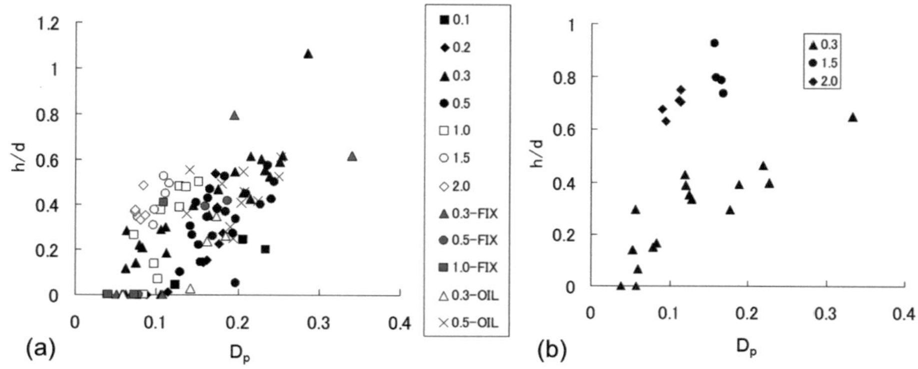

Figure 25.
Relations between Dp and ridge height rate h/d: (a) SUS304 plate and (b) C3604 plate [12]. Numbers, "FIX," "OIL" in boxes indicate ball diameter, using fixed ball, oil wetted conditions, respectively.

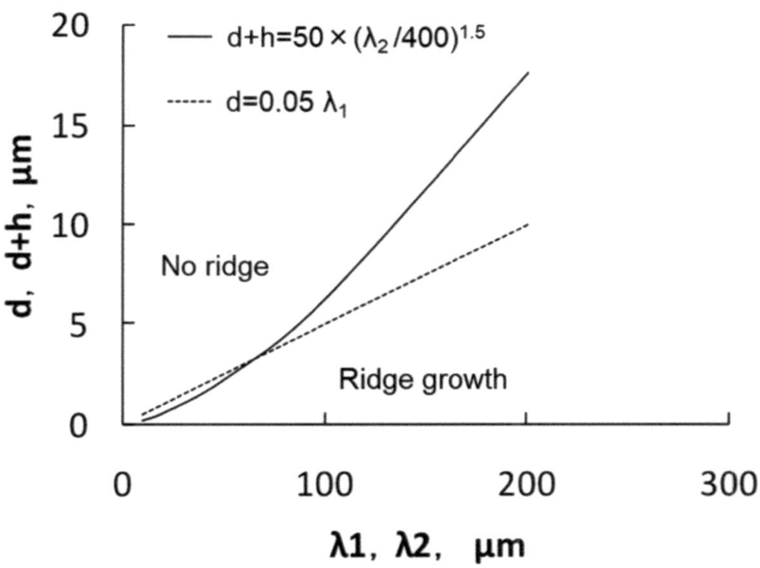

Figure 26.
Relations between calculated width and depth of groove [12].

(e.g., adding effective lubricant additives). These sounds obvious, but these may not have been understood well without understanding the phenomena above.

Considering these mechanisms above would offer some suggestions for reducing wear damages, however, the applicability of these mechanisms (such as for non-metallic materials) further need to be verified.

Notes

Most of the content of this chapter is based on Refs. [1, 2, 11, 12].

Author details

Naofumi Hiraoka
Institute of Technologists, Saitama, Japan

*Address all correspondence to: hiraoka@iot.ac.jp

References

[1] Hiraoka N. Time-dependent mild to severe wear transition in oscillation motion of cylindrical sliding pairs under boundary lubrication. Transactions of ASME, Journal of Tribology. 2002;**124**:822-828

[2] Hiraoka N. A study on mild-to-severe wear transition due to inconformity of wear-induced shape. Wear. 2005;**258**:1531-1535

[3] Hanief M. Effect of surface roughness on wear rate during running-in of En31-steel: Model and experimental validation. Materials Letters. 2016;**176**:91-93

[4] Hu Y, Meng X, Xie Y. A computationally efficient mass-conservation-based, two-scale approach to modeling cylinder liner topography changes during running-in. Wear. 2017;**386-387**:139-156

[5] SMITH AF. The influence of surface oxidation and sliding speed on the unlubricated wear of 316 stainless steel at low load. Wear. 1985;**105**(2):91-107

[6] Garbar II. Gradation of oxidational wear of metals. Tribology International. 2002;**35**(11):749-755

[7] Korkut MH. Microstructure and wear behavior of Al2024\SiFe and Al2024\SiFe\Al2O3 composites. Tribology International. 2003;**36**(3):169-180

[8] Feser T, Stoyanov P, Mohr F, Dienwiebel M. The running-in mechanisms of binary brass studied by in-situ topography measurements. Wear. 2013;**303**(1-2):465-472

[9] Perez E, Tanaka M, Jibiki T. Wear of stainless steels—Cause and transition of wear of martensitic stainless steel. Marine Engineering. 2013;**48**(5):662-669

[10] Hayashi K, Hirasata K, Kamenaka Y, Augita K. Friction and wear of cast iron under high sliding speed and high contact pressure. 2nd report. The condition of the transition from mild wear to severe wear. Transactions of JSME C. 1997;**63**(616):4322-4327

[11] Hiraoka N, Matsumoto H. Characteristic scales of wear track profiles generated by pin-on-disk wear tests. Tribology Online. 2008;**3**(3):205-210

[12] Hiraoka N, Yamane E. A study on the mechanism of generating wear track grooves. Tribology Letters. 2011;**41**:479-484

[13] Nakano T, Hiratsuka K, Sasada T. Fractal analysis of worn surface and wear particles. Journal of Analytical Science and Technology. 1990;**35**:151-154

[14] Iliuc I. Plate-like wear particle formation in a lubricated ball-on-plate friction pair. Tribology International. 1985;**18**:215-221

[15] Flodin A, Andersson S. Simulation of mild wear in helical gears. Wear. 2000;**241**:123-128

[16] Olofsson U, Andersson S. Simulation of mild wear in boundary lubricated spherical roller thrust bearings. Wear. 2000;**241**:180-185

[17] Ovist M. Numerical simulations of mild wear using updated geometry with different step size approaches. Wear. 2001;**249**:6-11

[18] Flodin A, Andersson S. A simplified model for wear prediction in helical gears. Wear. 2001;**249**:282-285

[19] Samuels B, Richards MN. The transition between mild to severe wear for boundary-lubricated steels.

Transactions of the ASME, Journal of Tribology. 1991;**113**:65-72

[20] Persson BN, Tartaglino U, Volokitin AI, Albohr O, Tosatt E. On the nature of surface roughness with application to contact mechanics, sealing, rubber friction and adhesion. Journal of Physics. Condensed Matter. 2005;**17**:R1-R62

[21] Sasada T, Norose S, Tomaru M, Mishina H. The intermittent transversal movement of the rubbing surfaces by interposed wear particles. Junkatsu. 1978;**23**:519-526

[22] Norose S, Sasada T. The mutual transfer of rubbing materials and the mixing structure of wear particles formed in lubricating oil. Junkatsu. 1979;**24**:226-230

[23] Hokkirigawa K, Kato K. An experimental and theoretical investigation of ploughing, cutting and wedge formation during abrasive wear. Tribology International. 1988;**21**:51-57

Chapter 15

Wear: A Serious Problem in Industry

Biswajit Swain, Subrat Bhuyan, Rameswar Behera, Soumya Sanjeeb Mohapatra and Ajit Behera

Abstract

Wear is the damaging, gradual removal or deformation of material at solid surfaces. Causes of wear can be mechanical or chemical. The study of wear and related processes is known as tribology. Abrasive wear alone has been estimated to cost 1–4% of the gross national product of industrialized nations. The current chapter focuses on types of wear phenomena observed in the industries (such as abrasive wear, adhesive wear, fretting wear, fatigue wear, erosive wear and corrosive wear), their mechanisms, application of surface coating for the protection of the surface from the industrial wear, types of surface coatings, thermal spray coating, types of thermal spray coating and its application in industry to protect the surface from wear. The detail information about the wear phenomena will help the industries to minimize their maintenance cost of the parts.

Keywords: wear, type of wear, tribology, surface coating, thermal spray coating

1. Introduction

Wear is the damaging, gradual removal or deformation of material at solid surfaces. Causes of wear can be mechanical also called as erosion or chemical also called as corrosion. Wear of metals occurs by plastic displacement of surface and near-surface material and by detachment of particles that form wear debris. In material science, wear is the erosion of material from a solid surface by the action of another solid. The study of the process of wear is the part of the theory of tribology. Wear in machine components, along with different cycles, for example, fatigue and creep, makes surfaces deteriorate, in the end prompting material degradation or loss of applicability. Subsequently, wear has enormous monetary significance as first mentioned in the Jost Report. Abrasive wear alone has been assessed to cost 1–4% of the gross national product of industrialized countries. Wear of metals happens by plastic dislodging of surface and close to surface material and by separation of particles that produce wear debris. The molecule size may change from millimeters to nanometers. This cycle may happen by contact with different metals, nonmetallic solids, streaming fluids, solid particles or fluid beads entrained in streaming gasses. The wear rate is influenced by components, for example, sort of stacking (e.g., stationary and active), kind of movement (e.g., gliding and continuing), surrounding temperature, and lubrication, specifically by the cycle of deposition and deterioration of the boundary lubrication layer. Contingent upon the tribosystem, diverse wear types and wear systems can be watched.

2. Industrial wear problems

The **Table 1** represents the various wear problems occur in the industries.

Sl no.	Industrial wear problems	Significant characteristic	Examples
1.	The wear of surfaces by hard particles in a stream of fluid	Erosion with one supply of erodent being continuously renewed in a gas or fluid	Valves controlling flow of crude oil laden with sand Gas pumping equipment
2.	The wear of surfaces by hard particles in a compliant bed of material	Abrasion, with supply of abrasive continuously renewed by movement of bed of material	Digger teeth. Rotors of powder mixes. Extrusion dies for bricks and tiles
3	Wear of metal surfaces in mutual rubbing contact, with abrasive particles present	Three body abrasion (solid abrasive-solid) with an ongoing supply of new abrasive particles	Pivot pins in construction machinery. Scraper blades in plaster mixing machines. Shaft seals for fluids containing abrasives
4.	The wear of metal components in rubbing contact with a sequence of other solid components	Adhesive wear and abrasion, but with one component in the wear process being continuously renewed	Tools used in manufacture, such as punching and pressing tools, sintering dies and cutter blades
5.	The wear of pairs of metal components in mutual and repeated rubbing contact	Adhesive wear, but with a wear rate that can be very variable depending on the detailed operating conditions	Piston rings and cylinder liners. Coupling teeth and splines. Fretting between machine components
6.	Component wear from rubbing contact between metals and non metals	Adhesive wear between two consistent components	Brakes and clutches. Dry rubbing bearings. Artificial hip joints

Table 1.
Examples of industrial wear problems.

3. Types of wear

3.1 Abrasive Wear

Removal of material by the mechanical action of an abrasive is known as abrasive wear (**Figure 1**). Abrasives are substances which are usually harder than the abraded surface and have an angular profile. Examples: sand particles between contact surfaces, the damage of crankshaft journals in reciprocating compressors. Abrasive wear is ordinarily ordered by the kind of contact and the contact condition. The sort of contact decides the method of abrasive wear. The two methods of abrasive wear are known as two-body and three-body abrasive wear. Two-body wear happens when the sand or hard particles eliminate material from the contrary surface. The basic similarity is that of material being eliminated or dislodged by a cutting or plowing activity. Three-body wear happens when the particles are not constrained, and are allowed to roll and slide down a surface. The contact condition decides if the wear is delegated open or shut. An open contact condition happens when the surfaces are adequately uprooted to be free of each other. There are various components which influence abrasive wear and therefore the way of material removal. A small number of unique components have been proposed to illustrate the way where the outer material is eliminated.

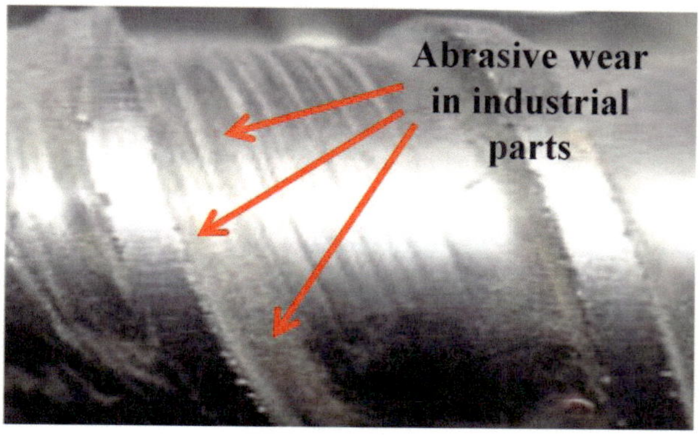

Figure 1.
Abrasive wear of industrial parts.

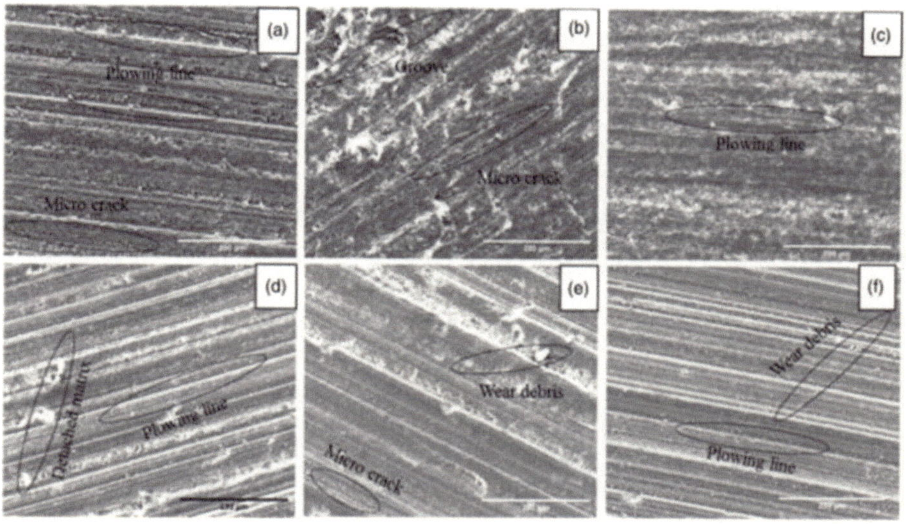

Figure 2.
SEM image of worn surface after abrasive wear test of (a) neat epoxy, (b) 0.5 wt% CNF, (c) 1.0 wt% CNF, (d) 1.5 wt% CNF, (e) 2.0 wt% CNF and (f) 2.5 wt% CNF [1].

Khanam et al. [1] investigated about the abrasive wear resistance of CNF enforced epoxy nanocomposites at different percentage of CNF concentration. It has been observed (**Figure 2**) that the neat epoxy composite revealed deep plowing line, many microcracks and surface covered with wear debris of detached matrix over the entire worn surface.

3.2 Adhesive wear

When one surface slides over the other interaction between the high spots produces occasional particles of wear debris. Mild adhesion is the expulsion of films, for example, oxides at a lower rate. Severe adhesion is the evacuation of metal because of tearing, breaking, and liquefying of metallic intersections (**Figure 3**). This prompts scraping or annoying of the surfaces and even seizure.

Adhesive wear can be found between surfaces during frictional contact and by and large alludes to undesirable dislodging and connection of wear debris and material mixes starting with one surface then onto the next. Two glue wear types can be recognized.

1. Adhesive wear is brought about by relative movement, “direct contact” and plastic deformation which make wear debris and material transfer starting with one surface then onto the next.

2. Cohesive adhesive load, holds two surfaces together despite the fact that they are isolated by a quantifiable separation, with or with no real exchange of material. By and large, glue wear happens when two bodies slide over or are squeezed into one another, which advance material exchange. This can be depicted as plastic distortion of little pieces inside the surface layers. The asperity or minute high focuses (surface roughness) found on each surface influence the seriousness of how sections of oxides are pulled off and added to the next surface, mostly because of solid adhesive force between atoms [2] yet in addition because of collection of vitality in the plastic zone between the severities during relative movement.

Yunxia et al. [3] investigated about the adhesive wear phenomena of aero-hydraulic spool valves and the investigation revealed the trimming and transformation of outer material due to the shear fracture of the bonded areas (**Figure 4**). It has been also claimed that the above mentioned work is an evidence of the adhesion wear process between spool and valve sleeve.

3.3 Fatigue wear

Surfaces can wear by fatigue when they are subject to fluctuating loads. High surface stresses cause cracks to spread into the material, and when two or more

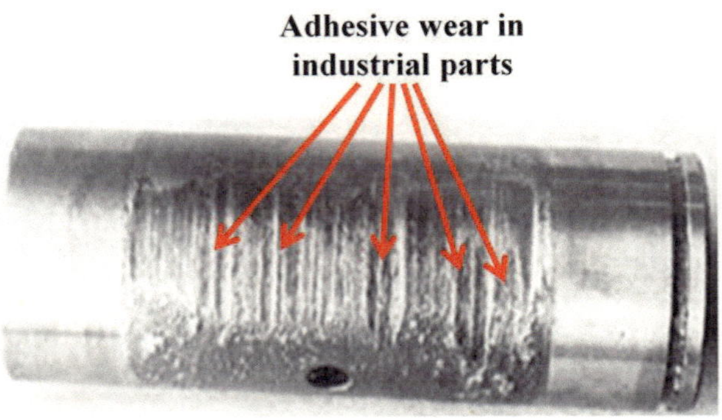

Figure 3.
Adhesive wear in industries.

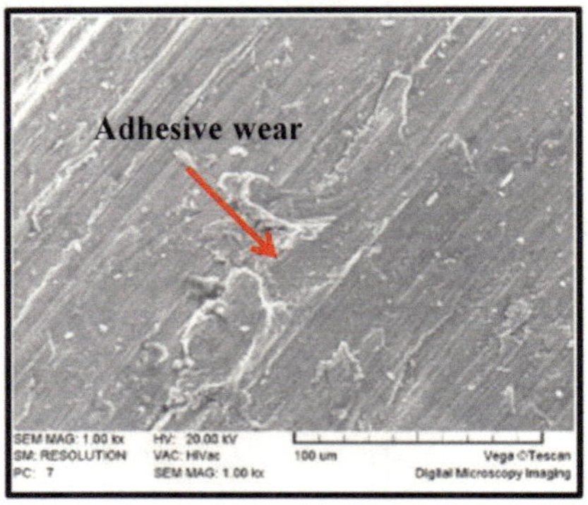

Figure 4.
SEM morphology of adhesive wear surface of spool shoulder [3].

of these cracks become joined together large loose particles are formed. Thermal Surface Fatigue occurs when high repetitive stresses are generated through the heating caused by the contact of the two contacting components which result in cracking of the surface and the loss of small chunks of material. Surface fatigue is a cycle where the outside of a material is debilitated by cyclic stacking, which is one sort of broad material weariness (**Figure 5**). Fatigue wear is developed when the wear particles are confined by cyclic split development of microcracks on a superficial level. These microcracks are either shallow splits or subsurface breaks.

Mao et al. [4] investigated the fatigue wear phenomena of the gear and in his investigation he found out that the main reason of fatigue wear is the high stress concentration. The pitting failure due to the stress concentration is illustrated in **Figure 5**.

3.4 Fretting wear

Fretting occurs where two contacting surfaces, often nominally at rest, undergo minute oscillatory tangential relative motion (**Figure 6**). Small particles of metal are removed from the surface and then oxidized. Typically occurs in bearings although the surfaces are hardened to compensate this problem and also can occur with cracks in the surface (fretting fatigue). This carries the higher risk of the two as can lead to failure of the bearings. Fretting wear is the rehashed recurrent scouring between two surfaces. Over some stretch of time fretting this will eliminate material from one or the two planes in contact. It happens normally in orientation, albeit most headers have their surfaces hardened to oppose the issue. Another issue happens when splits in either surface are made, known as fretting fatigue. It is the more genuine of the two marvels since it can prompt disastrous disappointment of the bearing. A related issue happens when the little particles eliminated by wear are oxidized in air. The oxides are generally harder than the fundamental metal, so wear quickens as the harder particles rub the metal surfaces further. Fretting corrosion acts similarly, particularly when water is available. Unprotected bearings on enormous structures like bridges can endure genuine debasement in conduct, particularly when salt is utilized during winter to deice the highways conveyed by the bridges. The issue of fretting corrosion was associated with the Silver Bridge misfortune and the Mianus River Bridge mishap.

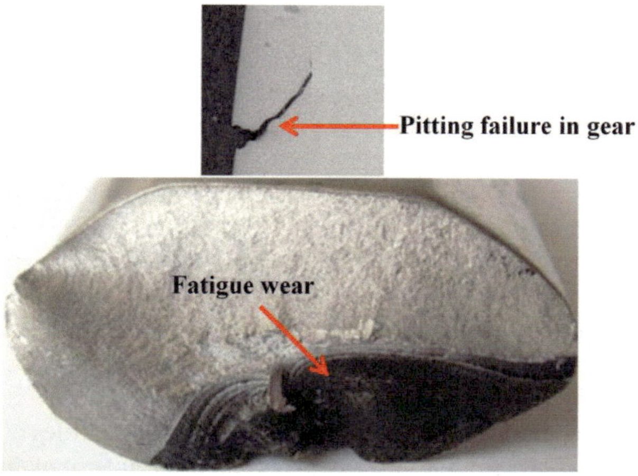

Figure 5.
Fatigue and pitting wear in industrial parts.

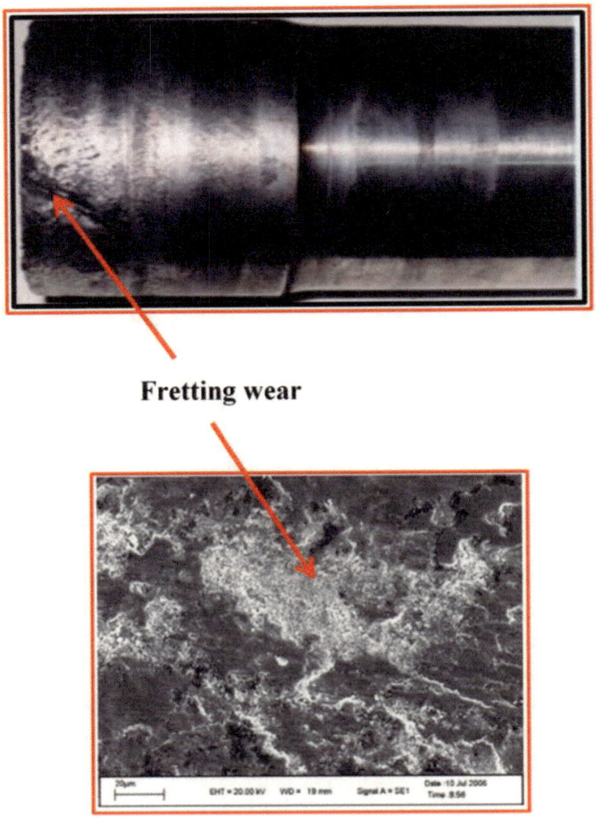

Figure 6.
Fretting wear in industrial parts.

Akhtar et al. [5] revealed in his research that the surface after 300 N testing have very heavy plowing of the steel matrix (**Figure 6**). At higher loads microplowing is very severe and causes the rapid removal of the material from the surface of the composite.

3.5 Erosive wear

Erosive wear is loss of material from a solid surface due to relative motion in contact with a fluid which contains solid particles impingement by a flow of sand, or collapsing vapor bubbles (**Figure 7**). Erosive wear closely depends on the material properties of the particles, such as hardness, impact velocity, shape and impingement angle. Example: A common example is the erosive wear associated with the movement of slurries through piping and pumping equipment. Erosive wear can be characterized as an amazingly short sliding movement and is executed inside a brief timeframe stretch. Erosive wear is brought about by the effect of particles of solid or fluid against the surface. The affecting particles steadily eliminate material from the surface through rehashed deformation and cutting mechanisms. It is a broadly experienced system in industry. Because of the idea of the passing on measure, funneling frameworks are inclined to wear when rough particles must be moved. The pace of erosive wear depends upon various elements. The material qualities of the particles, for example, their shape, hardness, impact speed and impingement edge are essential factors alongside the properties of the surface being disintegrated. The impingement point is one of the most significant factors. For ductile materials, the greatest wear rate

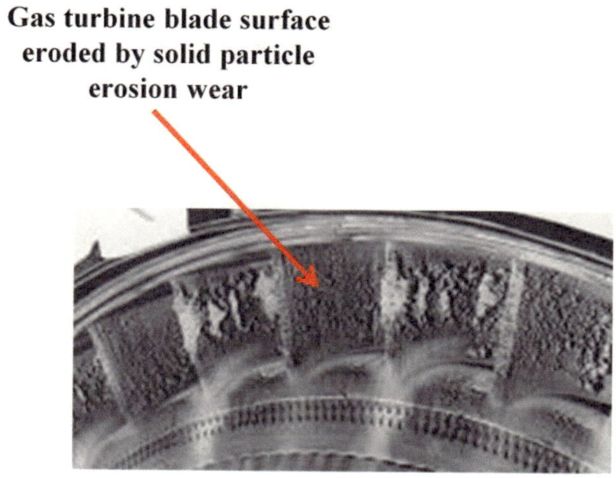

Figure 7.
Erosion of compressor blades in gas turbine engine.

is discovered roughly at 30° impingement angle, while for brittle materials the most extreme wear rate happens when the impingement angle is normal to the surface.

Swain et al. [6] investigated about the erosion behavior of the plasma sprayed NITINOL coating. In this work, the surface was eroded by 45 and 90° impingement angle of erodent. The wear mechanisms can be observed from the **Figure 8**. The surface impinged at 45° impingement angle (**Figure 8(a)–(c)**) having crater formation, chip formation and cutting grooves mechanisms. Whereas, the eroded surface at 90° impingement angle (**Figure 8(d)–(f)**) having crater formation, plastic deformation and lip formation mechanisms.

3.6 Corrosive and oxidation wear

Corrosion and oxidation wear happens both in oily and dry contacts (**Figure 9**). The essential reason are chemical reactions between weared surface and the eroding medium. Wear brought about by a synergistic activity of tribological stresses and

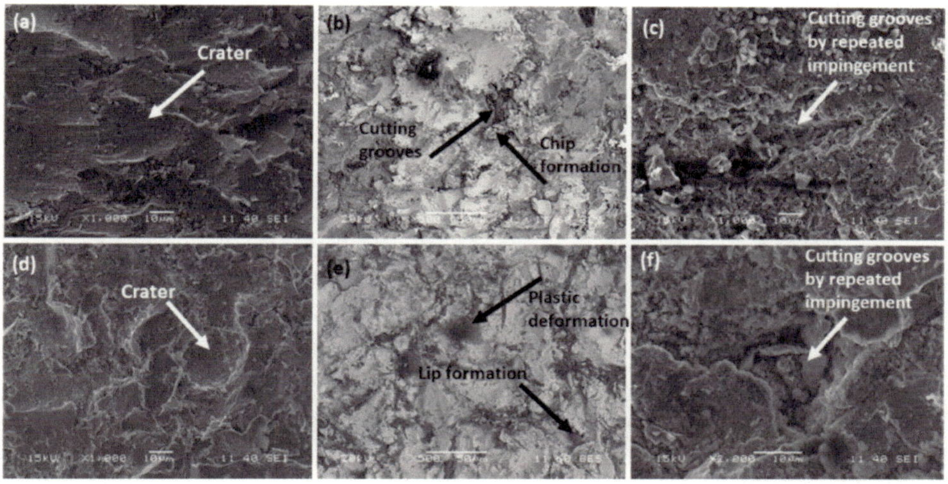

Figure 8.
SEM morphologies of eroded surface at (a), (b), (c) 45° and (d), (e), (f) 90° impingement angle [6].

Figure 9.
Corrosive and oxidation wear of structural members of industries.

consumption is likewise called tribocorrosion. Corrosive wear is otherwise called chemical wear. Corrosive wear is an assault on a material surface inside its condition. Corrosive wear can be either is wet or dry, contingent upon the sort of condition present for a specific response. Generally, wet erosion happens in an answer, for example, water, with some disintegrated species in it, which makes an acidic situation and response over the surface. Dry erosion is predominantly obstructed by the presence of dry gases, for example, characteristic air and nitrogen, etc. Since nature assumes an enormous function in corrosion wear, material choice is fundamental and ought to be the concentration before planning a segment. In erosion wear, corrosion and wear are two free instruments; if the demonstrations happen independently, the condition might be more basic than the consolidated impact of both. In presence of oil on a superficial level, consumption will be uniform all through the surface. On the off chance that limits of precious stone materials are defenseless to consumption rather than inside material, it is known as intergranular erosion. Pitting brought about by impingement of particles on the material surfaces produces pits and openings on the surfaces, which is difficult to perceive on a superficial level. Subsurface corrosion is disconnected particles that exist underneath the eroding material, essentially because of the response of constituents with the defused medium.

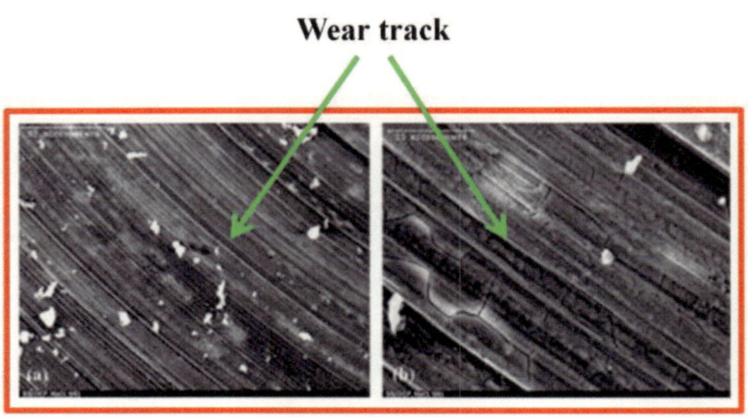

Figure 10.
SEM images of the wear track on an unprotected sample in the NaCl solution: (a) wear track, (b) a closer view at the wear track.

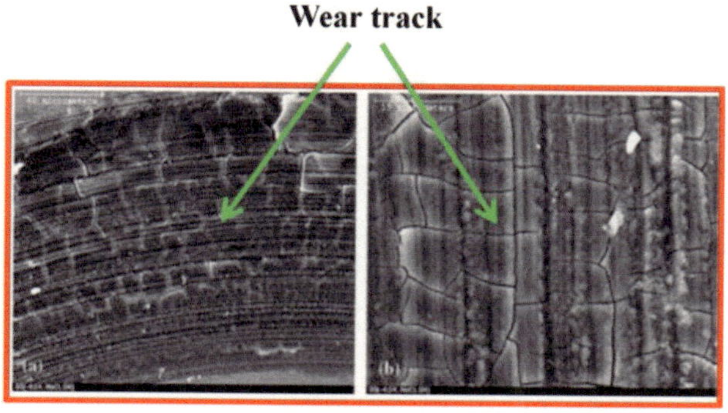

Figure 11.
SEM images of the wear track on a cathodically protected (−0.50 V) sample in the NaCl solution: (a) wear track, (b) a closer view at the wear track [7].

Akonko et al. [7] investigated the corrosive wear phenomena on both the protected and unprotected samples in the NaCl solution under a force of 5 N and found that the worn surface of a non-protected sample (**Figure 10**) indicated less cracks than those of cathodically protected (**Figure 11**). This indicates that the cathodic protection caused hydrogen embrittlement, and this has further boosted by stress, therefore caused more wear.

4. Wear mechanisms

4.1 Adhesive wear

The sort of mechanism (**Figure 12**) and the abundancy of surface fascination fluctuates between various materials yet are enhanced by an expansion in the thickness of "surface energy". Most solids will stick on contact somewhat. Nonetheless, oxidation films, oils and contaminants normally happening for the most part stifle attachment, and unconstrained exothermic chemical reactions between surfaces by and large produce a substance with low vitality status in the retained species.

Adhesive wear can prompt an expansion in harshness and the production of projections (i.e., protuberances) over the first surface. In modern assembling, this is alluded to as irking, which inevitably penetrates the oxidized surface layer and

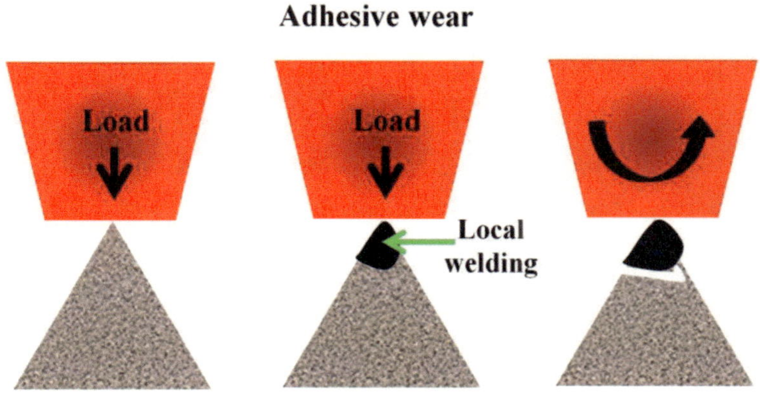

Figure 12.
Adhesive wear mechanism.

interfaces with the fundamental mass material, improving the opportunities for a more grounded bond and plastic flow around the knot.

A model for the wear volume for cement wear, V, can be portrayed by:

$$V = K \times WL / Hv$$

Where, *'W'* represents load, *'K'* is the wear coefficient, *'L'* represents the sliding distance, and *'Hv'* is the hardness.

4.2 Abrasive wear

The system of material expulsion in abrasive wear is essentially equivalent to machining and grinding during an assembling cycle (**Figure 13**). At the beginning of wear, the hard severities or particles enter into the milder surface under the typical contact tension. The wear trash regularly has a type of micro cutting chips.

A few methods have been suggested to foresee the volume misfortune in abrasive wear. A least difficult one includes the scratching of materials by angular shaped hard particles (indenter). Under an applied heap of P, the hard molecule enters the material surface to a profundity of h which is straightly relative to the applied burden (P) and conversely corresponding to the hardness (H) of the surface being scraped. As sliding happens, the molecule will furrow (cut) the surface delivering a depression, with the material initially ready being eliminated as wear flotsam and jetsam. On the off chance that the sliding distance (L) and the wear volume (V) can be written as:

$$V = k.\frac{PL}{H}$$

Here, 'k' is wear coefficient partially reflecting the effects of geometries, and properties of the particles (or asperities), and partly reflecting the influences of additional factors such as sliding speed, and lubrication environments.

4.3 Fatigue wear

Two mechanisms (**Figure 14**) of fatigue wear are recognized: high-and low-cycle fatigue. In high-cycle fatigue, the quantity of cycles before fatigue is high,

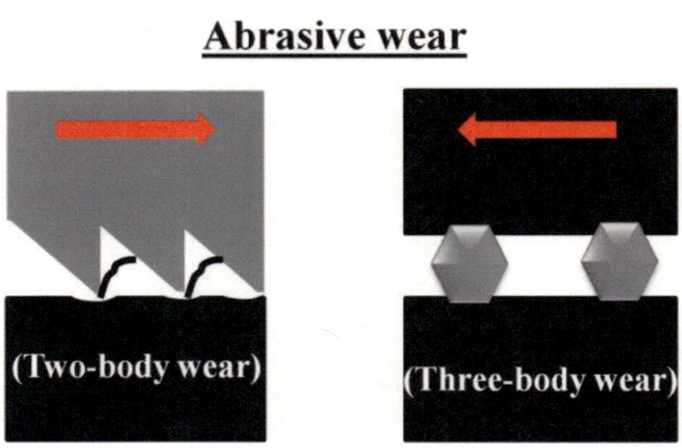

Figure 13.
Abrasive wear mechanism.

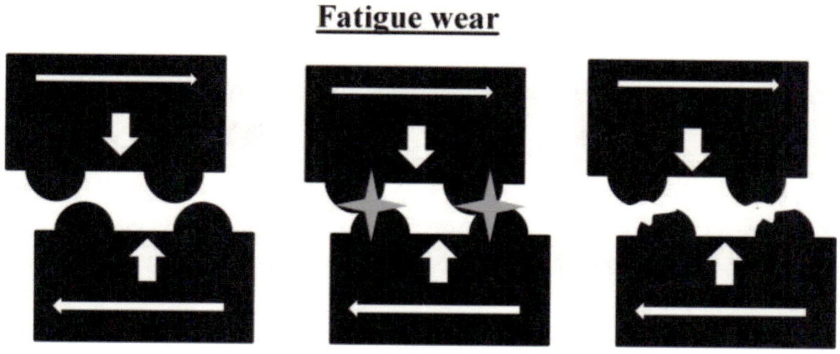

Figure 14.
Fatigue wear mechanism.

so the part life is generally long. The splits for this situation are created because of prior miniature imperfections in the material, near which the nearby pressure may surpass the yield esteem, despite the fact that ostensibly the naturally visible contact is in the flexible system. Gathering of plastic strain around inhomogeneities is an antecedent for commencement of a split. In the low-cycle fatigue, the quantity of cycles before disappointment is low, so the part bombs quick. In this mode, pliancy is prompted each cycle and the wear molecule is produced throughout aggregated cycles. The wear garbage is not produced at the principal cycles, yet just the shallow furrows because of plastic misshapening are framed, as talked about in. After a basic number of cycles, the plastic strain surpasses a basic worth and the crack happens. There are the three phases in break proliferation: split inception, development and post-basic stage, when the calamitous disappointment happens. The vast majority of the lifetime of the part is involved by the primary stage, with the spans of introductory splits around 2–3 μm and lower.

4.4 Fretting wear

Cyclic motion between contacting surfaces is the essential ingredient in all types of fretting wear. It is a combination process that requires interaction of two surfaces, and exposed to minor amplitude of oscillations.

According to the material properties of surfaces, adhesive, two-body abrasion and/or solid particles may produce wear debris. Wear particles detach and become comminuted (crushed) and the wear mechanism (**Figure 15**) changes to three-body abrasion when the work-hardened debris starts removing metal from the surfaces.

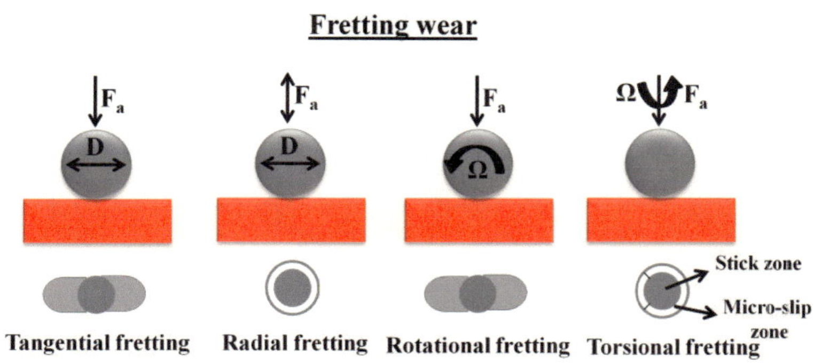

Figure 15.
Fretting wear mechanism.

Fretting wear arises as a result of the following order of events:

- The normal load causes asperities to stick and the tangential oscillatory motion shaves the asperities and produces wear debris that stores.
- The surviving (harder) asperities eventually act on the smooth surfaces causing them to undergo plastic deformation, create voids, propagate cracks and shear off sheets of particles which also gather in depressed areas of the surfaces.
- Once the particles have accumulated sufficiently to span the gap among the surfaces, abrasion wear follows and the wear zone extents laterally.
- As adhesion, delamination, and abrasion wear lasts, wear debris can no longer be contained in the primary zone and it outflows into surrounding valleys.
- Because the maximum stress is at the center, the geometry becomes curved, micropits form and these coalesce into larger and deeper pits. Finally, depending on the displacement of the tangential motion, worm tracks or even big cracks can be produced in one or both surfaces.

4.5 Corrosive and oxidation wear

Metal surface is normally covered with a layer of oxide, which could restrict metal-to-metal interaction, and therefore evading the development of adhesion and reducing the tendency of adhesive wear. In this connection, oxide is a favorable factor in reducing wear rate of metallic materials. However, whether such beneficial effect can be realized or not, is intensely reliant on the material properties and on contact conditions. When the hardness of the metal underlying an oxide layer is low, or when the contact load is relatively higher, the metal beneath the oxide layer will deform plastically, and asperities in the rigid surface will penetrate through the thin oxide layer, leading to the normal metal-to-metal contact. In such case, wear by abrasion or adhesion will occur depending on the mechanical properties and chemical properties of the contacting metals. The beneficial effect of oxide is minimal and wear rate is generally high. On the other hand, when the underlying metal is hard enough to support the oxide film, such as on a surface engineered hard surface, a process known as oxidation wear (**Figure 16**) will occur.

It needs to be mentioned that during sliding, the increased surface temperature promoted by frictional heating, and the less activation energy of oxide formation caused by plastic deformation, can increase the oxidation rate. Thus, rapid oxidation can be achieved, and the oxide layer can grow thicker during sliding than that

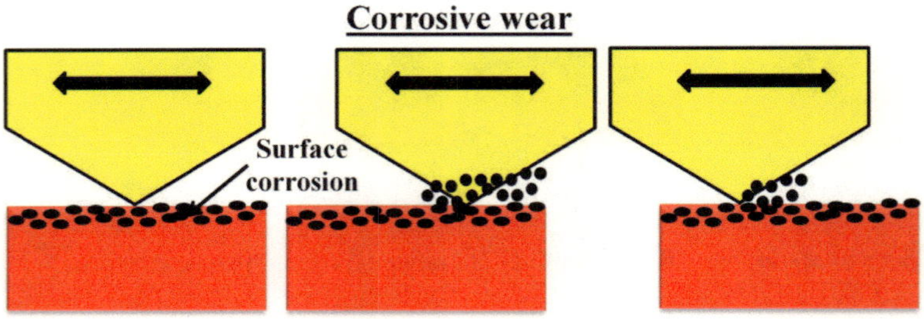

Figure 16.
Corrosive wear mechanism.

under static conditions. This ensures the fresh metal is rapidly covered with a new layer of oxide after the original oxide film was worn away. Oxidation wear will not happen in vacuum or in inert atmosphere, since re-oxidation is not possible. Oxidation is a minor form of wear. When the predominant wear mechanism is changed from abrasive or adhesive to oxidation wear, degree of wear can be reduced by some orders of magnitude.

5. Use of surface coating for the protection from wear

As wear is a surface or near surface phenomenon it has long been realized that the wear resistance of a component can be improved by providing a surface of different composition from the bulk material. After a brief introductory chapter wear phenomena and the properties required from a coating are addressed. Coating processes provide protection to a specific part or area of a structure exposed to harsh and corrosive environments in different fields ranging from aerospace and the automotive industry to tiny biomedical devices and implants inside the human body.

6. Types of surface coating

6.1 Physical vapor deposition (PVD) coating

PVD process is well-known for offering corrosion, wear resistance, and thin protective films on the surface of the materials that are exposed to corrosive media, and its applications range from decorative objects to industrial parts. The benefit of this technique is that the mechanical, corrosion, and esthetic properties of the coating layers could be adjusted on demand. Generally, PVD is a method that occurs in a high vacuum, and the solid/liquid materials transfer to a vapor phase followed by a metal vapor condensation, which creates a solid and thick film. The common types of PVD methods are sputtering and evaporation. Since the coating layers created by PVD are thin in nature, there is always a need for multilayered coatings while the materials selection should be considered carefully.

6.2 Chemical vapor deposition (CVD) coating

Another type of vapor deposition is called CVD. This process undergoes a high vacuum and is widely used in the semiconductors industry providing a solid, high quality, and a high resistance coating layer on any substrate. CVD can be used for mechanical parts in continuous interaction, which need protection for corrosion and wear. In this method, the substrate, known as a wafer, would be exposed to a set of volatile material precursors where a chemical reaction creates a deposition layer on the surface of the material. However, some by products of these chemical reactions, which are removed by continuous airflow of the vacuum pump, can stay in the chamber.

6.3 Micro-arc oxidation (MAO) coating

MAO method is known as a flexible method of coating concerning the composition of coatings. In general, MAO utilizes a high voltage difference between anode and cathode to generate micro-arcs as plasma channels. When these arcs hit the substrate, they melt a portion of the surface, depending on the intensity of the micro-arcs. Simultaneously, plasma networks discharge their pressure, which supports the deposition of coating materials in the working electrolyte on the substrate surface. The existing oxygen in the electrolyte causes a chemical reaction resulting oxidation and the oxides gets deposited on the surface of the substrate. The adaptability of this process lies in the flexibility of combining preferred elements and compounds as a solute in the working electrolyte. With MAO, the most common substrate materials are Al, Mg, Ti, and their alloys.

6.4 Electro deposition coating

Electro deposition of materials is considered a type of protection utilizing the deposition of metallic ions on a coating substrate. In this method, a difference in potential between anode and cathode poles causes an ion transfer in the unit cell. After a while, a coating layer forms on the submerged sample by getting ions from the other electrode. The common group of metals that have been intensively studied includes, but is not limited to, Ni-P, Ag/Pd, Cu/Ag, Cu/Ni, and Co/Pt.

6.5 Sol-gel coating

Sol–gel synthesis is used to obtain coatings that can modify the surfaces of metals to avoid corrosion or to enhance the biocompatibility and bioactivity of metals and their alloys that are of biomedical interest. Anticorrosion coatings composed of smart coatings and self-healing coatings will be described.

6.6 Thermal spray coating

Thermal spray coating is a general term for a series of processes that utilize a plasma, electric, or chemical combustion heat source to melt a set of designed materials and spray the melt on the surface in order to produce a protective layer. These are reliable types of corrosion- and wear-resistant coatings. In this process, a heat source or plasma, heats up the coating materials to a fully molten or semi-molten phase and sprays them on the substrate material with a high velocity jet.

Thermal spraying dates back to the early 1900s when Dr. Schoop [1] first carried out experiments in which molten metal were atomized by a stream of high-pressure gas and propelled on to a surface. The Schoop process consisted of a crucible filled with molten metal while the propellant, hot compressed air, provided enough pressure to break up the molten metal, creating a spray jet. This system was quite rudimentary and inefficient. Following Schoop's work some improvements to the process were introduced. But the disadvantages of the process is that, it was only useful for low-melting-temperature metals, that the molten metal caused severe corrosion and that it was not possible to establish a continuous process, were enough to stop further progress.

Schoop then focused his efforts in another direction and in 1912 the first device for spraying metal wires was produced. The principle of this process is simple; a wire was fed into a combustion flame which melted the tip of the wire and then compressed air surrounding the flame atomized the molten metal and drove the tiny droplets on to a substrate to form a coating. In addition to improvements to nozzle and gun design along with the wire feed drive rolls, the basic principle of the process is the same today. This procedure is called flame spraying (FS) and covers an enormous group of thermal spray techniques which use powder, wires or rods.

A completely new concept in thermal spraying was introduced by Schoop in 1914 when he used electricity to melt the feedstock material. The most advanced equipment made by Schoop was quite similar to current electric arc spraying. This method is based on producing an electric arc among two wires of conducting materials, which are fed together inside the gun. This arc is created at the tip of the wires and a jet of compressed air propels the molten metal to the substrate.

The concept of powder FS was introduced by F. Schori in the early 1930s, when a metallic powder was fed into a flame by the Venturi effect. The coating powder was heated in the nozzle and the exhaust gases (oxygen and acetylene) propelled the droplets. Improvements to the process incorporated in modern guns include an inert compressed gas that pressurizes the combustion chamber and results in rise in particle velocity.

The main problem associated with these early techniques was feedstock material. They all used a low-melting-point material, which leads to limited applications. Years passed, and the demand for high-temperature-resistant materials increased, until in the 1950s new systems that would boost the thermal spray market appeared. Firstly a modification of wire FS, the ceramic rod FS technique, which could use stabilized zirconias and aluminas appeared. However, it was the development, in about 1955, of the detonation gun (D-Gun) and atmospheric plasma spraying (APS) in about 1960 that proved to be the watershed as regards thermal spray applications.

The thickness achieved in thermal spray coating techniques can range from 20 micron to several milli meters which are significantly higher than the thickness offered by electroplating, CVD, or PVD processes. In addition, the materials that can be used as feedstock of thermal spray coatings range from refractory metals and metallic alloys to ceramics, plastics, and composites and can easily cover a relatively high surface area of a substrate. Therefore the current chapter will mostly focus on this coating process.

7. Types of thermal spray processes

There are various types of thermal spray coating processes introduced by the researchers (**Figure 17**).

7.1 Powder flame spraying

In powder flame spraying, the feedstock material is injected to the plume for heating and melted by the heating zone. After melting the molten particles are propagated towards the substrate surface. Then the molten particles are deposited on the substrate surface or pre-deposited splat to form a coating (**Figure 18**). The molten particles are ejected by the flame spray gun. The only difference between powder flame spraying and wire flame spraying is the feedstock material.

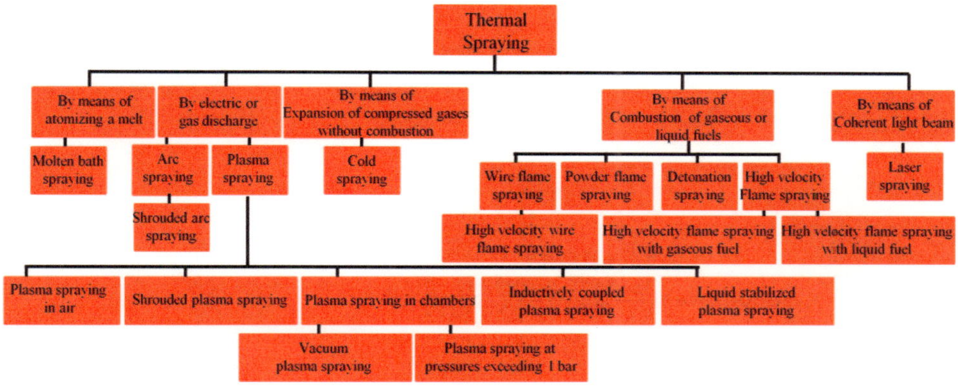

Figure 17.
Classification of thermal spray coating process.

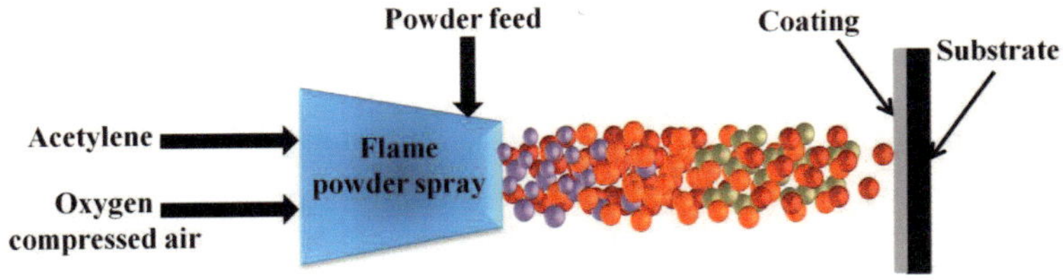

Figure 18.
Powder flame spray technology.

7.2 Wire or rod flame spraying

In rod type flame spray, the rod or wire is allowed to the heating zone where it melts and propagated by the plume towards the substrate to form coating (**Figure 19**). The feedstock rod may be a conventional rod or wire or manufactured by powder metallurgy process (sintering or binding). The melted particles are propelled towards the substrate, strike the surface at high velocity and flatten and form coating with a high adhesion strength with previously formed splat and substrate.

7.3 Detonation flame spraying

By the term "detonation" is meant a very rapid combustion in which the flame front moves at velocities higher than the velocity of sound in the unburned gases, and therefore characterized as supersonic velocities. A precisely measured quantity of the combustion mixture consisting of oxygen and acetylene is fed through a tubular barrel closed at one end. In order to prevent the possible back firing a blanket of nitrogen gas is allowed to cover the gas inlets. Simultaneously, a predetermined quantity of the coating powder is injected into the combustion chamber. The gas mixture inside the chamber is ignited by a simple spark plug. The gas mixture combustion generates plume which melt the particles to form coating (**Figure 20**).

7.4 High velocity oxy-fuel spraying

High velocity powder flame spraying was developed about 1981 and comprises a continuous combustion procedure that produces exit gas velocities estimated to be 4000–5000 feet per second. This is accomplished by burning a fuel gas (usually propylene) with oxygen under high pressure (60–90 psi) in an internal combustion chamber. Hot exhaust gases are discharged from the combustion chamber through exhaust ports and thereafter expanded into an extending nozzle. Powder is fed axially into this nozzle and confined by the exhaust gas stream until it exits in a thin high speed jet to produce coatings which are much denser than those produced with conventional or standard powder flame spraying techniques (**Figure 21**).

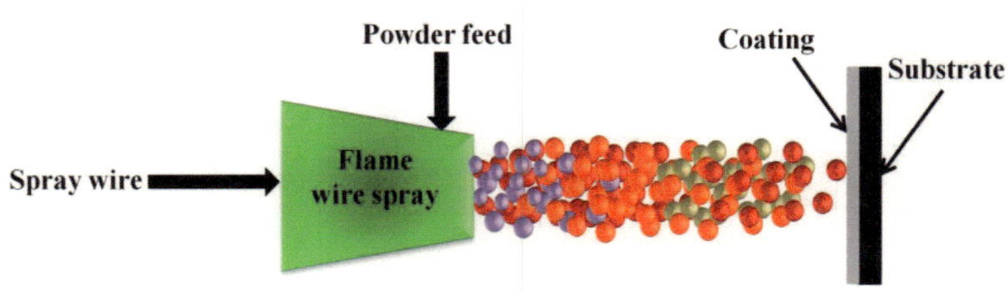

Figure 19.
Flame wire spray technology.

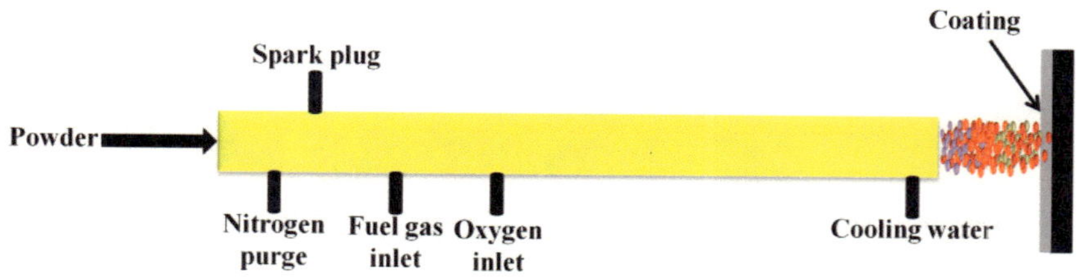

Figure 20.
Detonation flame spray technology.

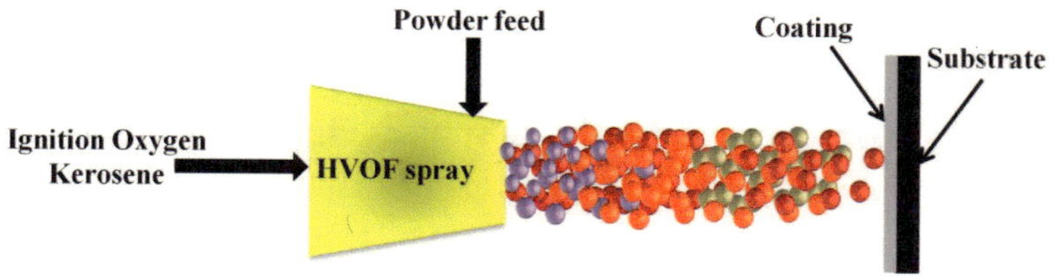

Figure 21.
HVOF spray technology.

7.5 Wire arc spraying

Twin wire arc spraying is the most economical thermal spraying process. In this type of spraying process the heating and melting occur when two oppositely charged wires are fed together in a way that arc is generated at their intersection (**Figure 22**). Once struck, the arc continuously melts the wires, and compressed air blown directly behind the point of contact, atomizes and projects the molten droplets, which sticks to the substrate to form a coating. Arc fluctuations due to periodic removal of molten droplets from the electrode tips have strong effects on melting and coating properties such as porosity, microstructure and oxide content.

7.6 Plasma spraying process

Plasma spraying is a flexible and low-cost method to manufacture coating and bulk materials. The first idea of a plasma spray process was patented in 1909 in Germany, and the first structural plasma installation appeared in the 1960's, as the product of two American companies Plasmadyne and Union Carbide. A gas, usually argon, but occasionally including nitrogen, hydrogen, or helium, is allowed to flow between a tungsten cathode and a water-cooled copper anode. The cathode is placed in the cylindrical nozzle and the cylindrical nozzle is the anode. An electric arc is initiated between the two electrodes using a high frequency discharge and then

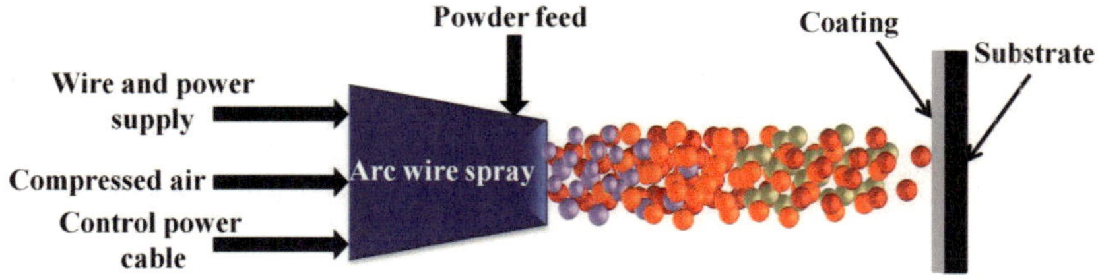

Figure 22.
Wire arc spray technology.

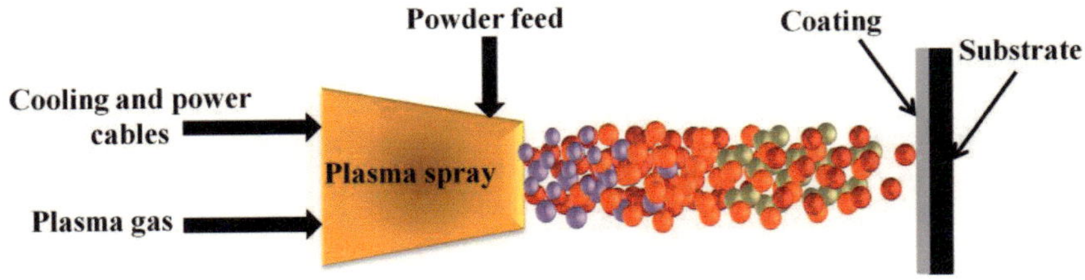

Figure 23.
Plasma spray coating technology.

sustained using dc power. The plasma is generated by the ionization of gas by the arc. The feedstock materials injected through the gun nozzle into the plasma plume, where it is melted and propagated to the substrates (**Figure 23**) [2, 6, 8–12].

8. Application of thermal spray technology for the protection from wear

A large range of industrial parts get advantage from thermal spraying, whether it is a portion of the manufacturing processes or as reclamations or re-engineering techniques. Some materials are used for minute role applications and others are sprayed by the tonne. Every application utilizes an amalgamation of procedure and material to give in the desired profit.

Reclamation and re-engineering of a wide range of rotating and moving parts from machines of all kinds, including: vehicles of both railways and highways, ships, aerospace, printing industries, paper industries, chemical industries, food industries, mining, earthmovers, machine tools, landing gear (chrome replacement) and any apparatus which is subject to wear, erosion or corrosion. This is done using either arc spray, flame spray or HVOF systems to spray steels, nickel alloys, carbides, stainless alloys, bronzes, copper and many other materials. New components which benefit from the enhanced surface properties that thermal spraying provides, include: Gate and ball valves, rock drilling bits, and down hole tools, print rollers, fluid seals, aerospace combustion chambers, turbine blades. Thermal sprayed coatings are used on a vast range of components which operate in harsh surroundings where, erosion, wear, corrosion or high temperature reduce part life. Part life is significantly prolonged due to thermal spray processes.

9. Conclusions

- Wear kinds of adhesive, abrasive, fatigue, and tribochemical wear are presented and their wear instruments are clarified with wear models in this chapter. In real wear of triboelements, a portion of these wear types are included simultaneously, and significant wear type changes at times starting with one then onto the next during running because of wear itself.

- Then again, wear is delicate to the difference in different framework boundaries, for example, mass, shape, stiffness, material properties, and condition. In light of such multiparameter affectability of wear, quantitative expectation of wear rate is still a long way from the real world.

- It gets significant, thusly, to perceive the significant wear type and its ordinary wear mechanisms according to framework boundaries.

- Surface covering improves the life of the segment and lessens the expense of substitution. The motivation behind surface innovation is to deliver practically compelling surfaces. A wide scope of coatings can improve the consumption, disintegration and wear obstruction of materials.

- We can reason that thermal spray coating is one of the most significant strategies of the surface change strategy. This examination was a push to give essential data with respect to a portion of the fundamental thermal spray strategies among which HVOF covering measure is most appropriate.

- By utilizing HVOF spray method uniform covering thickness, nonstop layer of covering and high hardness can be acquired. It has more favorable circumstances over the high quality, hardness, porosity, wear and erosion when contrasted with different cycle.

Author details

Biswajit Swain[1*], Subrat Bhuyan[2], Rameswar Behera[1], Soumya Sanjeeb Mohapatra[3] and Ajit Behera[1]

1 Department of Metallurgical and Materials Engineering, National Institute of Technology Rourkela, Rourkela, India

2 Department of Mechanical Engineering, G H Raisoni Institute of Engineering and Technology, Pune, India

3 Department of Chemical Engineering, National Institute of Technology Rourkela, Rourkela, India

*Address all correspondence to: biswajitnitrkl@gmail.com

References

[1] Khanam A, Mordina B, Tiwari RK. Statistical evaluation of the effect of carbon nanofibre content on tribological properties of epoxy nanocomposites. J Compos Mater. 2015;49(20): 2497-507.

[2] Swain B, Bajpai S, Behera A. Microstructural evolution of NITINOL and their species formed by atmospheric plasma spraying. Surf Topogr Metrol Prop [Internet]. 2018;7(1):015006. Available from: https://doi.org/10.1088/2051-672X/aaf30e

[3] Yunxia C, Wenjun G, Rui K. Coupling behavior between adhesive and abrasive wear mechanism of aero-hydraulic spool valves. Chinese J Aeronaut [Internet]. 2016;29(4): 1119-31. Available from: http://dx.doi.org/10.1016/j.cja.2016.01.001

[4] Mao K. Gear tooth contact analysis and its application in the reduction of fatigue wear. Wear. 2007;262(11-12):1281-8.

[5] Akhtar F, Guo SJ. Microstructure, mechanical and fretting wear properties of TiC-stainless steel composites. Mater Charact. 2008;59(1):84-90.

[6] Swain B, Mallick P, Bhuyan SK, Mohapatra SS, Mishra SC, Behera A. Mechanical Properties of NiTi Plasma Spray Coating. J Therm Spray Technol [Internet]. 2020 Apr 16;29(4):741-55. Available from: http://link.springer.com/10.1007/s11666-020-01017-6

[7] Akonko S, Li DY, Ziomek-Moroz M. Effects of cathodic protection on corrosive wear of 304 stainless steel. Tribol Lett. 2005;18(3):405-10.

[8] Swain B, Patnaik A, Bhuyan SK, Barik KN, Sethi SK, Samal S, et al. Solid particle erosion wear on plasma sprayed mild steel and copper surface. In: Materials Today: Proceedings. 2018.

[9] Biswajit Swain, Swadhin Patel, Priyabrata Mallick, Soumya Sanjeeb Mohapatra AB. Solid particle erosion wear of plasma sprayed NiTi alloy used for aerospace applications. In: ITSC 2019—Proceedings of the International Thermal Spray Conference. 2019. p. 346-51.

[10] Mallick P, Behera B, Patel SK, Swain B, Roshan R, Behera A. Plasma spray parameters to optimize the properties of abrasion coating used in axial flow compressors of aero-engines to maintain blade tip clearance. Mater Today Proc [Internet]. 2020 May; Available from: https://linkinghub.elsevier.com/retrieve/pii/S2214785320328492

[11] Kumar B, Soumya S, Mohapatra S, Power G. Sensitivity of Process Parameters in Atmospheric Plasma Spray Coating. J Therm Spray Eng. 2018;1(1):1-6.

[12] Swain B, Mallick P, Patel S, Roshan R, Mohapatra SS, Bhuyan S, et al. Failure analysis and materials development of gas turbine blades. Mater Today Proc [Internet]. 2020 Apr; Available from: https://linkinghub.elsevier.com/retrieve/pii/S2214785320316497

Chapter 16

Atomistic Simulation of Severely Adhesive Wear on a Rough Aluminum Substrate

Jun Zhong

Abstract

In this Chapter, a severely adhesive wear on a rough aluminum (Al) substrate is simulated by molecular dynamics (MD) under a high velocity impact of a hard-asperity (a hard-tip) with the Al-asperity. Multiple simulations include effects of four factors: the inter-asperity bonding, the geometry overlap between two asperities, the impact velocity between two asperities and the starting temperature of the Al-substrate. It is observed that the deformation mechanism on the Al-substrate would involve a local melting (from 1200 to 2500 K) which forms liquid type layers (amorphous textures) in the contact area between two asperities. Also, temperature profiles on the hard-tip and the Al-substrate is depicted. Moreover, a method in the Design of Experiments (DOE) is employed to interpret above all simulations. The DOE results indicate that the inter-asperity bonding and the geometry overlap between two asperities would substantially increase the wear rate (for about 53.56% and 67.29% contributions), while the starting temperature of the Al-substrate and the impact velocity between two asperities would play less important roles (about 10.30% and 6.61%) in raising the wear rate.

Keywords: the wear rate, molecular dynamics, the EAM potential, the Lennard-Jones potential, the design of experiment

1. Introduction

Surface wear of metals is very important to many industries such as automobile and aerospace etc. The wear rate often limits the lifetime and the durability of machinery parts, and thus leads to major economic losses. Wear phenomenon often involves the contacts between two or among more asperities on material surfaces, resulting in breaking old and forming new atomic bonds and plastic deformation in the contact areas. At the atomic scale, if only a few asperities come into contacts, the actual contact areas are very small when comparing with the macroscopic contact size. Thus local stresses in these areas can be exceptionally high, leading to high degrees of localized plastic deformation and heat generation, and even possibly local melting among asperities. In adhesive wear, aluminum (Al) wear is especially important because the Al is a relatively soft metal which is highly reactive with oxygen. Generally, a fresh (newly-formed) aluminum surface has little or no protection from oxides, and is less stable than the alumina (Al_2O_3). Therefore, excessive stresses and temperatures in the Al contact areas can provide an activation energy to initiate extremely exothermic reactions of fresh Al surfaces with oxides.

For example, during the rolling of an aluminum sheet by a steel roller, if excessive stresses are applied onto the sheet surface without lubricant coverings, the newly-formed (fresh) Al surface may easily react with oxides on the steel roller, resulting in the local melting of aluminum to bond onto the roller surface, a so-called severely adhesive wear which causes the catastrophic breakage of the steel roller.

During past six decades, studies on the Al wear under a dry-sliding (no lubricants) constraint have revealed that the large plastic strain would occur near subsurfaces on the Al-substrate when the wear process took place. Some experimental observations suggested that the Al wear rate be inversely proportional to the Al hardness because the higher Al hardness usually led to less plastic deformation on the Al-substrate [1]. Therefore, the understandings of the wear process may provide some valuable information for mechanisms of friction, lubrication and adhesion at the nano-scale [2–7]. Even if, during asperity contacts at the atomic-scale, it was still difficult to observe wear mechanisms in nano-seconds.

In theoretical study, molecular dynamics (MD) may simulate the nano-scale phenomena in a very short time. Thus during past four decades, advances in the MD simulations have helped researchers understand atomic mechanisms which brought two kinds of materials into contact. For examples, Landman *et al* used the MD simulations to observe the hard-tip (Ni) jump-to-contact, the plastic yielding, the adhesion to induce the atomic flow and the slip generation in the Au-substrate [8, 9]. Plimpton *et al* used the MD simulations to observe the nucleation of partial dislocation loops occurring within the contact areas where a displacement-controlled hard sphere indented into a gold (Au) substrate surface. They found that the dislocation loops would grow rapidly into the substrate, but emerge at the surface edges, and then dislocation slips may produce complicated structures in the substrate [10–12]. Tanaka *et al* ran the MD simulations to observe that, during the two—/three-body sliding contacts, an amorphous phase transformation would take place on the silicon substrate, i.e., the deformation on the silicon substrate would fall into adhering and plowing, but no wear regimes [13, 14]. Mendelev *et al* employed the MD simulations to observe that, during a flat ruthenium (Ru) slab downward into a gold (Au) substrate with a single asperity, the Au was very ductile at 150 and 300 K, while the Ru showed the considerably less plasticity at 300 and 600 K [15]. In our former works, we have ran several MD simulations on the Al deformation at the nano-scale, including the wear [16] and the nano-indentation on the Al-substrates [17]. However, our these studies adopted a very low strength of inter-asperity bonding between the Al-substrate and a hard-tip, in which we found that, even if there was a large plastic deformation on the Al-substrate, no Al atoms were removed from the Al-substrate if the inter-asperity bonding was below a critical value.

To summarize, although many interesting MD simulations for deformation and wear on metal surfaces have been discussed at the atomic scale, there have not yet any MD simulations to focus on investigating a severely adhesive wear between two sliding surfaces. In the actual manufacturing, this kind of wear would occur during the rolling of aluminum sheet and many other forming processes. So in this chapter, such the wear will be discussed by the MD simulations to find out what may occur when a soft Al-asperity on an Al slab is contacted by another asperity on a hard tool surface when these two surfaces are sliding relative to another. In our MD investigation, multiple MD simulations were conducted by varying some constraint factors: the inter-asperity bonding, the geometry overlap between two asperities, the relative impact velocity between two asperities and the starting temperature of the Al-substrate etc. In details, the Al-substrate was simulated by an EAM potential [18–20], while the hard-asperity was simulated by a strong Lennard-Jones (L-J) potential which may serve as a model for hard tool surfaces (with their oxides).

Moreover, different L-J potentials from weak to strong values may describe the bonding strength between the Al- and the hard-asperities, which may reveal an effect of the inter-asperity interactions on the severely adhesive wear. In Section 2, we describe our methodology. In Section 3, we list the simulation procedures. In Section 4, we assess the results of our MD simulations. Finally, in Section 5, we summarize the results.

2. Methodology

2.1 Molecular dynamics

Molecular dynamics (MD) is a methodology which depicts motions of a many-particle system using classical Newton's equations. It has been over 40 years since the first MD's application to a hard sphere system by Alder and Wainwright [21, 22]. The MD method is particularly useful for studying dynamical properties of materials, and help researchers extract physical-insights from the modeling works.

In the MD simulations, for a finite size system, classical trajectories of particles in real space are traced by solving the Lagrange equations numerically: when choosing the Cartesian coordinates, these equations would become Newton's equations

$$m_i \ddot{\mathbf{r}}_i = \mathbf{F}_i = \dot{\mathbf{P}}_i = -\frac{\partial V_{tot}}{\partial \mathbf{r}_i}, \tag{1}$$

where V_{tot} is the total potential energy of the system, m_i is the mass of particle i, $\boldsymbol{P}_i$ is the momentum of particle i, and $\boldsymbol{F}_i$ is the total inter-particle force acting on particle i, and the "•" denotes the first order of the time derivative. In practice, the convenience of using the MD simulation would mainly depend upon some particularly arithmetic algorithms to solve Eq. (1).

2.2 Algorithms and statistical ensembles

A standard way to solve Eq. (1) was the finite-difference method: given a configuration of a system (positions and velocities of all particles) and other dynamical information at time t, the numerical integration would determine the new configuration of the system at a later time $t + \Delta t$ (Δt is the time step). Commonly used methods in the MD simulations were *the Verlet algorithm* [23], *the Leapfrog algorithm* [24], and *the Gear predictor–corrector* (*GPC*) *algorithm* [25]. An ideal algorithm should be simple, run fast, require little memory and permit using a long time step Δt to hold the whole system trajectories as true as possible, and preserve conservation laws of momentum and energy. According to these, *the Verlet algorithm* can be regarded as the most widely used algorithm. *The leapfrog algorithm* was essentially identical to *the Verlet algorithm*. *The GPC* method was usually more accurate as well as more complicated than others.

Commonly used statistical ensembles in the MD simulations were *the microcanonical ensemble* (*NVE*), *the canonical ensemble* (*NVT*), *the isothermal-isobaric ensemble* (*NPT*) and *the grand canonical ensemble* (μVT). Please note, thermodynamic variables in parenthesis for each ensemble were fixed during the MD simulations. Usually, the (*NVE*) ensemble was most convenient to realize since all equations of motion conserve the total energy E and the particle number N, and the constant volume V was fixed by using periodic boundary conditions. The (*NVT*) ensemble was very commonly used for practical calculations. However, it needed to control the temperature: T. So, there were a number of ways to realize it: (1) the simplest one was the velocity scaling method, which was simple but crude in the MD simulations [26]. (2) the Nosé-Hoover's method was more conceptually fundamental, but had an adjustable parameter of thermal inertia in practice [27]. For the (NPT) ensemble, Anderson [28] proposed an extended system method for a constant pressure MD. In this method, a term of "external piston" was added into the Lagrange functional for adjusting the value of external pressure. In another way, Parrinello and Rahman [29] proposed "a variable cell constant MD scheme" which allowed the simulation box to change its shape as well as its size. This method, so far, is the most general constant pressure method in the MD simulations. The (μVT) is very useful for an open system: in this ensemble, the total number of particles: N will not be conserved. Regarding this, see Ref. [30] for details.

2.3 Inter-particle potentials

Traditionally, the total potential energy of a system V_{tot} in the MD simulations was calculated using empirical potentials. There have been many kinds of models like pair-wise potentials: Lennard-Jones or Morse potentials; multi-body interactions [26], etc. In early 1980s, M.S. Daw and M.I. Baskes [18, 19] proposed a model for semi-empirical potential: *the embedded-atom method* (EAM) which reflected the many-body effect in materials. Generally, this model was written as two following terms

$$E_{\text{tot}} = \sum_i F_i(\rho_i) + \frac{1}{2} \sum_{i,j(\neq i)} \varphi\left(r_{ij}\right), \tag{2}$$

where r_{ij} is the distance between atoms i and j. In Eq. (2), on the right-hand side, the first term reflects a many-body effect in nature, meaning to embed ionic-cores into electron gas, so the $F_i(\rho_i)$ was termed the embedding function; the second term represents the screening Coulombic repulsion among ionic-cores, which is a sum of pair-wise interactions. The total atomic-charge density ρ_i at the position $\boldsymbol{r}_i$ can be expressed as

$$\rho_i = \sum_{j(\neq i)} f_j\left(r_{ij}\right), \tag{3}$$

where the ρ_i was approximate to a sum of individual atomic-charge density f_j around site i neighbors, which was opposed to true charge distributions that resulted from the self-consistent *ab-initio* calculations. During past four decades, there have been several fitting techniques to successfully constitute the EAM model, among which a particularly reliable method created by Adams *et al* was developed by fitting to a large database of density functional calculated forces and experimental data, a so-called "*the force-matching method*," see Ref. [20].

In this work, since we were primarily interested in the wear deformation on an Al-substrate, we chose a reliable aluminum EAM potential developed by Liu, Ercolessi and Adams through the force-matching method [31]. This potential has proven reliable for many types of bulk and surface simulations [20, 31, 32]. Also, since little deformation was assumed to occur in the contact tool surface (steel roller surface), we used a simple L-J potential with a very high bond strength to describe the hard-tip (the LJ-tip, hard-asperity). Similarly, another simple L-J potential was applied to the interaction between an Al-asperity and the hard-tip, in which the strength of inter-asperity bonding would vary from weak to strong in order, so as to simulate a wide range of generic systems. The L-J potential was adopted in a simple pair-wise formula as shown below

Bond strength	ε (Å)	σ (Å)
LJ-tip	1.10	2.85
an Al-saperity with a LJ-tip	0.20 (weak)	2.85
	0.65 (medium)	2.85
	1.00 (strong)	2.85

Table 1.
Parameters of the Lennard-Jones potentials for interactions between an Al-asperity and a hard-tip (LJ-tip).

$$V(r) = 4\varepsilon\left[\left(\frac{\sigma}{r}\right)^{12} - \left(\frac{\sigma}{r}\right)^{6}\right], \tag{4}$$

where r was the distance between two atoms, ε and σ were parameters corresponding to the bond strength and the bond length between two atoms, respectively. **Table 1** provides relevant fitting parameters of those L-J potentials.

2.4 An implement simulation package: LAMMPS

The Large-scale Atomic/Molecular Massively Parallel Simulator (The LAMMPS) is a molecular dynamics package from the Sandia National Laboratories (see http://lammps.sandia.gov/), to make use of the MPI for parallel communication. The parallel algorithm for the LAMMPS was a force-decomposition method in which a subset of pair-wise force computations was assigned to each processor. To improve the computational efficiency, the LAMMPS used neighbor lists to keep tracking particles nearby, which were optimized for systems with particles that were repulsive at short distances, so that the local density of particles never became too dense. The LAMMPS is a free software with open-source, distributed under the terms of the GNU General Public License. Till now, the LAMMPS is particularly efficient (in a parallel computing sense) to systems whose particles fill a 3D rectangular box with approximately uniform density.

3. Simulation procedures

3.1 Geometry constructions

Figure 1 shows schematic models for the MD simulations. They consisted of a hemispherical hard-tip (the LJ-tip) with 4440 L-J atoms, plus one slab with a sinusoidal asperity totally with 45,299 Al atoms (the Al-substrate). **Table 2** provides the geometric information for the Al-substrate and the hard-tip at 0 K, both of which were perfect FCC single crystals.

3.2 Constraints in the MD simulations

These MD simulations were carried out through the LAMMPS which may apply the EAM and the L-J potentials to pure and alloy metal systems [10]. In each of these simulations, the lattice constant (a_0) of the Al bulk was determined at each desired temperature by the Hoover's thermostat method [30]. And then, the simulation system was constructed by an appropriate a_0 equilibrated for approximately 20,000 time steps (a time step = 0.001 ps) at one desired temperature. Bottom edge layers of the Al-substrate were fixed to prevent motions of the substrate during the whole MD simulations. Uppermost layer of the LJ-tip was constrained to move at a

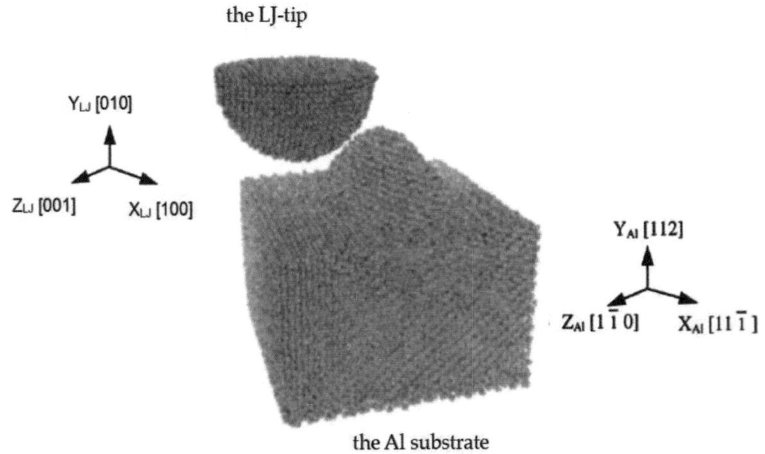

Figure 1.
The MD simulation models for the Al-substrate and the LJ-tip.

Component	Shape	Orientation	Lattice (0 K, Å)	Atom numbers
the LJ-tip	Hemisphere	FCC	4.009	4440
the Al-asperity	Sinusoid	FCC	4.032	4299
the Al-slab	Cubic	FCC	4.032	41,000

Table 2.
Geometric information in the MD simulations.

constant velocity along the x-direction (translational motion parallel to the Al slab surface), with no forces acting on this layer. All the MD simulations of asperity-asperity shear were performed at the constant energy (except for the fixed atoms), which allowed natural and realistic heat generation and diffusion due to formation of adhesive bonding and mechanical deformation in the contact area between two asperities.

3.3 Factors selected for design of experiments

Design of experiments (DOE) is a method which may be applied to analyses for experimental data, so that the relations of different selected factors (such as the strength of asperity–asperity bond) with response variables (such as the wear rate) can be determined. This method is widely applied in different fields to help researchers understand the factors which impact experimental processes most, so as to improve the desired experimental results [33].

During the MD simulations, the wear rate of aluminum was defined as the number of aluminum atoms removed from the Al-substrate, and was dependent upon many possibly significant factors. In order to get a better understanding of the wear process, some initial screening simulations were fulfilled to determine the most important factors and the range of these factors for investigation. Based upon the screening results, four factors were believed the most important ones: the inter-asperity bonding (D_I), the impact velocity (V_I) of the LJ-tip toward the Al-asperity along x-direction, the starting temperature of the MD system (T_I), and the geometry overlap between two asperities (O_I), see **Table 3**. The selected design technique was a so-called "*The 2^4 full factorial design*" with one central point for each factor (the mid-point of the range) in the Design Of Experiment (DOE). In this design, we

Factor	Variable	Data type	Low (−)	Central (0)	High (+)
A	T_I (K)	numerical	100	400	700
B	V_I (Å/ps)	numerical	1.00	5.50	10.00
C	O_I (Å)	numerical	1.5 a_0	2.5 a_0	3.5 a_0
D	D_I (eV)	numerical	0.20	0.60	1.00

Table 3.
Four selected factors and their levels in the DOE analyses.

did not consider any other replicates or blocking because they would not offer additionally valuable information due to the nature of computer simulations. So the chosen factors, their levels and their central points are listed in **Table 3**. In this Table, T_I = 100 K, 400 K and 700 K; $\boldsymbol{V}_I$ = 1.00 Å/ps, 5.50 Å/ps and 10.00 Å/ps; O_I = 1.5 a_0, 2.5 a_0 and 3.5 a_0; and D_I = 0.20 eV, 0.65 eV and 1.10 eV. The velocities ($\boldsymbol{V}_I$) were corresponding to 100 $\sim$ 1000 m/s, which were relevant to the surface edge of steel roller and some other forming processes; these high speeds were also chosen because they were more computationally efficient to the MD simulations. Factors that were held fixed in the DOE analyses: crystal defects (single crystals with no defects), crystal orientations, the LJ-tip moving direction, the MD model size and geometry, the LJ-tip geometry, and the contact load, etc. So totally, 17 MD simulations were carried out for different combinations of four selected factors, and with combinations chosen through the software of DesignExpertTM Version 6 [33].

3.4 Wear rate as a response variable

Since the wear rate is defined as the loss of materials from a rough substrate surface due to an interaction of the surface with its contacting environment, according to this, we chose the total number of Al atoms removed from the Al-asperity during asperity-asperity shear as the response variable.

4. Results and discussions

4.1 Wear rate

During a typical MD simulation of asperity-asperity shear, three different simulation stages took place: contact of two asperities, plowing of the LJ-tip through the Al-asperity, and final necking/fracture of the Al-asperity, as depicted in **Figure 2**. At the first stage, top of the LJ-tip impacted the Al-asperity, leading to the local heating, adhesion and plastic deformation at the contact area between two asperities, which was generally the region of maximum heat generation. At the second stage, the LJ-tip plowed through the Al-substrate, forming an Al neck to glide along a favorable slip system: . Finally, at the third stage, the Al neck was stretched and fractured, resulting in some Al atoms being removed from the Al-substrate. The extent of the Al necking was very sensitive to the $\boldsymbol{V}_I$ and the T_I. However, at the third stage, if the D_I reached below a critical value (approximately 0.04 eV), the removed Al atoms would slip off the LJ-tip surface. And then they would return to the Al residual substrate, resulting in no net removal of Al atoms [16].

Furthermore, occurrence of brittle or ductile fracture of the Al substrate during the MD simulations was observed to depend primarily upon the $\boldsymbol{V}_I$ and the T_I. That was, if the $\boldsymbol{V}_I$ was high but the T_I was low, there was very little or no elongation during the Al necking, and the final fracture of the Al neck was brittle. Conversely,

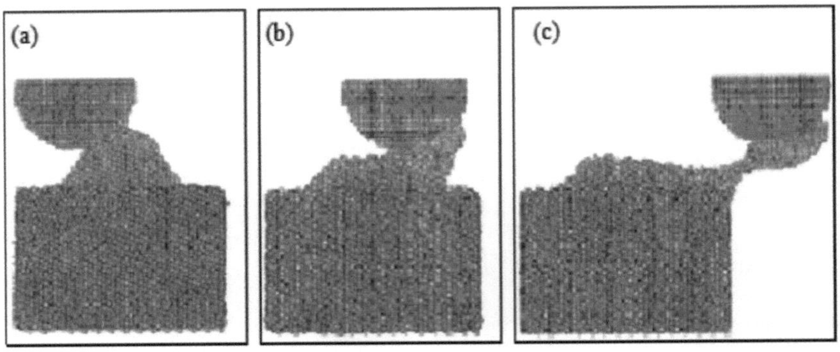

Figure 2.
Three stages of MD simulation for asperity-asperity shear: (a) the LJ-tip contact; (b) the LJ-tip plowing; (c) the Al necking and fracture.

if the $\boldsymbol{V}_{\mathrm{I}}$ was low while the T_{I} was high, the Al neck would be deformed and elongated more, and the final fracture of the Al neck was more ductile.

4.2 Effects of four selected factors

Outputs of the MD simulations indicated that the number of Al atoms removed from the Al-substrate (wear rate) varied from 262 to 3144, depending upon different factors in the MD simulations. Analyses for these data indicated that, increasing the D_{I} and the O_{I} would increase the wear rate mostly, increasing the T_{I} would just slightly increase the wear rate, whereas raising the $\boldsymbol{V}_{\mathrm{I}}$ exhibited a weak inverse correlation with the wear rate. In details, (1) if the D_{I} became stronger, two asperities would adhere together via solid–liquid welding in the contact area, and more bonding states would occur along with more heat generation to increase temperature profiles in the contact area. So, a stronger D_{I} may result in more removal of Al atoms. Due to thermal diffusion, temperature profiles in other areas of two asperities were also increased; (2) if enlarging the O_{I} which determined the contact area and hence controlled the volume of Al atoms being plowed by the LJ-tip, the deformation of the Al-substrate would increase. And then, it resulted in a higher wear rate. Komanduri *et al* also reported similar results in their studies [9]; (3) if increasing the T_{I}, dislocations in the Al-substrate can move and glide more easily through the elevated temperature region, resulting in large plastic deformation, i.e., the transformation from the brittle fracture (a small Al necking) at lower temperatures to the ductile fracture at higher temperatures can be easily observed; (4) it has found that the $\boldsymbol{V}_{\mathrm{I}}$ would play a minor inverse correlation role in the wear rate. This could be due to the increase of strain rate along with the rise of the $\boldsymbol{V}_{\mathrm{I}}$, which decreased the ductility of materials so that the wear process on the Al-substrate may undergo less plastic deformation before fracture. As discussed earlier, more plastic deformation at the stage of the LJ-tip plowing may enlarge the area of adhesion. Therefore, when less plastic deformation took place, fewer Al atoms were removed by the LJ-tip. Similar experiments on dry-sliding wear of Al-Si alloys indicated that the coefficient of wear in these alloy systems was highly dependent upon the disc speed: at a faster disc speed (0.356 m/s), the wear rate was found to be low or moderate; While at a slower speed (0.089 m/s), the wear rate increased dramatically [34]. So, it should be noticed that, in experiments when the $\boldsymbol{V}_{\mathrm{I}}$ increased, more asperity-asperity collisions could take place, whereas in our simulation, only two single asperities were simulated. Thus when comparing with experimental observations, one would have to consider both the damage per asperity and the increased number of two interacting asperities.

4.3 Plastic deformation during asperity-asperity shear

It was well known that, during the wear process, severe temperature fluctuations could result in major changes of mechanical properties. That was, during a higher temperature region, moduli and yield strength of metals could usually decrease, so the localized plasticity would increase and the welding could even occur if a critical temperature was reached [35]. It was found that, near the contact area of two asperities, the Al neck was deformed amorphously under very high stresses and temperatures. In this area, linear motion of debris along the sliding direction was transformed into thermally random atomic motions. Therefore, the higher the V_I, the much higher the temperature profiles in the contact area became, so a thin and soft/liquid like layer may form. As a result, the deformed Al neck would behave like a viscous liquid under very high temperatures and stresses [36]. When at a very high V_I (= 10.0 Å/ps), amorphous plasticity became the major deformation mechanism. However, at a very low V_I (= 1.00 Å/ps) and other lower T_I, D_I and O_I, in addition to amorphous deformation at the contact area, dislocation cores were identified by using the evaluation of the centrosymmetry parameter [10], so they were found to emit into the Al-substrate from the high temperature region (amorphous deformation region) along a favorable system $(111)[10\bar{1}]$, see **Figure 3**.

4.4 Thermal analysis

Thermal distributions from four different MD simulations are shown in **Figures 4** and **5**, which give insight into heat generation and heat transfer during asperity-asperity shear. Temperature profiles of the Al-substrate were calculated by using the following equation:

$$\frac{3}{2} n k_B T_i = \frac{1}{2} \sum_k m_k v_k^2, \tag{5}$$

where m_k, T_i and v_k were mass, temperature, and velocity for atom i and k, respectively. k_B was the Boltzmann constant, and n was the number of atoms for the Al substrate and the LJ-tip within a sphere of about 27 Å in diamond at a point i considered [16]. For the moving LJ-tip and the Al atoms removed by the LJ-tip, the temperature profiles were calculated by

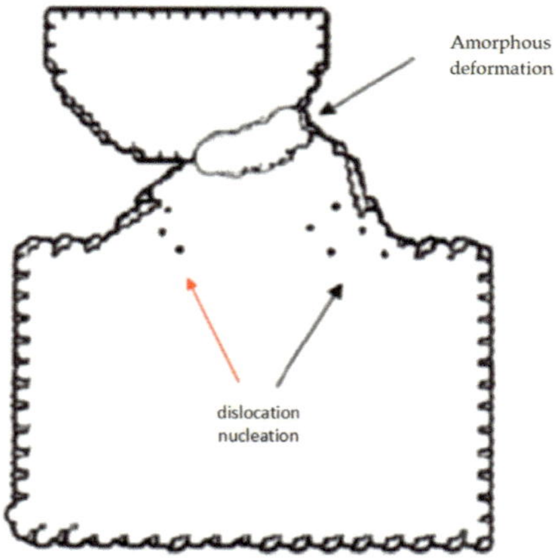

Figure 3.
Amorphous deformation of the Al neck and emission of dislocation cores in the Al-substrate.

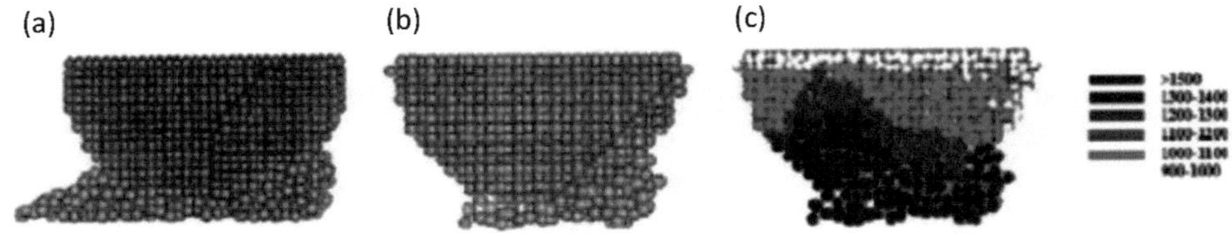

Figure 4.
(a). side view of the LJ-tip at the contact stage; (b) slide view of the LJ-tip at the plowing stage; (c) temperature profiles for a slice of the LJ-tip at the final stretching stage. (T_I = 700 K, O_I = 3.5 a_o, D_I = 0.20 eV and V_I = 10.0 Å/ps).

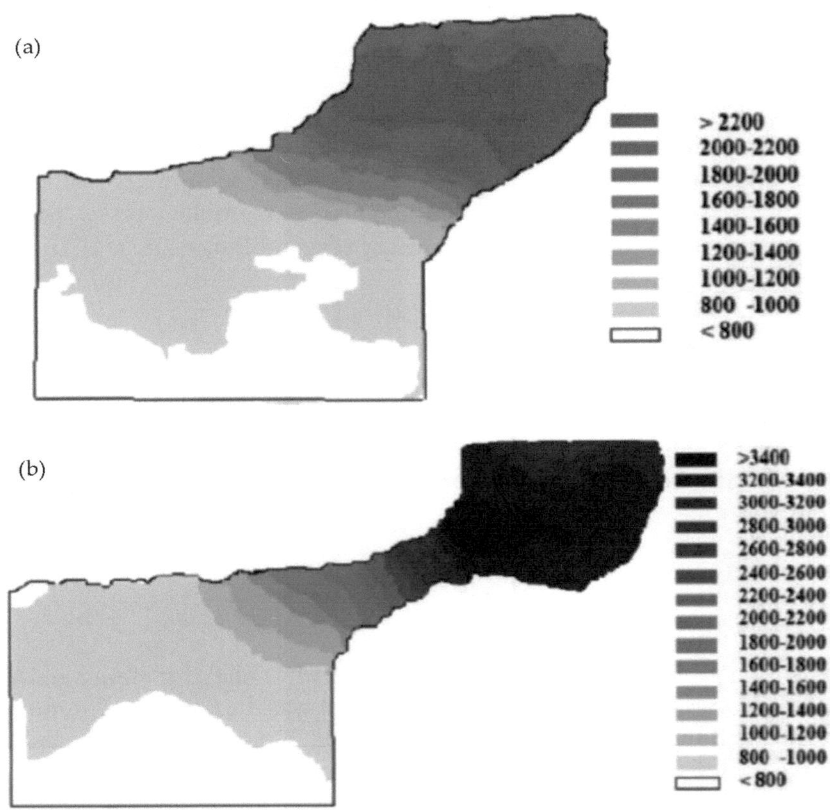

Figure 5.
Thermal analyses for two different MD simulations under T_I = 700 K, O_I = 3.5 a_o, D_I = 1.00 eV, and (a) $\mathbf{V}_I$ = 1.00 Å/ps and (b) $\mathbf{V}_I$ = 10.0 Å/ps.

$$\frac{3}{2}nk_BT_i = \frac{1}{2}\sum_k m_k(\mathrm{v}_k - \mathrm{V}_I)^2, \tag{6}$$

where V_I was the impact velocity of the LJ-tip along the x-direction.

For examples, **Figure 4** shows three selected views of narrow slice for the LJ-tip during the MD simulations along with relevant temperature profiles. In **Figure 4(a)** and **(b)**, atoms in the first layer on the Al-asperity were found to follow almost same stacking sites as those L-J atoms being located at the top of the LJ-tip when two asperities came into bond together. And then in **Figure 4(c)**, the order of subsequent layers in the Al-asperity became more random in high temperature region (about 1200 $\sim$ 1550 K): atoms in these layers moved randomly around the top of the Al-asperity (local melting) until they bonded to and transferred heat to the cooler LJ-tip.

Figure 5 shows the temperature profiles for two selected systems: T_I = 700 K, O_I = 3.5 a_0, D_I = 1.00 eV, and $\mathbf{V}_I$ = 1.00 Å/ps, 10.0 Å/ps, respectively. **Figure 6** shows the temperature profiles for two selected systems: T_I = 100 K, O_I = 3.5 a_0, D_I = 1.00 eV, and $\mathbf{V}_I$ = 10.0 Å/ps, 1.00 Å/ps, respectively. In these two Figures, (1) at the contact stage of the LJ-tip, the local heating occurred as new atomic bonds formed in the contact area between two asperities. Thus for the LJ-tip, the temperature gradient was positive from the contact area to its top layers; while for the Al-substrate, this positive gradient was found from the deformed Al neck to its far ends; (2) at the plowing stage of the LJ-tip, the heat diffused into the LJ-tip bulk from the contact area; while for the Al-asperity, the Al necking also produced more heat in the remnant surface where the Al necking root glided, so the heat diffused into the Al-substrate from the remnant surface, and hence increased the temperature profiles there.

In these two Figures, the maximum local temperature and the temperature gradient were quite different: the heat generation was much faster than the heat diffusion under a very high $\mathbf{V}_I$, so the $\mathbf{V}_I$ seemed to play a major role in generating the maximum local temperature. In addition, a higher local temperature may lead to a larger local softening of the Al-substrate, so the deformation was mostly limited to a narrow region, and hence materials were removed. It was noticed that the maximum local temperature was much higher than the starting temperature T_I, so the T_I seemed not to play a major role in removing the Al atoms. It should also be noticed that the maximum local temperature can briefly exceed the boiling point of the Al, and yet the local liquid was not boiling. The reason was that, the increase of

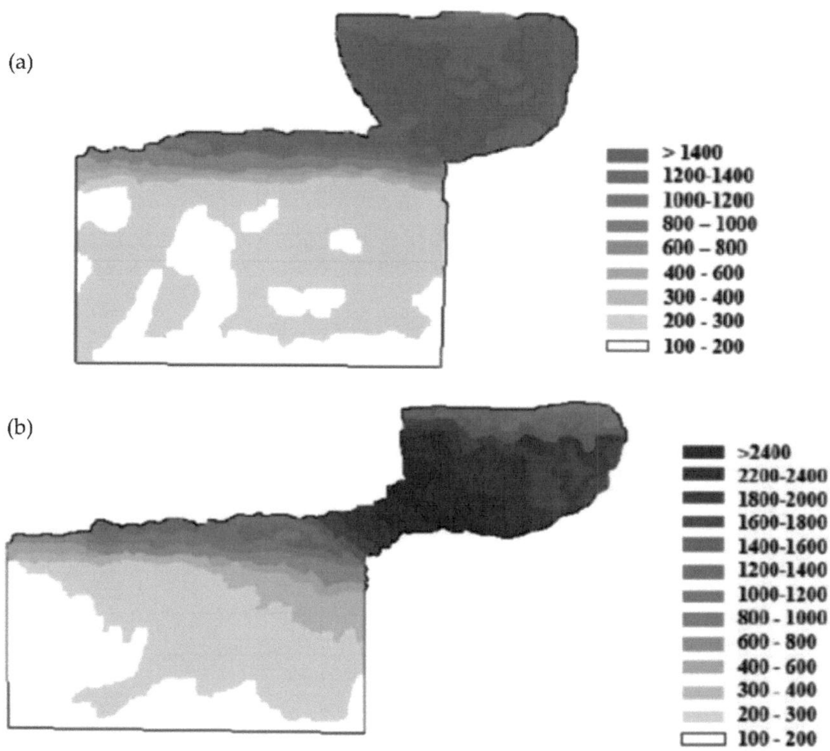

Figure 6.
Thermal analyses for two different MD simulations under T_I = 100 *K,* O_I = 3.5a_0*,* D_I = 1.00 *eV, and (a)* $\mathbf{V}_I$ = 1.00 *Å/ps and (b)* $\mathbf{V}_I$ = 10.0 *Å/ps.*

temperature was very brief (about 10–100 ps), while the activation energy for atoms to jump into the gas phase was much higher (approximately the cohesive energy), so the kinetics prevented any significant amount of evaporation from occurring.

4.5 Statistical analysis

Seventeen outputs for the MD simulations are listed in **Table 4**. The goal of the statistical analysis by the DOE is to determine the "best fit" equation to describe the wear rate as a function of four simulation variables (inter-asperity bonding, geometry overlap, impact velocity and starting temperature). It was assumed that the effect of each variable is additive, and there is no interaction between each of two variables (no cross term). So, this simplest model may work very well for the DOE analysis [16]. However, because the wear rate varied over a large range (the ratio of wear rate from its maximum to minimum was about 12), the wear rate must by transformed by using a natural log. Therefore, the equation for the wear rate in terms of the selected four factors was expressed as follows

$$Wear = \exp\left[5.08 + 0.000639 \times A - 0.03 \times B + 0.42 \times C + 0.93 \times D\right], \quad (7)$$

where the *Wear* (wear rate) was the number of Al atoms removed by the LJ-tip, *A* was the starting temperature (K), *B* was the impact velocity (Å/ps), *C* was the geometry overlap (Å), and *D* was the inter-asperity bonding (eV). Please note, Eq. (7) applied an exponential formula to describing the wear rate because of the large variation in the wear rate (see the detailed analyses of variance for the selected factors in **Table 5**).

T_I (K), *A*-term	V_I (Å/ps), *B*-term	O_I (Å), *C*-term	D_I (eV), *D*-term	Wear rate (atom number)
100	1.00	1.50 a_0	0.20	409
700	1.00	1.50 a_0	0.20	453
100	10.0	1.50 a_0	0.20	262
700	10.0	1.50 a_0	0.20	292
100	1.00	3.50 a_0	0.20	947
700	1.00	3.50 a_0	0.20	1231
100	10.0	3.50 a_0	0.20	842
700	10.0	3.50 a_0	0.20	982
100	1.00	1.50 a_0	1.00	711
700	1.00	1.50 a_0	1.00	1433
100	10.0	1.50 a_0	1.00	694
700	10.0	1.50 a_0	1.00	933
100	1.00	3.50 a_0	1.00	1346
700	1.00	3.50 a_0	1.00	3144
100	10.0	3.50 a_0	1.00	1402
700	10.0	3.50 a_0	1.00	1855
400	5.00	2.50 a_0	0.50	1244

Table 4.
Seventeen combinations of the MD simulations for the DOE analyses.

Source	Sum of square	DF	Mean square	F values	
Prob > F					
Model	6.33	4	1.58	34.44	< 0.0001
A	0.47	1	0.47	10.30	0.0075
B	0.30	1	0.30	6.61	0.0245
C	3.09	1	3.09	67.29	< 0.0001
D	2.46	1	2.46	53.56	< 0.0001
Residual	0.55	12	0.046		
Core Total	6.88	16			

Please note: A — T_I; B — V_I; C — O_I; D — D_I.

Table 5.
ANOVA for those selected factorial models provided by the design expert™ software.

It is more useful to describe the wear rate using the normalized variables (variables are normalized to a scale from −1 to +1, the low and high levels for each variable) because the magnitude of A, B, C and D coefficients allows one to easily determine the relative importance of each variable.

Through using the normalized variables, Eq. (7) can be rewritten as follows

$$Wear = \exp\left[6.80 + 0.19 \times A - 0.12 \times B + 0.42 \times C + 0.37 \times D\right]. \quad (8)$$

In Eq. (8), coefficient magnitudes of variables indicated that A (= +0.19) term had a small effect, B (= −0.12) term had a minor inverse correlation, while C (= +0.42) and D (= +0.37) terms had major effects on the wear rate. That was, the inter-asperity bonding (D) and the geometry overlap (C) had much more effects on the wear rate than the starting temperature (A) and the impact velocity (B).

5. Conclusions

In this work, a severely adhesive wear was investigated by simulating asperity-asperity shear between a fast moving rigid LJ-tip toward an Al-asperity. Molecular dynamics simulations were conducted for 17 different combinations of four variables: the starting temperature, the relative velocity, the geometry overlap and the inter-asperity bonding between two asperities. It was found that the wear process occurred in three stages: the contact, the plowing, and the necking/fracture on the aluminum substrate. Thermal analyses indicated that the heat generated during the MD simulations stemmed from the adhesive reaction in the contact area between two asperities, and then in the remnant surface on the Al residual substrate where the Al necking root glided. Bond formation and mechanical deformation during asperity-asperity shear may result in large increases of local temperature in the contact area (1200 ∼ 2500 K), so the primary mechanism of deformation on the Al-substrate was due to amorphous plasticity and local melting. Generations and motions of dislocation cores were observed under milder conditions where little melting occurred. A method: *The 2^4 full factorial design* in the Design Of Experiment was adopted in analyzing effects of those four variables (factors) on the wear process. Analysis results indicated that, the inter-asperity bonding and the geometry overlap between two asperities would play much more important roles in the wear process than the starting temperature and the impact velocity.

Acknowledgments

This work was sponsored by the National Science Foundation (Grant no.: DMR9619353), USA; the General Program of The National Nature Science Foundation of China (Grant no.: 11972348), the Joint Fund Program of The National Nature Science Foundation of China (Grant no.: U1738108), and the Hebei Provincial Key Laboratory of Thermal Protection Materials (SZX2020038, in preparation), China.

Author details

Jun Zhong
North China Institute of Aerospace Engineering, Langfang 065000, P.R. China

*Address all correspondence to: settings83@hotmail.com

References

[1] Wert J.J, Singerman S.A, Galdwell S. G, Quarles R.A. The role of stacking fault energy and induced residual stresses on the sliding wear of aluminum bronze. Wear. 1983; **91**: 253–267. DOI: https://doi.org/10.1016/0043-1648(83)90071-6

[2] Syed S.A, Wahl K.J, Colton R.J. Nanoindentation and contact stiffness measurement using force modulation with a capacitive load-displacement transducer. Review of Scientific Instruments. 1999; **70**: 2408–2413. DOI: https://doi.org/10.1063/1.1149770

[3] Engelder J.T, Scholz C.H. The role of asperity indentation and ploughing in rock friction—II: Influence of relative hardness and normal load. International Journal of Rock Mechanics and Mining Sciences & Geomechanics Abstracts. 1976; **13**: 155–163. DOI: https://doi.org/10.1016/0148-9062(76)90820-2

[4] Putman C and Kaneko R. Experimental observation of single-asperity friction at the atomic scale. Thin Solid Film. 1996; **273**: 317–321. DOI: https://doi.org/10.1016/0040-6090(95)06795-7

[5] Maw W, Stevens F, Langford S.C, Dickinson J.T. Single asperity tribochemical wear of silicon nitride studied by atomic force microscopy. Journal of Applied Physics. 2002; **22**: 5103–5109. DOI: https://doi.org/10.1063/1.1510595

[6] Xue X, Polycarpou A.A, Phinney LM. Measurement and Modeling of Adhesion Energy Between Two Rough Microelectromechanical System (MEMS) Surfaces. Journal of Adhesion Science and Technology. 2008; **22**: 429–455. DOI: https://doi.org/10.1163/156856108X305570

[7] Mate C.M, McClelland G.M, Erlandsson R, Chiang S. Atomic-scale friction of a tungsten tip on a graphite surface. Physical Review Letters. 1987; **59**: 1942–1945. DOI: https://doi.org/10.1103/PhysRevLett.59.1942

[8] Landman U, Luedtke W.D, Nancy A, Colton R.J. Atomistic Mechanisms and Dynamics of Adhesion, Nanoindentation, and Fracture. Science. 1990; **248**: 454–461. DOI: 10.1126/science.248.4954.454

[9] Komanduri R, Chandrasekaran N, Raff L.M. MD simulation of indentation and scratching of single crystal aluminum. Wear. 2000; **240**: 113–143. https://doi.org/10.1016/S0043-1648(00)00358-6

[10] Kelchner C.L, Plimpton S.J, Hamilton J.C. Dislocation nucleation and defect structure during surface indentation. Physical Review B. 1998; **58**: 11085. https://doi.org/10.1103/PhysRevB.58.11085

[11] Saraev D, Miller R.E. Atomistic simulation of nanoindentation into copper multilayers. Modelling and Simulation in Materials Science and Engineering. 2005; **13**: 1089–1100. DOI: https://doi.org/10.1088/0965-0393/13/7/006

[12] Zimmerman J.A, Kelchner C.L, Hamilton J.C, Foiles S.M. Surface Step Effects on Nanoindentation. Physical Review Letters. 2001; **87**: 165507. DOI: https://doi.org/10.1103/PhysRevLett.87.165507

[13] Zhang L, Tanaka H. Towards a deeper understanding of wear and friction on the atomic scale—a molecular dynamics analysis. Wear. 1997; **211**: 44–53. DOI: https://doi.org/10.1016/S0043-1648(97)00073-2

[14] Zhang L, Tanaka H. Atomic scale deformation in silicon monocrystals

induced by two-body and three-body contact sliding. Tribology International. 1998; **31**: 425–433. DOI: https://doi.org/10.1016/S0301-679X(98)00064-4

[15] Fortini A, Mendelev M.I, Buldyrev S, Srolovitz D. Asperity contacts at the nanoscale: Comparison of Ru and Au. Journal of Applied Physics. 2008; **104**: 074320. DOI: https://doi.org/10.1063/1.2991301

[16] Zhong J, Adams J.B, Hector L.G. Jr. Molecular dynamics simulations of asperity shear in aluminum. Journal of Applied Physics. 2003; **43**: 4306–4314. DOI: https://doi.org/10.1063/1.1558966

[17] Yu H, Adams J.B, Hector L.G. Jr. Molecular dynamics simulation of high-speed nanoindentation. Modelling and Simulation in Materials Science and Engineering. 2002; **10**: 319–330. DOI: https://doi.org/10.1088/0965-0393/10/3/305

[18] Daw M.S, Baskes M.I. Semiempirical, Quantum Mechanical Calculation of Hydrogen Embrittlement in Metals. Physical Review Letters. 1983; **50**: 1285–1288. DOI: https://doi.org/10.1103/PhysRevLett.50.1285

[19] Daw M.S, Baskes M.I. Embedded-atom method: Derivation and application to impurities, surfaces, and other defects in metals. Physical Review B. 1984; **29**: 6443–6453. DOI: https://doi.org/10.1103/PhysRevB.29.6443

[20] Ercolessi F, Adams J.B. Interatomic Potentials from First-Principles Calculations: The Force-Matching Method. Europhysics Letters. 1994; **26**: 583–588. DOI: https://doi.org/10.1209/0295-5075/26/8/005

[21] Alder B.J, Wainwright T.E. Studies in Molecular Dynamics. I. General Method. The Journal of Chemical Physics. 1959; **31**: 459–466. DOI: https://doi.org/10.1063/1.1730376

[22] Alder B.J, Wainwright T.E. Studies in Molecular Dynamics. II. Behavior of a Small Number of Elastic Spheres. The Journal of Chemical Physics. 1960; **33**: 1439–1451. DOI: https://doi.org/10.1063/1.1731425

[23] Verlet L. Computer "Experiments" on Classical Fluids. I. Thermodynamical Properties of Lennard-Jones Molecules. Physical Review. 1967; **159**: 98–103. DOI: https://doi.org/10.1103/PhysRev.159.98

[24] Beeman D. Some multistep methods for use in molecular dynamics calculations. Journal of Computational Physics. 1976; **20**: 130–139. DOI: https://doi.org/10.1016/0021-9991(76)90059-0

[25] Gear G.W. *Numerical Initial Value Problems in Ordinary Differential Equations* (Prentice-Hall Inc., Englewood Cliffs, New Jersey, 1971). ISBN-13: 978–0136266068

[26] Heermann D.W. *Computer Simulation Method in Theoretical Physics* (Springer Verlag, Berlin, 1990). ISBN: 978–3–540-52210-2

[27] Nosé S. A Molecular Dynamics Method for Simulations in the Canonical Ensemble. Molecular Physics. 1984; **52**: 255–268. DOI: 10.1080/00268978400101201

[28] Anderson H.C. Molecular dynamics simulations at constant pressure and/or temperature. The Journal of Chemical Physics. 1980; **72**: 2384–2393. DOI: https://doi.org/10.1063/1.439486

[29] Parrinello M, Rahman A. Polymorphic transitions in single crystals: A new molecular dynamics method. Journal of Applied Physics. 1981; **52**: 7182–7190. DOI: https://doi.org/10.1063/1.328693

[30] Hoover W.G. *Computational Statistical Mechanics* (Elsevier Science, 1991). ISBN: ISBN-13: 978–0444881922

[31] Liu X-Y, Ercolessi F, Adams J.B. Aluminium interatomic potential from density functional theory calculations with improved stacking fault energy. Modelling and Simulation in Materials Science and Engineering. 2004; **12**: 665–670. DOI: https://doi.org/10.1088/0965-0393/12/4/007

[32] Liu X-Y, Adams J.B, Ercolessi F, Moriarty J.A. EAM potential for magnesium from quantum mechanical forces. Modelling and Simulation in Materials Science and Engineering. 1996; **4**: 293–304. DOI: https://doi.org/10.1088/0965-0393/4/3/004

[33] Montgomery D.C. Design and Analysis of Experiment (8th ed., Wiley, 2012). ISBN-13: 978–1118146927

[34] Lasa L, Rodriguez-Ibabe J.M. Effect of composition and processing route on the wear behaviour of Al–Si alloys. Scripta Materialia. 2002; **46**: 477–481. DOI: https://doi.org/10.1016/S1359-6462(02)00020-9

[35] Markov D, Kelly D. Mechanisms of adhesion-initiated catastrophic wear: pure sliding. Wear. 2000; **239**: 189–210. DOI: https://doi.org/10.1016/S0043-1648(99)00373-7

[36] Ling F.F, Edward Saibel E. Thermal aspects of galling of dry metallic surfaces in sliding contact. Wear. 1957; **1**: 80–91. DOI: https://doi.org/10.1016/0043-1648(57)90002-9

Chapter 17

Fretting Wear Performance of PVD Thin Films

Brahim Tlili

Abstract

Nowadays, most surface treatments are realized through vapor deposition techniques as thin hard coatings to guarantee high surface hardness, low friction coefficient, and improve wear resistance. Several experimental investigations have led to the development of multilayer coatings in preference to the traditional TiN coating. In the current chapter, research was conducted on the fretting wear of (TiAlCN/TiAlN/TiAl) and (TiAlZrN/TiAlN/TiAl) multilayer coatings deposited by reactive DC (magnetron sputtering) of Ti-Al and Ti-Al-Zr alloys on AISI4140 steel. Fretting wear tests (20,000 cycles at 5 Hz) were conducted in ambient conditions, where the interaction between normal load and displacement amplitude determined the fretting regime. The influence of the normal load and displacement amplitude on the coefficients of instantaneous coefficient of friction and stabilized coefficient of friction is different in the two multilayer, coated steels. The PVD coating (TiAlZrN/TiAlN/TiAl) reduces the friction. The worn volume of coated AISI4140 steel is sensitive to normal load and displacement amplitude. The relation between worn volume and cumulative dissipated energy was established for the two coated steels. The energetic fretting wear coefficients were also determined. A multilayer (TiAlZrN/TiAlN/TiAl) coating has a low energetic wear coefficient.

Keywords: fretting, wear, PVD, coatings, hardness

1. Introduction

The hard coatings have been used extensively to increase the wear resistance under loading conditions in fretting wear. Various methods of coating deposition have found industrial application. One successful approach is the physical vapor deposition (PVD) technique. Presently, development of TiAlN-based multilayer and nanocomposite coatings are the most important trends in the hard coating industry [1]. The ternary coating has better resistance to oxidation and poor adherence [2]. In order to remedy these imperfections, additional constituents such as carbon or zirconium are recommended. In order to characterize the quality of a coating, evaluation of a number of properties is essential. Hardness, friction coefficient, roughness, wears, and oxidation resistance have primary importance. The addition of carbon to TiAlN provides a quaternary coating, TiAlCN, which has better resistance to oxidation, improved mechanical properties, and has good adherence with high tribological capacities [3, 4]. However, the addition of zirconium to TiAlN provides a quaternary coating, TiAlZrN. Although some studies are available on TiAlZrN layers and their behavior in terms of wear [2], data on resistance to fretting wear are not available. First, according to the intensity of the imposed parameters (normal force, amplitude of displacement, and frequency), fretting maps were established to determine the sliding conditions and the type of damage according to the fretting conditions. The running condition fretting maps indicate when total or partial sliding conditions occur, and the material response fretting maps quantify the damages (wear, cracking). Actually, the partial slip regime (PSR) is associated with cracking as a result of a fatigue phenomenon, whereas the gross slip regime (GSR) leads to wear by debris formation [5]. The mechanical models of contact including elastoplasticity are used to delimit the boundary between the partial slip and gross slip [6]. Crack propagation is studied using efficient fatigue criteria [7, 8], while wear is studied by more or less empirical multiple quantitative approaches [9]. Second, in this step, we have to determine the tribological properties of quaternary coatings and correlate the worn volume according to the parameters of loading in fretting wear. Finally, the comprehension of fretting behavior remains sensitive as soon as the mechanical, thermal, and physical-chemical interactions [occurring in a contact associated with the role played by the interfacial layer generated by detachment of the particles during friction (third body)] need to be taken into account; when wear particles have been liberated from the surface, some of them may attach to the counterface to form a transfer layer and significantly change the tribological properties of the counterface (like forming a new counterface) [10–13]. The wear analysis performed with imposed displacements or tangential forces is based on an energetic approach [6, 14]. The wear volume increases linearly with the dissipated energy within the contact.

The PVD surface treatment process makes it possible to produce nanocomposite coatings with a metal matrix. During this process, the particles are accelerated at very high speed through a flow of inert propellant gas under pressure and are preheated to a temperature below the melting temperature of the materials to be sprayed [1–3]. Due to the very high speeds, the coatings thus formed are very dense [2]. Obtaining large thicknesses (several micrometers), accompanied or not by superficial oxidation, with densities close to the volume of base substrate have advantages over deposits produced by other thermal processes. The cold spraying process is also a repair method used either on new parts under development or on parts under maintenance. Certain layers are also a means of protection against corrosion, for example, for components of aeronautical parts. Reinforcement phases can be filled layers, the composition of which also varies depending on the material of the matrix [2, 4]. Solid lubricants can also be incorporated [5]. Coatings with solid lubricants are used for technical parts such as bearings or rings to reduce friction.

The tribological circuit in general describes the different flow rates of the third body capable of being activated in an elementary contact in two dimensions [6, 7]. The internal source flow corresponds to the detachment of particles, due to surface tribological transformations (TTSs) [8, 9], cracking, and adhesion. It leads to the formation of the third natural body. The external source flow comes from the introduction of a third artificial body into the contact. The internal flow is the flow of the third body which circulates between the first bodies. The external flow is the flow of the third body which escapes from the contact. It is divided into a recirculation flow and a wear flow. The recirculation flow consists of the third body which, when reintroduced into contact, driven for example by one of the first bodies, will again contribute to speed accommodation. On the other hand, the wear flow is made up of the third body which, when definitively ejected from contact, will never again participate in speed accommodation. The debits will be estimated on a relative qualitative scale from very low (*) to very high (******).

The rheology of the third solid body is characterized relatively on the basis of its cohesion and its ductility [7]. These two properties are determined from the morphology and texture of the third body obtained by observations and analyses after tests. The cohesion translates a third body more or less compacted or pulverulent. Ductility reflects the ease with which the third body spreads in contact.

In this chapter, the multilayer coatings' behavior under various loading parameters in the fretting wear is presented.

2. Experimental procedures

2.1 Substrate

Specimens used in this work were machined from steel trade AISI4140 hardness 420 HV0.05, with surfaces of (10 × 10) mm^2 and (10 × 12) mm^2. These samples were divided into two sets; the first set was coated by TiAlCN/TiAlN/TiAl and the second round by TiAlZrN/TiAlN/TiAl. Before PVD treatment, all samples were cleaned and polished with trichloroethylene, acetone, and alcohol in an ultrasonic cleaner.

2.2 Deposition

The multilayer coatings were deposited on the steel AISI4140 by the process of spraying DC magnetron mode, using target compounds TiAl (50°/°Ti, 50°/°Al) [3] and Al-Ti-Zr (19°/°Ti, 21°/° to Al, 10°/°Zr, 50°/°N) of high purity (99.9999%).

The samples were mounted on a continued rotating satellite inside the vacuum plasma chamber. The atmosphere was chosen to produce successively under a layer of TiAl, a buffer layer of TiAlN, and then a layer of TiAlCN or TiAlZrN. Indeed, the thin films of TiAl and TiAlN aim to improve the adhesion layers TiAlCN [2, 3] and TiAlZrN [10] with steel. The optimum conditions for filing as the bias voltage targets, temperature, and time of deposition were determined and optimized in previous studies; **Table 1** shows in detail the optimal conditions for verification. In addition, Argon, acetylene, and nitrogen gas which are of very high purity (99.999%) were introduced into the vacuum chamber. The basic pressure in the room was 5.10^{-5} Pa, which grew to 0.1 Pa for the deposition of desired layers. The distance between the target and the substrate surface was 35 mm. Before the deposition, the surface of the substrate was cleaned by argon ion bombardment at the end to eliminate all these antagonists.

2.3 Characterization of the coatings

The multilayer coatings prepared in this work and their mechanical properties are shown in **Table 2**.

Unit	Traget power (K_w)	Bias (V)	Temper (°C)	Rotation velocity (rpm)	Time (mn)	Ar	Gas flow N_2 SCCM	C_2H_2
TiAl	6	-300	250	0.7	22	400	–	–
TiAlN	7.5	-200	150	0.6	17	770–800	90–150	–
TiAlCN	7.5	-200	100	0.6	42	950	150–135	20–135
TiAlZrN	7.5	-200	100	0.6	42	950	150–135	20–135

Table 1.
Deposition conditions.

	Hardness (GPa)	Young's modulus (GPa)
TiAlCN/TiAlN/TiAl [3]	15	260
TiAlZrN/TiAlN/TiAl	28	310
AISI4140 steel	42	210

Table 2.
Mechanical properties of coating layer and substrate.

The fretting wear behavior of multilayer TiAlCN/TiAlN/TiAl has been carried out in previous studies [3]. Indeed, a special focus in this chapter was given to the characterization and the study of the behavior of fretting wear coating of multilayer TiAlZrN/TiAlN/TiAl. The total thickness of the coating is determined by the electronic scanning microscope (SEM) after a major fracture in vertical section, followed an analysis of the chemical composition by energy dispersive spectroscopy (EDS). The morphology of the coating TiAlZrN is being reviewed by an atomic force microscope (AFM). The microstructural characterization of this coating is investigated by X-ray diffraction (Philips X'pert system). Scans were carried out in the grazing angle mode with an incident beam angle of 3° and the normal θ-2θ method classically used in the same situation. Young's modulus and hardness were measured by nanoindentation tests with a nanoindenter MTS-XP. The indentation was performed using a triangular Berkovich diamond pyramid. SEM and AFM observations allowed determination of the global coating thickness and the morphology of the surface, respectively. The multilayered TiAlZrN/TiAlN/TiAl coatings had a mean thickness of 800 nm, distributed in three layers as shown in **Figure 1**. The surface was globally uniform with some domes and tiny craters spread all over the area **Figure 2**.

Dimensional measurements showed that the domes had a mean diameter of about 25 nm and the craters a maximum depth of 16 nm. The crystallographic structure and orientation of the coatings were determined by X-ray diffraction. Phase identification for multilayer coatings TiAlZrN/TiAlN/TiAl revealed the presence of reflection peaks corresponding to stripes (100) and exhibited a weak intensity peak at 2θ = 2.83° (44.74 Å) (**Figure 3**). It can be seen that the as-coated state already has a crystalline structure of $AlZr_3$. The presence of crystallographic structure is linked to the columnar morphology of the layers observed by SEM on a cross section **Figure 4**. The chemical composition of multilayered coatings is shown in **Figure 5**. The measurements of nanoindentation carried out on a depth exceeding the thickness of the coating made it possible to determine the average hardness and average Young modulus in this multilayer TiAlZrN/TiAlN/TiAl. The results are presented in **Table 2**.

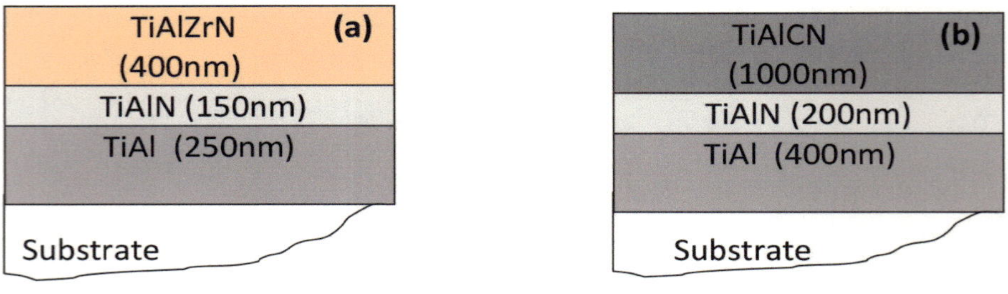

Figure 1.
Distribution and thickness of the layers in the multilayers coating: (a) TiAlZrN/TiAlN/TiAl and (b)TiAlCN/TiAlN/TiAl.

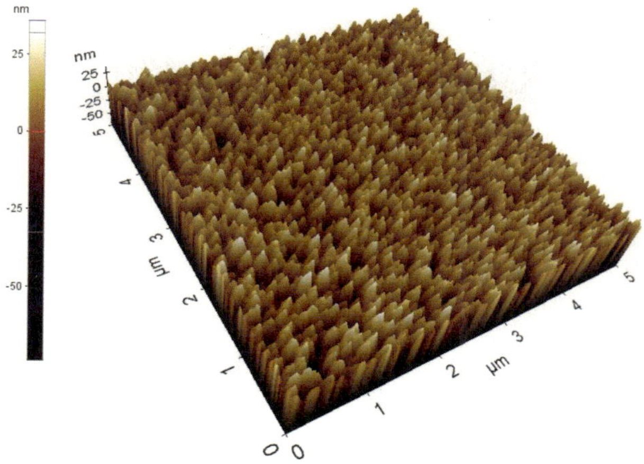

Figure 2.
AFM morphologies of the surface layer of TiAlZrN.

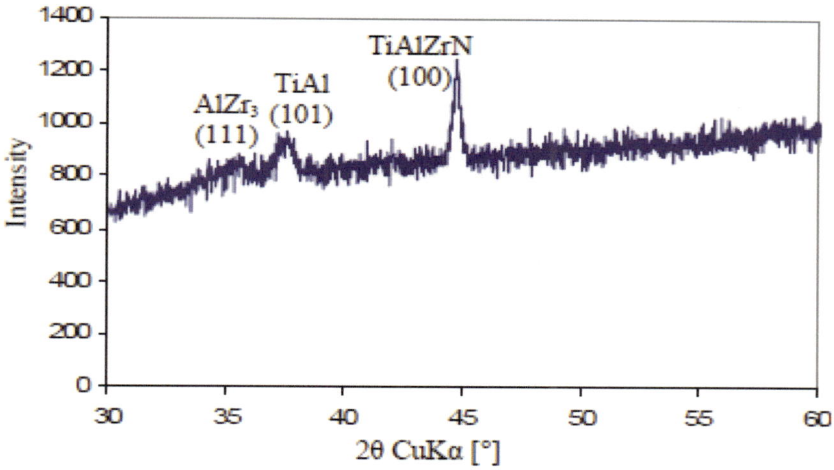

Figure 3.
XRD diffractogram of the TiAlZrN/TiAlN/TiAl multilayer.

2.4 Fretting tests

Fretting wear, which is considered a nuisance in several branches of industry such as aeronautics, nuclear industry, etc., refers to the degradation of the contact surface resulting from wear, which requires an overhaul or replacement of the machine components. It is defined as the wear that takes place during a low amplitude oscillatory movement between two apparently immobile solids under a load normal to the contact surface. Such a phenomenon is observed especially in assemblies subjected to vibrations. To simulate the industrial process, a test bench was developed within the laboratory and subsequently installed on a fatigue machine (MTS).

The fretting tests were carried out on an MTS tension compression hydraulic machine. A sphere-on plane configuration was employed as shown in **Figure 6(a)**. The counter-body was a polycrystalline alumina ball with 24 mm diameter, a Young's modulus of 310 GPa, and a hardness of 2300 Hv0.05. The flat coating alloy (10 mm × 10 mm × 12 mm) specimens were manufactured from a cast bar of

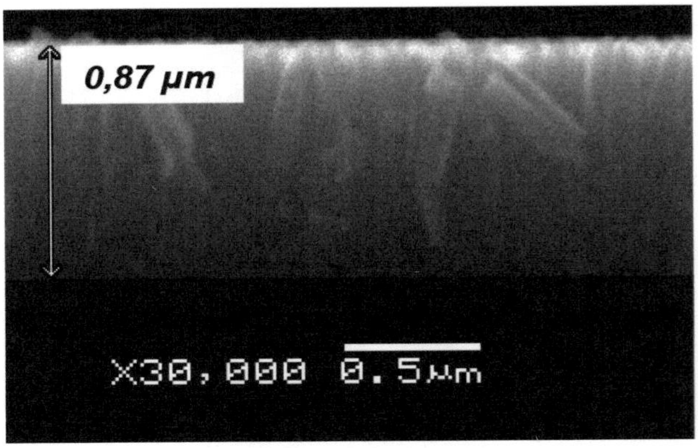

Figure 4.
Scanning electron micrograph of a cross section of multilayered TiAlZrN/TiAlN/TiAl coating as made by fractography.

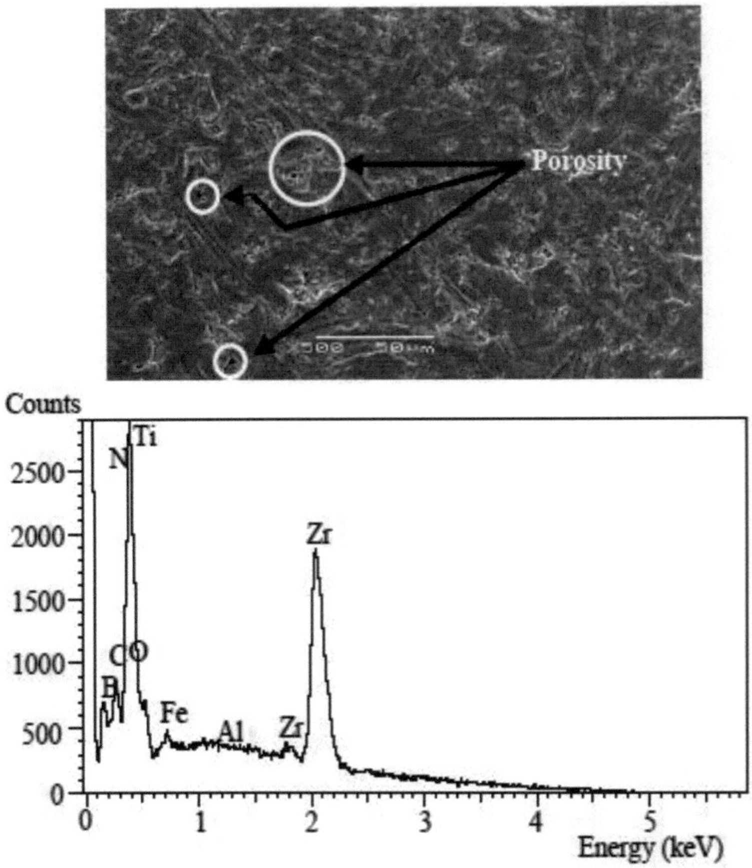

Figure 5.
Energy dispersive spectroscopy (EDS) of multilayered TiAlZrN/TiAlN/TiAl coatings.

AISI4140 steel. During the test, the instantaneous displacement, the normal force, and the tangential force were monitored and recorded for every cycle.

The fretting tests were conducted with displacement control, using an extensometer as a displacement transducer. When it is possible to plot the hysteresis loops of the fretting-wear stresses from the instantaneous measurement of the tangential force as a

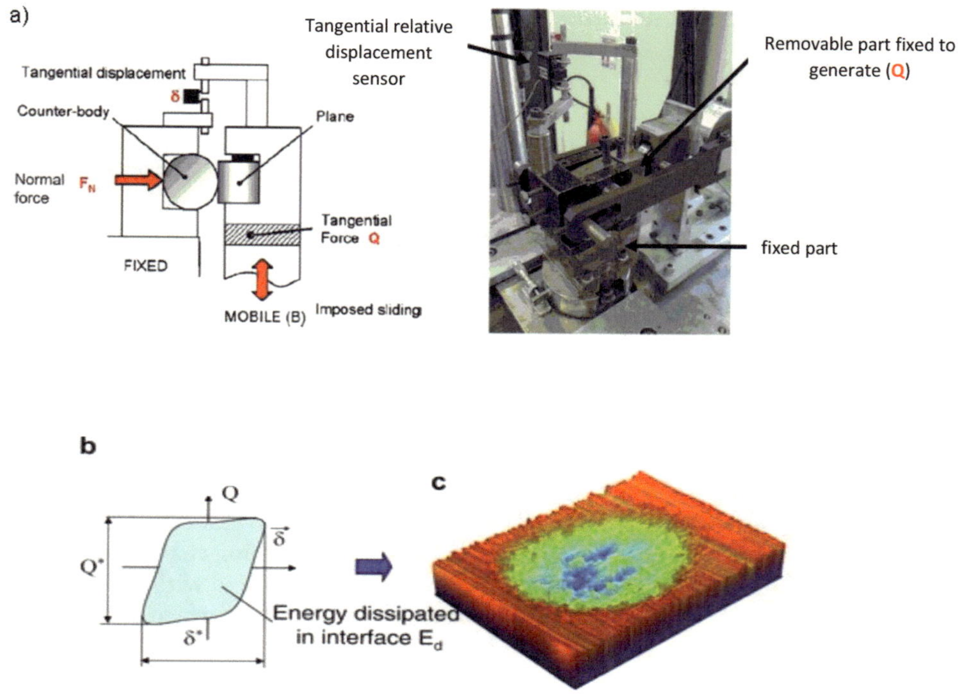

Figure 6.
Illustration of the experimental fretting wear approach: (a) fretting log of one fretting cycle and test bench, (b) full slip fretting loop, and (c) 3D wear trace, obtained by optical profilometer.

function of the sliding distance for normal force values between 50 and 750 N and the sliding ranges from ± 20 μm to ± 100 μm [3, 5, 15–17]. The area of the fretting loop corresponds to the dissipated energy during the fretting cycle (Ed), whereas the residual opening of the cycle (i.e., the residual displacement when Q = 0) is related to the full sliding amplitude (δg) (**Figure 6(b)**). All tests were performed at 20×10^3 cycles and the frequency was set at 5 Hz [3, 5, 15, 18].

The fretting tests were conducted in dry conditions at an ambient temperature of 25°C and a relative humidity of 60%. Prior to the fretting test, the specimen and counter-body were cleaned with acetone and alcohol. Hundred to 2000 fretting cycles were performed. The tests were stopped in two different ways: once the displacement was stopped abruptly after the last fretting cycle, and once the displacement amplitude was reduced to zero during several cycles. However, to quantify the wear volume, a specific 3D surface profilometry methodology, fully developed in reference, [15] is applied. It consists of determining the wear volume below the reference surface (V^-) and the transfer volume above the reference surface (V^+). This analysis is performed both on sphere and plane fretting scars. A system wear volume is then deduced from the following relationship:

$$V_{system} = \left(V^- - V^+\right)\text{plane} + \left(V^- - V^+\right)\text{orb} \quad (1)$$

However, other conditions have been tested to investigate the effect of certain parameters suspected of having a dominant role in the wear mechanism. These conditions will be given in due course. Furthermore, optical observations are coupled with scanning electron microscopy to examine posttest fretting scars.

3. Test results and discussion

3.1 Fretting wear

Preliminary work was carried out to determine the fretting running map of PVD multilayered coatings in contact with alumina sphere. The tangential force (Q) and displacement (δ) amplitudes are determined for each cycle, and each sliding rate is reported on a 2D map of the fretting displacement and friction force. After a certain number of cycles, the partial slip regime (PSR) is manifested as a change in the hysteresis loop form, whereas the gross slip regime (GSR) maintains the buckle form with a variation of tangential force [3]. Running condition fretting maps (RCFMs) can then be determined from this map [15]. **Figure 7** shows the boundary lines of both sliding rates for different multilayered TiAlCN/TiAlN/TiAl and TiAlZrN/TiAlN/TiAl coatings. It can be seen that the gross slip regime region of the PVD-coated AISI4140 steel is enlarged due to the presence of the TiAlZrN, then the TiAlCN layers. From a phenomenological consideration, the gross slip regime corresponds to wear, and in the partial slip regime, the wear is associated with cracking, and the contribution of TiAlZrN/TiAlN/TiAl-reducing coatings should be interpreted positively. Indeed, in such a situation, wear is favored with the cracking of the covered part, which makes it possible to sacrifice the surface in order to protect the volume of the part [3].

The TiAlZrN coating thus reduces the partial slip regime field, which is the most detrimental for fretting. However, sliding amplitudes are rather large and seen to be related more to the reciprocating condition. However, it is fundamental to relate the displacement value to the contact dimension. The boundary between the fretting and reciprocating conditions can be related to the ratio between the displacement amplitude and the contact radius, e = d/a [19]. It transpires that when e remains little than 1, a nonexposed surface exists and grosses slip fretting conditions prevail, whereas if e is above 1, the whole surface is exposed to the

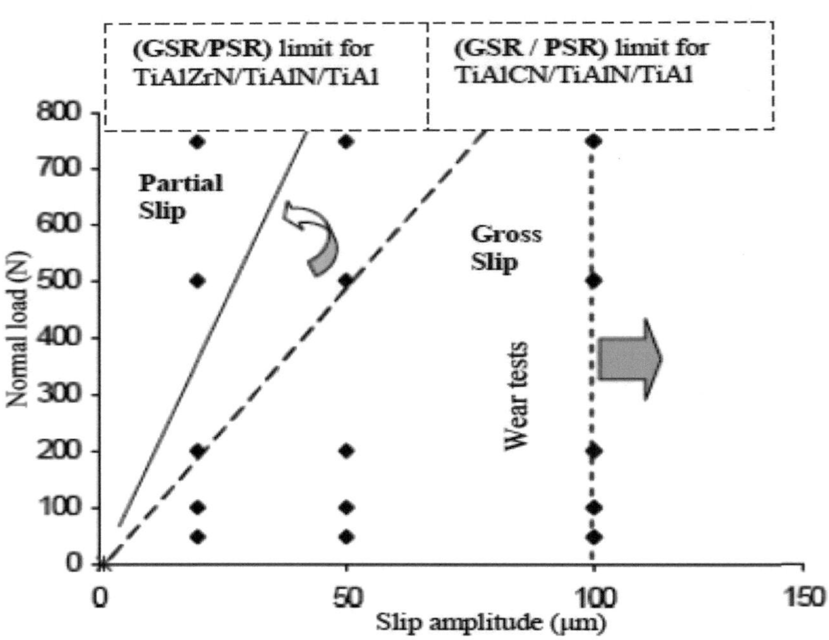

Figure 7.
Effect of multilayered coatings (TiAlZrN/TiAlN/TiAl) on the running condition.

ambient and the contact is under reciprocating conditions [18]. The maximum e value calculated for all performed test situations remains lower 0.5, which implies gross slip fretting conditions.

3.2 Tribological properties

Figure 8 shows the evolution of the coefficient of friction as a function of the total number of cycles for the two multilayered PVD coatings under same loading parameters. The first cycle systematically presents a low friction coefficient around 0.11, and the incipient low friction coefficients can be explained by the presence of surface oxides. During the test, the friction coefficient increases progressively toward a level known as the stabilized friction coefficient. Such a difference of friction behavior between the two antagonists (TiAlZrN and TiAlCN) is clearly illustrated in the graph of evolution of the friction versus the fretting cycles. It confirms the previous fretting cycle analysis and outlines the difference of friction kinetics between the two antagonists. The transition period is systematically longer in the presence of the TiAlZrN coating. In the case of the PVD layer, De wit [20] showed that the transition period corresponds to the formation of debris made of amorphous retiles and nanocrystallines. Beyond this transition, the amorphous phase is transformed into a crystalline phase and contributes to further wear. Depending on the loading condition, the film of TiAlCN or TiAlZrN can be eliminated, thus favoring a significant increase of the friction coefficient **Figure 9**. It explains the influence of the pressure and displacement amplitude on the evolution of the friction coefficient on a coated specimen, taking into account a slight surface degradation at low friction, which must be introduced into the friction cycle [21, 22]. It can be seen that by increasing the displacement amplitude or the pressure, the wear depth will grow faster and the elimination of surface porosity will be accelerated. As soon as the surface porosity and the aluminum oxide, which plays the role of a solid lubricant, are removed, low friction conditions can no longer be maintained and high metal/metal interactions with high friction coefficients in the range of 0.5–0.6 are observed, indicating that the PVD film has been breached. When a microarc oxidation coating was used, the fretting friction coefficient of modified PVD coatings alloy under higher loading condition remains as high as about 0.6; however,

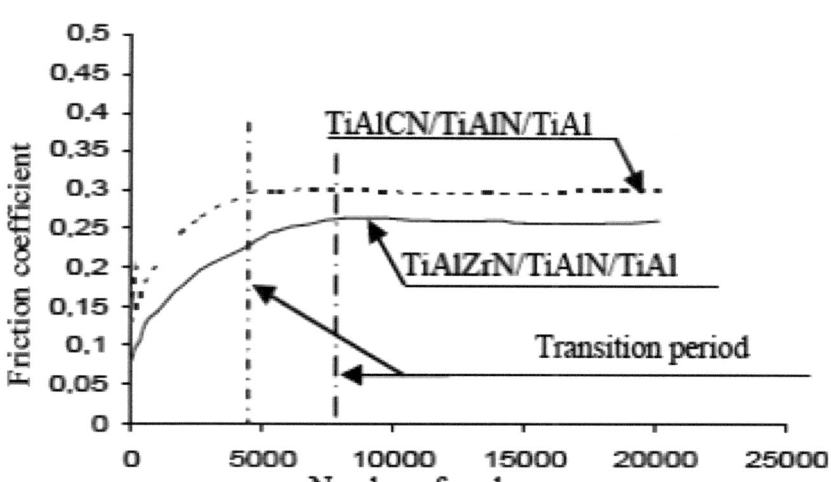

Figure 8.
Evolution of the friction coefficient for the TiAlZrN/TiAlN/TiAl and the TiAlZrN/TiAlN/TiAl (FN = 50 N, δ = 20 μm).

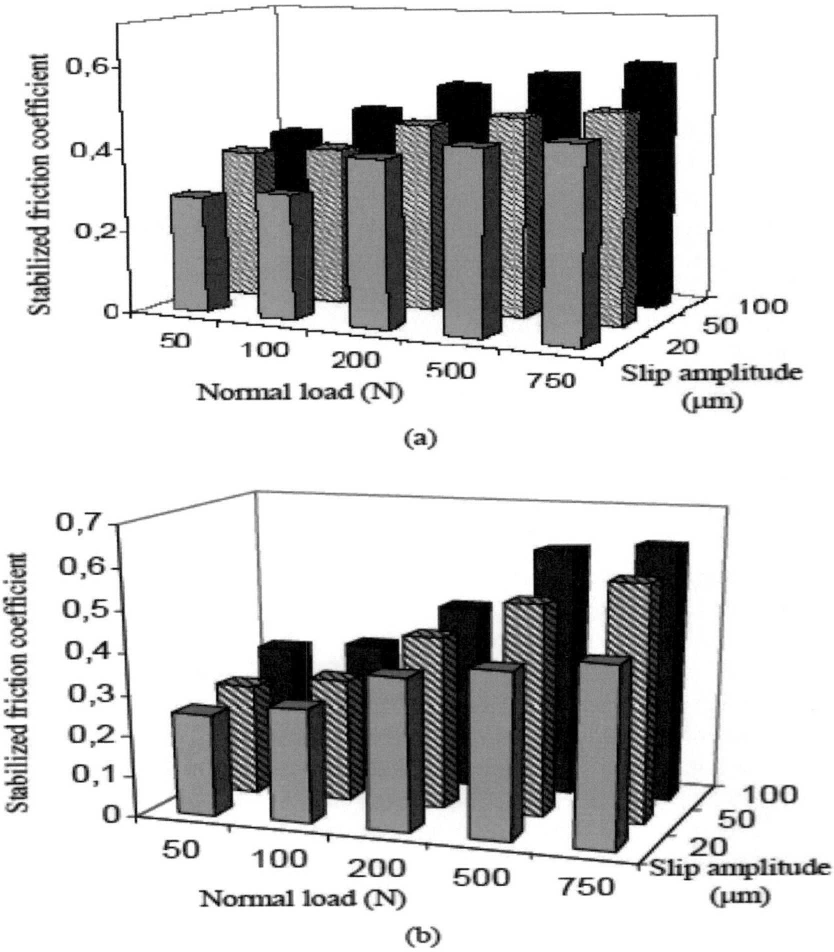

Figure 9.
Stabilized friction coefficient as a function of normal load and slip amplitude of PVD multilayered coating. (a) TiAlCN/TiAlN/TiAl [3] and (b) TiAlZrN/TiAlN/TiAl.

the cracking domain and severe adhesive nature was limited. Under low loading condition, the friction coefficient was significantly reduced and remained favorable stable at 0.4–0.3 in the long term, which indicates that the multilayered coatings lowered the shear and adhesive stresses between contact surfaces, consequently alleviating the possibility of initiation and propagation of cracks in the inner layer of multilayered coatings.

3.3 Wear properties

In every tribological application, the extent of the damage or surface deterioration is of interest. There are several methods of evaluating the wear volume/loss, which can be roughly classified into three methods: weight measurement, topographical analysis, and 2D analysis by means of empirical equations. In tribological research, where many specimens need to be analyzed, a simple and fast procedure is desirable for wear volume/loss determination. Moreover, the effect of different material combinations, slip amplitude, normal force, and the energy dissipated during sliding are also presented.

3.3.1 Prediction of the wear volume evolution

The linearity of the variation of fretting wear with normal load and displacement obtained by the two multilayered TiAlCN/TiAlN/TiAl- and TiAlZrN/TiAlN/TiAl-coated specimens are used to impose the same definition as Archard's equation, namely, a quasi-linear function. The wear volume evolution of the two coated steels is shown in **Figure 10**, as a function of slip amplitude and normal force. For all specimens, the beneficial effects of the coating on wear volume diminished with the increasing normal force and fretting stroke. The latter observation is consistent with the work of Santner et al. [23], who reported that TiN was much more effective at suppressing wear under sliding wear conditions. For the steel substrate, the coating TiAlCN or TiAlZrN had no good effect on the fretting wear for sliding amplitudes larger than 50 μm, regardless of the applied normal forces out of 500 N. The wear transition is attributed to the higher variation of both normal load and sliding amplitude. However, for all fretting wear tests, the behavior evolution of wear volume versus displacement is the same. This means in all tests, wear volumes remain constant and are not greatly influenced by the low normal load or sliding distance. In fact, it is only the wear amplitude which changes according to the displacement and high normal load. In every case, there is a constant wear volume which precedes the establishment of the high wear regime. Three suppositions can be made to synthesize all the results presented above. Wear volumes are similar for all loading conditions and consist of two phases. The first one corresponds to the elimination of the contamination layer and the native oxides. Thus, alumina to PVD coatings contact will be established. As a result, the adhesion phenomenon is favored as regards the miscibility antagonists by plastic deformation, which increases the micro-junction density by crushing asperities. Hence, adhesive wear appears by creating transfer of the softer material (PVD coatings) on the harder

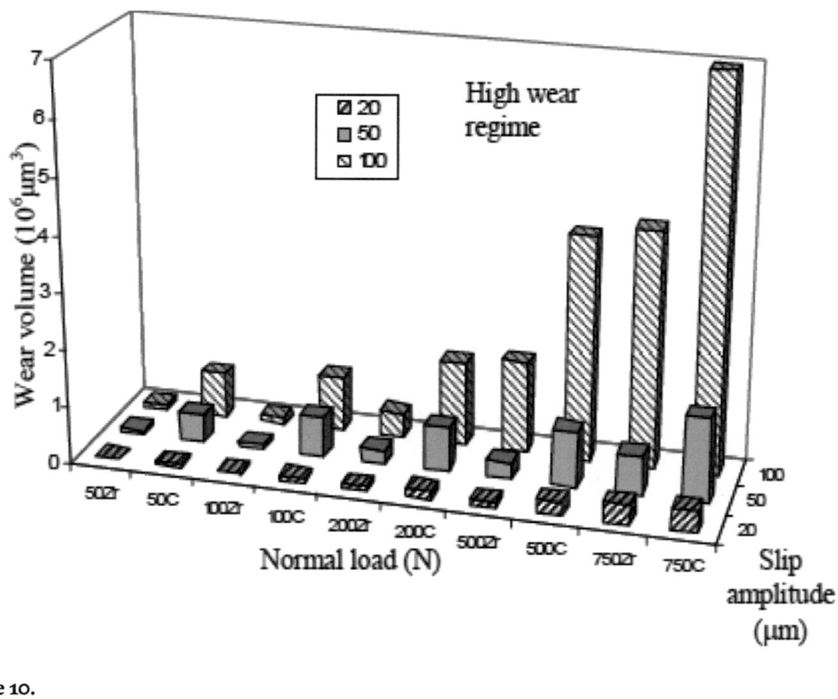

Figure 10.
Wear volume as a function of parameters loading multilayered coatings: (Zr) TiAlZrN/TiAlN/TiAl, (C) TiAlCN/TiAlN/TiAl.

material (alumina). This phase will be followed by a transitional stage: wear of transferred PVD coatings. The second phase (high wear regime) happens as soon as the trapped debris are oxidized. Gradually, wear is accentuated on both sides of contact (on PVD coatings and alumina ball). For larger displacements ($\delta > 50$ µm) and a high normal load, fatigue wear induced by multicracks has been detected **Figure 11**. One hypothesis is that the generated plastic strain leads to a brittle layer, which can be associated to the tribologically transformed structure (TTS) [24, 25]. Enduring cyclic loading, multicracks are activated which induce debris formation and wear. De Wit [16] showed that the transition period corresponds, in the case of the PVD layer, to the formation of debris made up of amorphous rutiles and nanocrystallines. Beyond this transition, the amorphous phase is transformed into a crystalline phase and contributes to further wear. SEM observations showed that the debris appeared on this TiAlCN/TiAlN/TiAl multilayer coating during the first cycles, in the form of particles less than 1 µm in size [3].

The main wear mechanism of the TiAlZrN-coated AISI4140 steel was the brittle cracks of TiAlZrN, before destroying the coating. After destroying the TiAlZrN coating, the main wear mechanism was cracks due to a low normal force and high slip amplitude and cumulated plastic flow on the edge of the fretting scar and adhesive and abrasive wear at high slip amplitude. Oxidation was observed on most the worn surface.

Moreover, the effect of surface roughness before deposition on the wear behavior of multilayer coatings deposited on an AISI 4041 steel substrate. After deposition, the surface roughness of the coating was approximately half of the original substrate surface roughness. While the frictional behavior was not apparently affected, the wear rate of the coatings increased significantly with increase in the substrate surface roughness. Wear rate increased rapidly when the substrate surface roughness exceeded Ra 1 µm. Above this substrate roughness, the dominant wear mechanism also changed from adhesion to chip/flake formation and fragmentation of the coatings. Chipping/flaking of the coatings initially occurred mainly at the tops of asperities of the surface texture. The Archard's specific wear

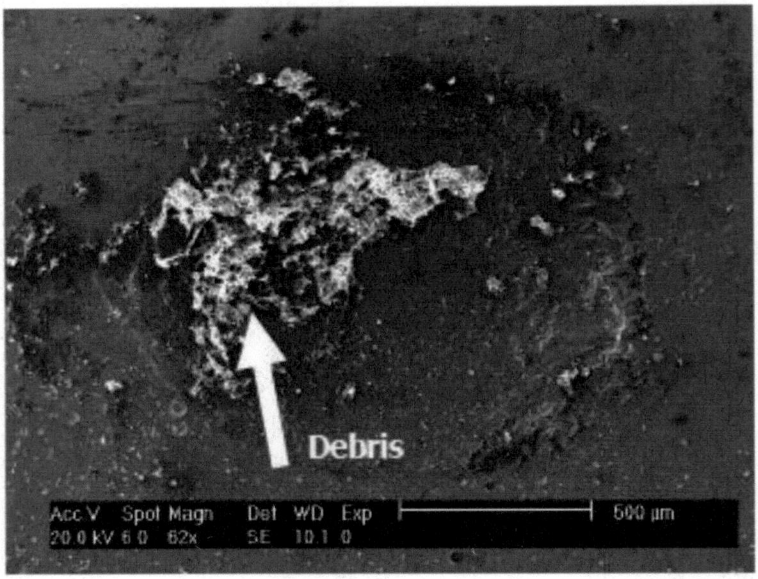

Figure 11.
Fretting wear scar morphology of multilayered TiAlZrN/TiAlN/TiAl coating (gross slip regime), FN = 200 N. δ = ±100 µm.

rate increased with the increase in total load for coatings on the rough substrate surfaces; however, this was almost invariant with the increase in load for coatings on the smoother substrate surfaces, nominally following the Archard's wear law. Contact pressure distributions over the real area of contact between the ball and the rough coating surfaces have been analyzed by applying the elastic foundation model of contact mechanics. It has been shown that the contact pressures increase significantly with the increase in surface roughness of the coatings. Plastic yielding is highly possible in the coatings deposited on rough substrate surfaces above Ra 1 µm. The observed apparent effect of surface roughness on wear and wear mechanism transitions of the multilayer coatings can be explained according to the contact mechanics analysis results.

3.3.2 Prediction energy wear coefficient

Investigations at various loads and slip amplitudes confirm that there exists a correlation between the wear volume extension of the TiAlZrN/TiAlN/TiAl multilayer coating and dissipated energy. For all the different loading conditions previously defined in Section 2.4, **Figure 12** shows the rates of the wear volume as a function of the dissipated energy. The used volume is measured using a 3D optical profilometer, and the dissipated energy is estimated directly by the area of the fretting loop for each cycle. As a separate form of this behavior, when the energy approach is applied, all of the test parameters are represented by the one and only linear equation from which a single overall energy wear coefficient (α = slope of the curve) can be detreminated for each antagonist. In fact, the energy coefficient represents the slope of the straight line which connects the lost volume and the energy dissipated by friction: $V_{Lost} = \alpha E_d$. In the case under consideration, $\alpha_{TiAlZrN} = 104\ \mu m^3/J$, and TiAlCN coating provides an energetic wear coefficient of $\alpha_{TiAlCN} = 23.103\ \mu m^3/J$ [3].

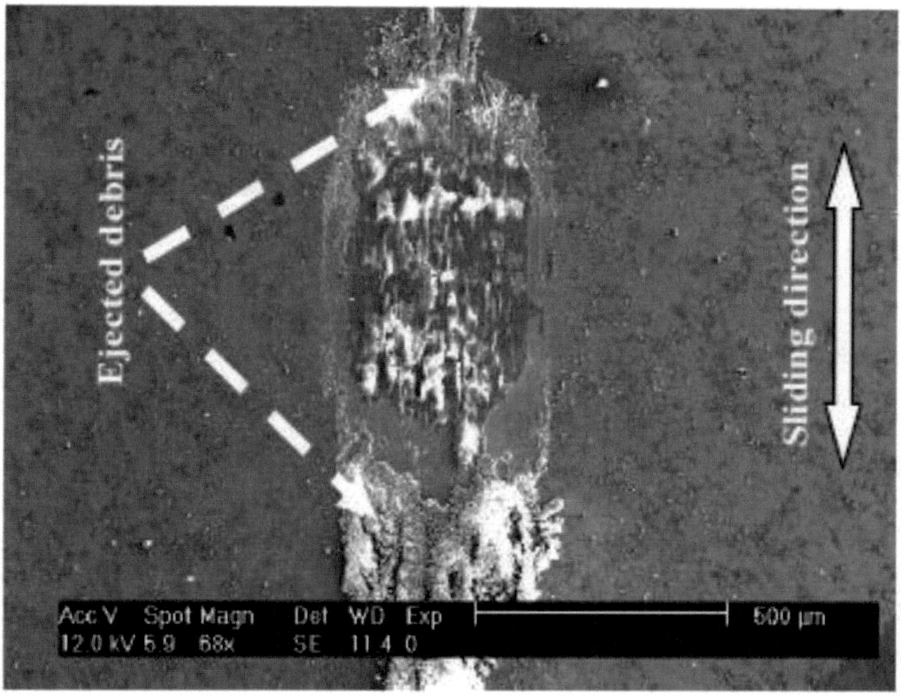

Figure 12.
Fretting wear scar morphology of multilayer TiAlZrN/TiAlN/TiAl coating (partial slip regime).

This result shows that the TiAlZrN coating improves its higher capacity to wear in fretting. The principal factors favoring this tendency are generally related to the presence of compression residual stress, the decrease of friction coefficient, the increase of superficial hardness, and the roughness effect of the surface [20]. Without measuring the residual stress, the contribution of the reduction of friction coefficient and the increase in hardness are confirmed in this study. One can notice as in **Figure 13** that hard coatings quickly give rise to particle detachment and prevent the partial slip regime by accommodating the displacement in the powder bed, and favor debris formation. A large amount of debris has been observed during fretting tests, and coatings are damaged by material transfer due to adhesion and/or abrasion. It can be seen that a small amount of debris remains within the contact area, while a large amount of debris is ejected outside and located near to the border of the fretting scar. However, fretting scars have three characteristic regions, which are clearly visible for the multilayered TiAlZrN/TiAlN/TiAl coating; central convex region with slight abrasion traces, outer annular region covered with debris, and transition region where the abrasion traces parallel to the sliding direction can be easily identified. **Figure 14** compares the energetic wear coefficients with those reported in the literature [3, 20, 21]. Magnetron-sputtered TiAlZrN coatings in the as-deposited condition possess a better fretting wear resistance (104 $\mu m^3/J$) than TiAlCN coatings (23 $\mu m^3/J$) for tests performed in ambient air. According to energetic considerations, the multilayered TiAlZrN coatings improve the resistance to fretting wear by a factor 6.5, compared with non-coated steel [3]. But the TiAlCN multilayer has a lower performance as it improves the resistance only by a factor of 2.8. This result is acceptable since the addition of Al, Zr forms stable oxides, especially Zr, which forms a very thin oxide layer similar to Al_2O_3 [2, 21, 22, 26] that strongly influences the energetic wear coefficient and provides good tribological properties [2].

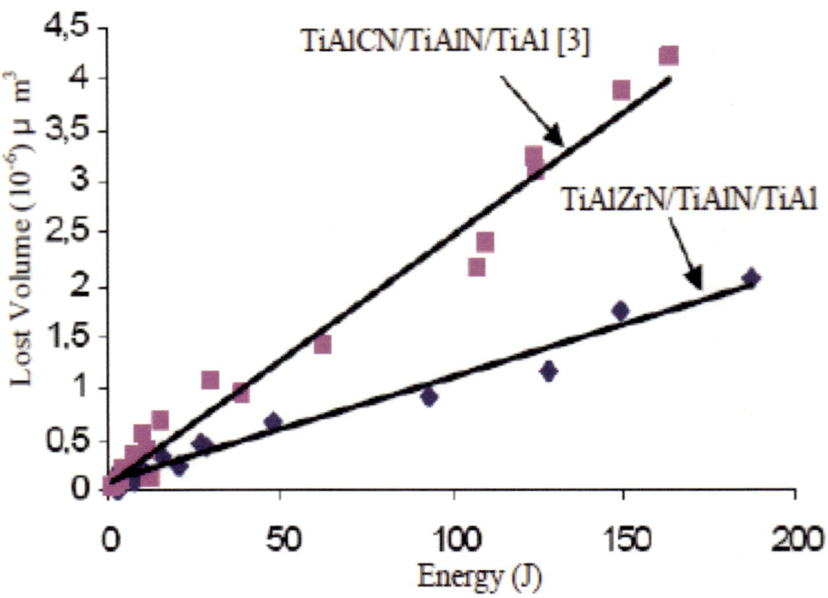

Figure 13.
Wear volume as a function of cumulated dissipated energy.

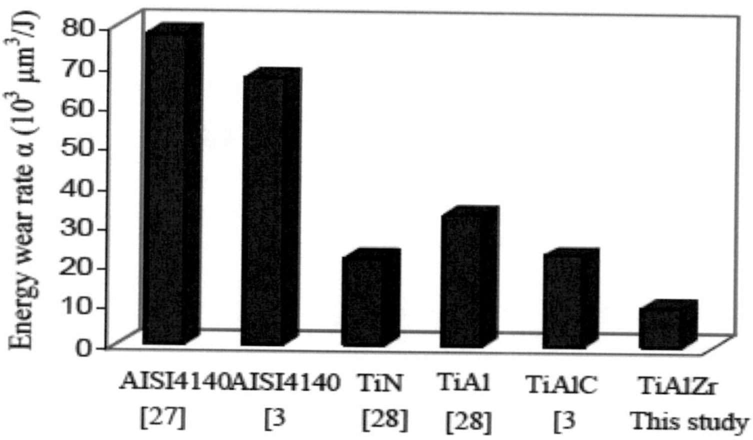

Figure 14.
Energy wear rates of different substrates and hard coatings in fretting wear tests [27].

4. Conclusion

In this work, we reviewed the deposition parameters and properties of titanium and titanium-aluminum-based quaternary coatings. It is found that the effect of the individual as well as multiple alloying elements are manifested in further modifying the properties of (Ti, Al) N coatings. Research and developments on simple binary and ternary coatings in previous studies are discussed. This was followed by the investigations on quaternary multicomponent coatings (TiAlZrN/TiAlN/TiAl and TiAlCN/TiAlN/TiAl); the behaviors in these coatings in fretting wear are compared. The main conclusions are the following. The local card solicitation obtained for the two layers of quaternary studied shows the delimitation of both sliding rates. It is shown that the gross slip regime region (GSR) of the coated AISI4140 steel is extended by the presence of the TiAlZrN layer in comparison with the TiAlZrN layer. These results are being very useful in the tribology. Indeed, the GSR is always associated to wear, but the partial slip regime (PSR) is accompanied to a crack that can be disastrous and leads to the failure. To reduce the instantaneous friction coefficient and stabilized friction coefficient, it is necessary to choose the coating-based zirconium coating instead of the coating-based carbon. In general, the stability of the coefficient of friction observed under the fretting conditions tested for a PVD coating was linked to the presence of a third multilayer body: a first layer consisting of particles mainly of submicron size; a second, discontinuous lamellar layer; and a third layer consisting of particles having a microstructure with ultrafine grains (refinement). In steady state, the third body formed easily remains trapped in contact. The third body, less than 50 μm thick, protects the surface (reduced wear) and is stable.

However, the gap between these two coatings is governed by the amplitude of the loading parameters. The thin coatings formed by physical vapor deposition may provide an initial friction reduction but it is not sufficiently durable for this application. However, the beneficial effect of the PVD coatings on fretting wear diminishes with increasing normal force and decreasing fretting stroke. In this chapter, the worn volume of the two quaternary layers is very influenced by loading conditions. Therefore, special attention should be given to better distinguish the effect of zirconium. Adding Zr improves the wear resistance of Al-Ti-N coating. Zr stabilizes Ti-Al-N lattice and also forms a very thin stable oxide layer similar to Al_2O_3. These two effects together enhance the wear resistance of the Ti-Al-Zr-N coatings. The stability of the energy approach is used to predict the wear kinetics in a volume of tribo-systems: Wv = α × Ed Σ, where, α is the energetic coefficient, Wv is the worn volume, and Ed is the dissipated cumulated energy. The multilayered TiAlZrN/TiAlN/TiAl coatings improve the resistance to the fretting wear of AISI4140 steel by twice as much as the multilayered TiAlCN/TiAlN/TiAl coatings.

In the light of various experimental investigations, the study of the wear mechanisms of the different films was carried out. It follows that the advantages of a multilayer coating over a single-layer coating are mainly presented in three points:

- multilayer coatings are multifunctional,
- good resistance to wear, and
- reduce friction.

Acknowledgements

The authors thank T. Gendre and Co. (Waterman S.A., France) for providing the coatings.

Author details

Brahim Tlili
LR-11-ES19, Laboratory of Applied Mechanics and Engineering (LR-MAI), National School of Engineers of Tunis, University of Tunis El Manar, Tunis, Tunisia

*Address all correspondence to: tlilii_brahim@yahoo.fr

References

[1] Nanastructured coatings for high performance tools (PLATIT). Werkzeug Technik. March 2003. p. 77

[2] PalDey S, Deevi SC. Single layer and multilayer wear resistant coatings of (Ti, Al) N: A review. Materials Science and Engineering A. 2003;**342**:58-79

[3] Ben Cheikh Larbi A, Tlili B. Surface and Coatings Technology. 2006;**257**(8972):167-168

[4] Lii D-F, Huang J-L, Lin MH-H. Surface and Coatings Technology. 1998;**99**:197

[5] Fouvry S, Kapsa P, Vincent L. A global methodology to quantify fretting damage. In: Mutch Y, Kiyon SE, Hoeppner DW, editors. Fretting Fatigue: Advances in Basic Understanding and Applications, STP 1425. West Conshohocken, PA: American Society for Testing and Materials; 2003. pp. 17-32

[6] Fouvry S, Kapsa P, Vincent L. Wear. 1995;**185**:35-46

[7] Fouvry S, ELLeuch K, Simeon G. Journal of Strain Analysis. 2000;**37**(6):703-716

[8] Proudhon H, Fouvry S, Yantio GR. International Journal of Fatigue. 2006;**28**:707-713

[9] Meng HC, Ludema KC. Wear. 1995;**181-183**:443-457

[10] Lugscheider E, Knotek O, Barimani C, Leyendecker T, Lemmer O, Wenke R. Surface and Coatings Technology. 1999;**112**:146-151

[11] Godet M. The third body approach. Wear. 1984;**100**:325-437

[12] Descartes S, Desrayaud C, Nicollini E, Berthier Y. Wear. 2005;**258**:1081-1090

[13] Berthier Y, Vincent L, Godet M. Wear. 1998;**125**:25-38

[14] Mohrbacher H, Blanpain B, Celis JP, Roos JR, Stals L, Van Stappen M. Wear. 1995;**188**:130-137

[15] Dubourg MC, Chateauminois A, Villechaise B. Tribology International. 2003;**36**:109-119

[16] Dewit D, Blanpain B, Froyen L, Celis JP. Wear. 1998;**217**:215-224

[17] Bowden FP, Tabor D. Friction and Lubrification of Solids. Vol. 1. London: Oxford University Press; 1964

[18] Elleuch K, Fouvry S. Wear. 2002;**253**:662-672

[19] Santner E, Klaffke D, Kocker GMZ. Wear. 1995;**190**:204-211

[20] Holmberg K, Matthewst A, Ronkainen H. Tribology International. 1998;**31**(1-3):107-120

[21] Knotek O, Münz WD, Leyendecker T. Journal of Vaccum Science and Technology A. 1987;**5**(4):2173-2179

[22] Knotek O, Böhmer M, Leyendecker T, Jungblut F. Materials Science and Engineering. 1988;**A105-106**:481-488

[23] Fouvry S, Kapsa P, Vincent L. Wear. 1996;**200**(1-2):186-205

[24] Sauger E, Fouvry S, Ponsonnet L, Kapsa P, Martin JM, Vincent L. Wear. 2000;**245**:39-52

[25] Sauger E, Ponsonnet L, Martin JM, Vincent L. Tribology International. 2000;**33**:743-750

[26] Varenberg M, Halperin G, Etsion I. Wear. 2002;**252**:902

[27] Prakash B, Ftikos C, Celis J-P. Surface Coatings and Technology. 2002;**154**:182